云和降水的遥感测量

[美]康斯坦丁·安德罗纳什　著
阮　征　崔　晔　李浩然　译
李浩然　校

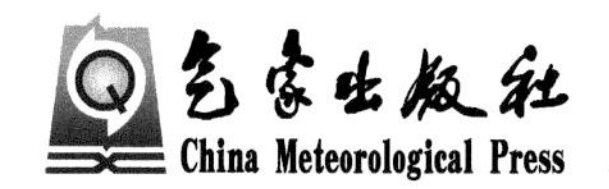

内容简介

本书旨在介绍云和降水遥感的一系列前沿研究课题，分为两部分：第一卷，地基遥感，包含了关于天气雷达及其在龙卷、风暴和雷暴探测、表征和预报中应用的文章；第二卷，星基遥感，涵盖了降水的卫星遥感探测，并以重大项目为例进行说明，包括热带降雨测量任务（Tropical Rainfall Measuring Mission，TRMM）和全球降水测量（Global Precipitation Measurement，GPM）项目，以及在卫星遥感中使用云雷达进行云探测和表征的最新进展。

云和降水遥感的论题非常宽泛，本书带来了在多个尺度上提升精细遥感探测技术的巨大力量和潜力的证据。这些方法具有显著的社会效益，并将极大地有助于解答关于云系统在地球科学中的作用的复杂问题。本书面向本科生、研究生，大气科学、气象学、环境科学、水文学和遥感方向的研究人员，以及涉及天气雷达和云雷达技术和应用的工程师。

图书在版编目（CIP）数据

云和降水的遥感测量 / （美）康斯坦丁·安德罗纳什著；阮征，崔晔，李浩然译. -- 北京：气象出版社，2023.2（2024.4重印）
书名原文：Remote Sensing of Clouds and Precipitation
ISBN 978-7-5029-7933-1

Ⅰ. ①云… Ⅱ. ①康… ②阮… ③崔… ④李… Ⅲ. ①卫星遥感－应用－降水－大气物理学－研究 Ⅳ. ①P426.61

中国国家版本馆CIP数据核字(2023)第034681号
北京市版权局著作权合同登记：图字 01-2023-0252

云和降水的遥感测量
YUN HE JIANGSHUI DE YAOGAN CELIANG

出版发行：气象出版社
地　　址：北京市海淀区中关村南大街 46 号　　**邮政编码**：100081
电　　话：010-68407112(总编室)　010-68408042(发行部)
网　　址：http://www.qxcbs.com　　**E-mail**：qxcbs@cma.gov.cn
责任编辑：刘瑞婷　　**终　　审**：张　斌
责任校对：张硕杰　　**责任技编**：赵相宁
封面设计：艺点设计
印　　刷：北京建宏印刷有限公司
开　　本：787 mm×1092 mm　1/16　　**印　　张**：15
字　　数：375 千字
版　　次：2023 年 2 月第 1 版　　**印　　次**：2024 年 4 月第 2 次印刷
定　　价：180.00 元

译者前言

随着云和降水在天气和气候及其应用中所发挥的重要作用被逐渐认识，云降水微物理研究逐渐深入并进入快速发展阶段，云和降水的遥感测量也迎来了机遇与挑战。近年来国内垂直探测系统发展迅速，该领域国外已经开展了多年的研究，这些研究成果与使用经验对国内开展相关研究具有非常重要的借鉴意义。

针对这一需要，译者选择将《Remote Sensing of Clouds and Precipitation》一书引入国内并进行翻译，该书包含地基遥感卷和星基遥感卷两部分内容，分别从地基雷达遥感及星基卫星遥感不同探测视角，对云和降水的遥感测量进行阐述。第一卷"地基遥感"介绍了云雷达最新进展及其应用，也包含了天气雷达在龙卷、风暴和雷暴的探测特征和预报中应用的研究进展。第二卷"星基遥感"介绍卫星遥感中云雷达探测的最新进展，特别针对利用遥感数据提高数值天气预报系统和定量降水估计的改进进行了阐述。

本译著的出版得到了国家重点研发计划"超大城市垂直综合气象观测技术研究及试验（2017YFC1501700）"项目的资助。该项目的研究目的之一旨在引领国内云降水垂直探测技术及应用发展，自2018年执行以来，致力于发展我国地基云降水垂直探测雷达，并组织了连续多年的云和降水垂直观测试验，形成了观测系统标校、观测数据质控处理及云降水微物理参数反演分析等一系列研究成果，但我国在该领域的研究与应用还任重道远，因此有必要对国外的研究进展进行系统性介绍，对标国际研究前沿，对我国在该领域的探测技术及研究发展进行精准定位。考虑到国内星基和地基垂直探测系统的快速发展，广大科研和业务人员迫切需要全面、系统地了解国外云和降水遥感的系列前沿研究成果，我们将此书引入并翻译，希望本译著对从事该领域研究及应用的相关人员有所帮助，也非常期待此书能够为我国新型云降水垂直探测系统的发展与应用贡献微薄之力。

本书的翻译工作分工如下：第一卷"地基遥感"由崔晔、李浩然翻译，阮征校对；第二卷"星基遥感"由阮征翻译，崔晔校对；最后由李浩然对全部书稿进行终审校对、润色。在翻译过程中，中国气象局气象干部培训学院俞小鼎教授、国家卫星气象中心谷松岩研究员、气象探测中心姚聘副研究员对本书提供了非常专业的指导意见，也非常感谢气象出版社刘瑞婷编辑在此书出版过程中付出的极大热情和心血。

希望本书能够为读者带来专业上的收获、借鉴。

2022年11月于北京

原版前言

本书旨在介绍云和降水遥感的一系列前沿研究课题，是在进一步深化对云和降水在天气、气候中扮演角色及应用的认识背景下完成的。云影响着大气辐射平衡和地球水循环。为了减少当前天气预报和气候预测中的不确定性，需要获得准确的云观测信息以改进其在数值模式中的参数化方案。在进一步认识云的角色方面，云降水遥感面临突出的挑战和机遇。这些在本书中都有所阐述。

本书分为两部分，涵盖了很多的主题。第一卷“地基遥感”包含天气雷达基础及其在探测、分析和预报龙卷、风暴和雷暴中的应用，同时包括面向气象和水文应用的雷达系统的最新进展，其中一个章节介绍了云雷达及其应用。第二卷“星基遥感”涵盖了降水的卫星遥感，并介绍了包括热带降雨测量任务（Tropical Rainfall Measuring Mission，TRMM）和全球降水测量（Global Precipitation Measurement，GPM）在内的重大项目，以及在卫星遥感中利用云雷达开展云探测和分析的最新进展。这一部分展现了通过云探测卫星 CloudSat、云-气溶胶激光雷达与红外探路者卫星 CALIPSO 观测的云结构，以及它们对未来卫星项目的启示。因为水汽在气候系统和云降水过程中扮演的重要角色，书中包含了利用卫星微波传感器进行水汽探测的一个章节，另有一章讨论了在遥感探测支持下数值模式中云模拟的进展。近年来的研究表明，遥感数据可以有效提高数值天气预报和定量降水估计的表现。

虽然云和降水遥感的论题非常广泛，本书从多个尺度上提供证据阐释了提升精细遥感探测技术的巨大能力和潜力。这些方法具有显著的社会效益，并将为解决云在地球科学中扮演的角色这一复杂问题做出重大贡献。本书面向高年级本科生、研究生，来自大气科学、气象学、环境科学、水文学和遥感领域的研究人员，以及天气雷达、云雷达技术和应用领域的工程师。

Constantin Andronache

于美国波士顿马萨诸塞州

2017 年 10 月

致谢

这本书的出版离不开所有作者积极热情的帮助，感谢所有作者和支持他们的机构使本书的出版成为可能。特别感谢 Scott Collis 博士（美国阿贡国家实验室）、Susanne Crewell 博士（德国科隆大学地球物理与气象研究所）、Raquel Evaristo 博士（德国波恩大学气象研究所）、Patrick N. Gatlin 博士（美国国家航空航天局）、Karen Kosiba 博士（美国灾害性天气研究中心）、Hirohiko Masunaga 博士（日本名古屋大学空间-地球研究所）、Angela Rowe 博士（美国华盛顿大学大气科学系）在评阅过程提出的许多建议。感谢美国地球物理联合会、欧洲地球科学联合会和美国气象学会最近举行的一系列关于遥感及其应用的令人振奋的会议。感谢 John Wiley & Sons 出版公司允许我们使用在他们杂志上发表的一组插图。最后，感谢施普林格（Springer）出版社的 Zachary Romano、Susan Westendorf、Aaron Schiller、Kalaiselvi Ramalingam 和 John Ram Kumar 在本项目执行期间提供的热情帮助。

作 者

ElisaAdirosi 大气科学与气候研究所，意大利国家研究委员会（意大利罗马）。

FilipeAires 巴黎天文台（法国巴黎）。

ConstantinAndronache 波士顿学院（美国马萨诸塞州栗树山）。

LucaBaldini 大气科学与气候研究所，意大利国家研究委员会（意大利罗马）。

WesleyBerg 科罗拉多州立大学大气科学系（美国科罗拉多州柯林斯堡）。

DavidJ. Bodine 美国俄克拉何马大学诺曼分校（美国俄克拉何马州）。

DavidT. Bolvin 科学系统与应用公司（美国马里兰州拉纳姆）；美国宇航局戈达德太空飞行中心（美国马里兰州格林贝尔特）。

MirceaGrecu，摩根州立大学，戈达德地球科学技术研究中心（美国马里兰州巴尔的摩）；美国宇航局戈达德太空飞行中心（美国马里兰州格林贝尔特）。

GeorgeJ. Huffman 美国宇航局戈达德太空飞行中心（美国马里兰州格林贝尔特）。

TakamichiIguchi 美国宇航局戈达德太空飞行中心，中尺度大气过程实验室（美国马里兰州格林贝尔特）；马里兰大学帕克分校地球系统科学跨学科中心（美国马里兰州）。

ChrisKidd 马里兰大学帕克分校（美国马里兰州）；美国宇航局戈达德太空飞行中心（美国马里兰州格林贝尔特）。

DaliaB. Kirschbaum 美国宇航局戈达德太空飞行中心（美国马里兰州格林贝尔特）。

MatthewR. Kumjian 美国宾州州立大学帕克分校气象与大气科学系（美国宾夕法尼亚州）。

JamesM. Kurdzo 麻省理工学院林肯实验室（美国马萨诸塞州列克星敦）。

TakeshiMaesaka 国家地球科学与灾害防御研究所（日本茨城筑波）。

ToshihisaMatsui 美国宇航局戈达德太空飞行中心，中尺度大气过程实验室（美国马里兰州格林贝尔特）；马里兰大学帕克分校地球系统科学跨学科中心（美国马里兰州）。

MarioMontopoli 大气科学与气候研究所，意大利国家研究委员会（意大利罗马）。

HajimeOkamoto 九州大学应用力学研究所（日本福冈春日）。

WalterA. Petersen 美国航空航天局马歇尔太空飞行中心（美国阿拉巴马州亨茨维尔）。

NicolettaRoberto 大气科学与气候研究所，意大利国家研究委员会（意大利罗马）。

KaoriSato 九州大学应用力学研究所（日本福冈春日）。

GailSkofronick-Jackson 美国宇航局戈达德太空飞行中心（美国马里兰州格林贝尔特）。

YukariN. Takayabu 东京大学气候系统研究中心（日本千叶柏市）。

缩略词

AIR(Atmospheric Imaging Radar)　大气成像雷达
ARM(Atmospheric Radiation Measurement)　大气辐射测量
ARRC(Advanced Radar Research Center)　先进雷达研究中心
AVSET(Automated Volume Scan Evaluation and Termination)　自动体扫评估和终止
CALIOP(Cloud-Aerosol Lidar with Orthogonal Polarization)　正交偏振的云-气溶胶激光雷达
CALIPSO(Cloud-Aerosol Lidar and Infrared Pathfinder Satellite Observation)　云-气溶胶激光雷达与红外探路者卫星观测
CaPE(Convection and Precipitation/Electrification project)　对流和降水/电气化项目
CASA(Collaborative Adaptive Sensing of the Atmosphere)　大气协同自适应传感
CRM(Cloud Resolving Model)　云解析模型
CSWR(Center for Severe Weather Research)　灾害性天气研究中心
DMSP US(Defense Meteorological Satellite Program)　美国国防气象卫星计划
DOW(Doppler on Wheels)　机动多普勒雷达
ECMWF(European Centre for Medium-Range Weather Forecasts)　欧洲中期天气预报中心
EOS(Earth Observing System)　地球观测系统
EOV(Effective Field of View)　有效视场
GBVTD(Ground-Based Velocity Tracking Display)　地基速度跟踪显示
GMI(GPM Microwave Imager)　GPM 微波成像仪
GPM(Global Precipitation Measurement mission)　全球降水测量任务
GPM-CO(GPM Core Observatory)　GPM 核心观测台
GPROF(Goddard profiling algorithm)　戈达德廓线算法
G-SDSU(Goddard Satellite Data Simulator Unit)　戈达德卫星数据模拟器单元
HCS(Hydrometeor classification scheme)　水凝物分类方案
ISCCP(International Satellite Cloud Climatology Project)　国际卫星云气候学项目
JAXA(Japan Aerospace Exploration Agency)　日本宇宙航空研究开发机构
JMANHM(Japan Meteorological Agency Nonhydrostatic Model)　日本气象厅非静力模式
LANL(Los Alamos National Laboratory)　洛斯阿拉莫斯国家实验室

LEO(Low Earth Orbit)　近地轨道

LES(Large Eddy Simulations)　大涡模拟

LIS(Lightning Imaging Sensor)　闪电成像传感器

LPVEx(Light Precipitation Validation Experiment)　弱降水验证试验

MDA(Mesocyclone Detection Algorithm)　中气旋探测算法

MJO(Madden-Julian oscillation)　Madden-Julian 振荡

ML(Melting layer)　融化层

MODIS(Moderate-resolution imaging spectroradiometer)　中分辨率成像光谱仪

MPAR(Multi-function Phased Array Radar sensing experiments)　多功能相控阵雷达传感实验

NASA(National Aeronautics and Space Administration)　美国国家航空航天局,又称美国宇航局

NCAR(National Center for Atmospheric Research)　国家大气研究中心

NEXRAD(Next-Generation Radar)　新一代天气雷达

NIC(Non-inductive charging theory)　无感充电理论

NPOESS(National Polar-Orbiting Operational Environmental Satellite System)　国家极轨业务环境卫星系统

NSF(National Science Foundation)　国家科学基金会

NSSL(National Severe Storms Laboratory)　国家强风暴实验室

NWP(Numerical Weather Prediction)　数值天气预报

PAIR(Polarimetric Atmospheric Imaging Radar)　偏振大气成像雷达

PIA(Path-integrated attenuation)　路径积分衰减

PR(Precipitation Radar)　降水雷达

PSD(Particle Size Distribution)　粒度分布

QPE(Quantitative precipitation estimation)　定量降水估测

Radar(Radio detection and ranging)　雷达

RaXPol(Rapid X-band Polarimetric Radar)　快速 X 波段偏振雷达

RFGF(Rear-flank gust front)　后侧阵风锋

RHI(Range height indicator mode)　距离高度显示

ROC(Radar Operations Center)　雷达操作中心

ROTATE(Radar Observations of Tornadoes and Thunderstorms Experiment)　龙卷和雷暴的雷达观测试验

SAILS(Supplemental Adaptive Intra-Volume Low-Level Scans)　补偿自适应低层体扫

SMART-R(Shared Mobile Atmospheric Research and Training Radar)　共享式移动大气研究和训练雷达

SRT(Surface return technique)　地表回波技术

SSM/I(Special sensor microwave/imager)　特殊微波传感器/微波成像仪

STAR(Simultaneous transmit and receive mode)　同时发送和接收模式

TDA(Tornado detection algorithm)　龙卷监测算法

TDS(Tornado debris signature)　龙卷碎片特征

TDWR(Terminal Doppler Weather Radar)　多普勒天气雷达终端

TIP(Tornado Intercept Project)　龙卷拦截项目

TKE(Turbulence Kinetic Energy)　湍流动能

TMI(TRMM Microwave Imager)　TRMM 微波成像仪

TMPA(TRMM Multi-satellite Precipitation Analysis)　TRMM 多卫星降水分析

TOA(Time of arrival)　到达时间

TRMM(Tropical Rainfall Measuring Mission)　热带雨量测量任务

TVS(Tornado vortex signature)　龙卷涡旋特征

TWIRL(Tornadic Winds: In situ and Radar observations at Low levels)　龙卷:低空实地和雷达观测

VAD(Velocity azimuth display)　速度方位显示

VCP(Volume coverage pattern)　体扫模式

VIIRS(Visible Infrared Imaging Radiometer Suite)　可见红外成像辐射计套件

VORTEX(Verification of the Origins of Rotation in Tornadoes)　龙卷涡旋源地验证项目

WSR-88D(Weather Surveillance Radar 1988 Doppler)　1988 多普勒天气监测雷达

目　录

第一卷　地基遥感

第二卷 星基遥感

第一卷 地基遥感

01

第1章　引　　言

Constantin Andronache

1.1 综述

云对大气辐射平衡和水循环有重要影响。它们与入射的短波辐射和发出的长波辐射相互作用，从而影响地球的能量收支。云通过影响水汽输送和降水，在地球的水循环中也发挥着重要作用[15]。利用遥感方法观测、测量和预测降水是具有重大经济和社会影响的研究和业务活动[23,24,26,55]。天气雷达的发展及其在气象学中的应用，提升了多学科遥感探测的能力，更广泛的应用以及应对天气和气候监测和预测挑战的能力，反映在最近发表的许多论文中[2,3,10,14,33,38,39,45,46,57,59]。

本卷介绍了关于云和降水遥感的一系列前沿研究主题。其研究动机来自于认识云系统在天气、水文和气候研究与应用中的重要性。传统上，由于人们对降水及其对各种经济活动和日常生活的影响有直接的兴趣，降水云一直是人们关注的主要焦点。数值天气预报(NWP)和气候模拟方面的研究表明，也需要对非降水云特征进行描述和认识。因此，云雷达得到了发展，并广泛应用于固定地点、外场活动和卫星[25,54,60]。云雷达与其他遥感方法一起，为进一步描述云的结构特征提供了手段。目前正在努力建立一个综合测量系统，能够在全球尺度上连续地描述所有云(非降水云和降水云)的特征。这一目标涉及将地面和卫星传感器的探测数据结合起来，可以为全球大气模型提供急需的数据集。在实践中，许多测量系统都是为特定的应用而设计的，有特殊的限制，包括量程、覆盖范围、分辨率、精度和连续性，只使用地基探测会导致探测范围无法覆盖海洋或极地地区。如本卷所述，卫星提供有价值的探测数据，覆盖地球的大片区域。利用地面和卫星测量，并辅以数值模拟的协同方法，成为有希望解决当前挑战的一个途径。

本书第一卷包含了地基遥感的章节，重点是在风暴、雷暴和龙卷的监测中使用的天气雷达，也介绍了云雷达及其应用。第二卷讨论了被动和主动卫星遥感的应用，包括热带降雨测量任务(TRMM)和全球降水测量(GPM)项目，以说明天基遥感测量降水的方式。其他研究主题包括通过云卫星和云—气溶胶激光雷达和红外探路者卫星(CALIPSO)进行的云观测，以及在全球范围内云结构和微物理特征描述方面的进展。有1章讨论了利用微波辐射计对水汽进行遥感，以说明大气中水汽的重要性。另有1章讨论了在数值模式中使用遥感数据，聚焦于诸如云微物理方案、模式结果与遥感探测之间的比较等问题，并讨论用于改进云系统数值模拟的技术。在简要介绍本书的章节之前，我们回顾了在本卷中没有提及的云系统特性描述方面的一些突出进展。

美国宇航局(NASA)在地球观测系统(EOS)范围内进行的一系列任务中使用了大量的传感器，这些传感器有助于了解地球－大气系统。EOS旨在获得陆地、海洋和大气的全面观测，而本卷只关注与云和降水有关的内容。中分辨率成像光谱仪(MODIS)是地球科学调研中最常用的卫星遥感平台之一。它于1999年12月18日由美国宇航局的Terra卫星发射，随后于2002年5月4日由Aqua发射。该仪器的设计目的是提供对地球大气层、陆地和海洋的全球观测。MODIS利用从可见光(VIS)到红外(IR)的36个光谱通道，探测反射的太阳辐射和发射的热辐射。MODIS提供光谱和空间信息，用以反演云顶、云的光学和微物理特性，测得的云顶特征参数是白天和晚上的气压、温度和高度。云光学和微物理特征参数为：云光学厚度、粒子有效半径以及推导出的白天水云和冰云路径[44]。MODIS后续的一个具有类似功能的仪器是可见光和红外成像辐射计套件(VIIRS)。这些传感器是国家极地轨道环境卫星系列(NPOESS)和NPOESS筹备项目(NPP)的一部分。NPP旨在弥补Terra和Aqua任务与NPOESS系列之间的差距[45,46]。

美国能源部(DOE)大气辐射测量(ARM)计划对云和降水遥感作出了重大贡献，其重点是云在气候系统中的作用[37,56]。ARM建立了一套永久性地面观测站，并为许多外场任务提供了便利[51,58]。在许多外场试验的帮助下，ARM项目发展了地基遥感能力，从而获得了观测与水汽、气溶胶、云和辐射相关的大气过程的精确方法[50]。ARM站点探测和描述云系统特征的能力为在世界其他地方发展更多的观测站提供了基础。ARM对卫星云探测方案的验证以及对云特征的卫星反演做出了重大贡献。这些ARM活动是通过一系列先进的地面设备实现的，如毫米波云雷达、拉曼激光雷达以及被动微波辐射计和红外干涉仪[36,50]。

欧洲的一项重大进展是Cloudnet计划，该计划建立了一套标准的地面遥感仪器，能够提供云参数，可与当前运行的NWP模式进行比较[9,21]。继Cloudnet计划之后的进一步发展以及ARM能力和协作的扩展，已经产生了一种更全面的方法来监测不同地点的云系统，从而能够评估和改进高分辨率数值模式[17]。

1.2 地基遥感

本书第一卷阐述了天气雷达在探测、监测包括风暴、龙卷和雷暴等恶劣天气方面的应用实例。第1章专门讨论云雷达的使用及其应用。

第2章对作为云和降水遥感重要工具之一的双偏振多普勒天气雷达的概念原理进行了概述[10,11,14,27,28]。首先是单个非球形粒子的电磁散射基本原理，接着介绍粒子群散射和等效反射率因子的概念、多普勒原理和多普勒速度估计，同时介绍了雷达系统的基本概况。本章给出了双偏振雷达变量及其物理解释，双偏振多普勒天气雷达探测促进了对云和降水物理的新认识，并在探测和监测有害天气业务方面取得了重大进展。这些天气情况包括严重的对流风暴(冰雹、龙卷、破坏性大风、洪水)和冬季风暴。本章总结了众多偏振多普勒天气雷达的应用，强调提高对云和降水认识新的研究途径。雷达回波分类和定量降水估计算法在最近几十年也取得了实质性的进展。未来的发展将提高我们对云和降水中基本物理过程的理解,并通过数据同化技术将数值天气预报模式和偏振多普勒天气雷达观测相结合。

第3章首先讨论了雷达在龙卷研究中的作用。龙卷大爆发时会造成许多人死亡和重大破坏。每年,在美国和世界其他地区都有许多龙卷的报道。本章接着描述了龙卷的理论和模拟。基于理论、涡室实验和大涡数值模拟(LES),建立了龙卷流场结构的概念模型。下一部分介绍雷达探测龙卷的历史,从早期雷达到现在的移动多普勒雷达系统。WSR－88D是气象雷达一项杰出的发展,通常也被称为NEXRAD("下一代雷达"),在20世纪80年代末和90年代初,这种分布广泛、地基、定点的多普勒雷达覆盖了美国大陆的绝大多数地区。从历史的角度来看,一个重大的进展是利用雷达实现了严重龙卷的现场观测。接下来本章介绍了自20世纪70年代多普勒雷达问世以来,地面雷达观测取得的成就。这些成果包括龙卷探测、雷达分析技术、龙卷成因和动力学、龙卷的双偏振和快速扫描雷达观测[6－8,31,32]。最后,本章提出了几个未解决的龙卷研究领域和需要改进的雷达分析技术。

第4章探讨雷达在探测和表征雷暴方面的应用。天气雷达是识别和分析雷暴的重要工具,雷暴通常与强降水、闪电和强风有关[12,13,41,47]。对流雷暴中的闪电会影响航空和电力基础设施的管理。雷暴的严重程度、运动轨迹、粒子在雷暴云系统内的分布等特征都可以被揭示出来。本章描述如何利用不同的雷达探测方法来识别雷暴的基本特征。本章介绍了对流过程中雷达信号的特征。接着讨论了从双偏振雷达观测结果推断出的云起电过程。双偏振天气雷达可以探测和测量与雷暴中电活动有关的积雨云中的霰粒子。最后,作者介绍了雷达和闪电联合探测的应用。

第5章介绍了云雷达的进展及其在提供暴雨预警的地基网络中的应用。本章介绍了云雷达的功能需求,并根据天线扫描和性能提供了这些仪器的类型。随后分析了影响雷达反射率的物理条件,并考虑了大气的布拉格散射和透射率,指出Ka－波段和W波段适合云观测。最后,本章对使用双偏振扫描云雷达系统的观测进行了说明和讨论,特别是从其能够提供降水雷达无法提供云系统细节的角度进行了讨论。双偏振云雷达有望得到广泛应用,将为云微物理研究做出重大贡献[25,35]。

1.3 星基遥感

本书的第二部分介绍了通过TRMM和GPM计划进行的卫星降水测量。有一章专门介绍了CloudSat和CALIPSO探测的云特性,另一章讨论利用卫星微波传感器测量水汽,最后一章分析了利用遥感数据进行数值云模拟的进展。

第6章阐述了在热带雨量测量任务(TRMM)中探测和地面降水的雷达应用。这项由NASA和日本宇宙航空研究开发机构(JAXA)于1997年发起的联合任务一直运行到2015年4月。TRMM是为天气和气候研究而设计的,提供了热带和亚热带地区关键的降水测量[29,30]。TRMM生成的长期降雨产品,为热带气旋预报、数值天气预报、降水气候学以及广泛的社会应用作出了贡献[24]。本章描述了TRMM最重要的仪器,即降水雷达(PR)和

TRMM 微波成像仪（TMI），以及用于从其观测数据估算降水的算法。PR 为热带风暴物理学提供了新的见解，TMI 使测量大气中的水汽、云水和降雨强度成为可能。本章还讨论了从 TRMM 观测资料获得降水的挑战以及克服这些挑战的策略。本章有一节专门介绍 TRMM 多卫星降水分析（TMPA）的降水产品，该产品是在项目内开发的，旨在减少降水估计中的时间采样误差。本章还简要介绍了 TRMM 观测结果和产品在科学和社会应用中的使用方向和示例[19,34,61]。

第 7 章回顾了雷达在全球降水测量（GPM）计划中探测和地面降水的应用，GPM 计划是成功的 TRMM 计划的后续项目。GPM 是由 NASA 和 JAXA 发起的一项国际卫星任务，旨在扩大天基的全球降水测量[18]。GPM 核心天文台（GPM-CO，于 2014 年 2 月发射）携带 Ka/Ku 波段双频降水雷达（DPR）、多频（10～183 GHz）微波辐射计和 GPM 微波成像仪（GMI）[16]。GPM-CO 是降水系统的天文台，也是天基多频微波辐射计估测降水的校准参考[18]。该任务旨在为研究和社会化应用提供新一代星载全球降水产品[53]。GPM 以地球的水和能量循环为重点，提供近实时（NRT）的降水观测，用于监测恶劣天气事件、淡水资源和其他社会应用。GPM NRT 数据的可用性使天基降水观测在各种实际应用中得以使用[24]。本章介绍 GPM 任务设计、传感器特性描述、卫星间校准、反演方法、地面校准活动和应用。

第 8 章从使用主动传感器对云进行遥感的视角，重点介绍了 CloudSat、CALIPSO 和 EarthCARE。Cloudsat 和 CALIPSO 是 A-Train 卫星星座[55]的一部分，该卫星星座包括 NASA 的 Aqua 和 Aura 卫星，以及法国的 PARASOL 卫星。A-Train 地球观测方法可以增进我们对大气水文过程的理解。本章描述了 CloudSat 上搭载的云廓线雷达和 CALIPSO 上的激光雷达如何共同提供云宏观尺度的垂直结构及其微物理特性。为了评估云在气候系统中的作用，数值模型需要充分的观测信息，包括大气温度、气压、水汽、云的三维多层结构、云相态和微物理参数。本章介绍了云雷达和激光雷达的原理，用于确定云特征中有兴趣的物理参数算法，并讨论了各种类型云检测所面临的挑战[42]。

第 9 章介绍了利用微波观测获取晴朗和多云情况下海洋和陆地上空大气水汽廓线的方法，重点研究了一种利用微波频段卫星观测资料的水汽反演算法，采用神经网络方案，其中包括卫星观测的专用校准系统。在没有降水的晴朗和多云条件下，可以反演海洋和地表上空的水汽，同时反演大气相对湿度廓线和大气柱水汽总量。该算法已经开发并用于 AQUA（或 MetOp）平台上搭载的仪器 AMSR-E/HSB（或 AMSUA/MHS）。本章介绍了反演方法的原理和理论上反演的不确定性。然后，利用 AQUA（或 MetOp）平台上搭载的仪器 HSB/AMSRE（或 MHS/AMSU-A）的实测数据对算法进行了测试，结果与欧洲中期天气预报中心（ECMWF）分析和无线电探空数据进行了比较。相对于先验 ECMWF 分析，亮度温度的后验验证试验显示，反演结果总体上呈现积极的影响[1,4]。

第 10 章概述了与遥感测量相比较，近期在改进大气模式中云和降水微物理参数化方面所作的努力。大量的研究工作致力于在不同尺度的数值模式中表示云的微物理过程[20,22,40,43,48,49,52]。本章描述了典型云微物理参数化的结构，随后介绍了用于比较数值模式输出值与遥感探测值的两种常规方法。一种方法是将探测得到的物理量与模式模拟计算得到的相应物理量进行比较，该方法能够直接识别目标量误差，但由于遥感探测数据的不确定性，该方法存在潜在的缺陷。另一种方法是使用信号模拟器，使探测值与大气模式模拟值之间能够基于信号进行比较，避免了反演算法内固有的不确定性。本章讨论了这些方法的优点和挑战。

1.4 结束语

本卷中介绍的关于云和降水遥感的最新研究，以及每一章所选的参考文献，概述了目前使用遥感描述云系统特征的工作进展。理解云和降水在地球系统中的作用，对于短期天气预报、准确气候预测及其多种应用等当代活动越来越重要。为了应对这些挑战，有必要改进大气探测技术、数值模式，并使用协同方法来解决复杂的问题。遥感方法在这项工作中起着至关重要的作用。这些工作由许多不断发展的研究项目和一个充满活力的科学家和工程师团体来维持。《云和降水的遥感测量》一书提供了必要的信息，帮助读者了解云系统的遥感现状，以及它们对天气和气候监测和预测的影响和很多实际应用。

致谢 在此，我向所有为本书做出贡献的作者和评论家表示衷心的感谢。非常感谢施普林格(Springer)出版社在这个项目期间的持续帮助。

参考文献*

[1] Aires, F., F. Bernardo, and C. Prigent. 2013. Atmospheric water-vapour profiling from passive microwave sounders over ocean and land. Part I: methodology for theMegha-Tropiques mission. Quarterly Journal of the Royal Meteorological Society 139: 852-864. https://doi.org/10.1002/qj.1888.

[2] Andronache, C., ed. 2017. Mixed-Phase Clouds: Observations and Modeling. Amsterdam: Elsevier.

[3] Atlas, D., ed. 2015. Radar in Meteorology: Battan Memorial and 40th Anniversary Radar Meteorology Conference. New York: Springer.

[4] Bernardo, F., F. Aires, and C. Prigent. 2013. Atmospheric water-vapour profiling from passive microwave-sounders over ocean and land. Part II: validation using existing instruments. Quarterly Journal of the Royal Meteorological Society 139: 865-878. https://doi.org/10.1002/qj.1946.

[5] Bluestein, H. B. 2013. Severe Convective Storms and Tornadoes: Observations and Dynamics. Berlin/Heidelberg: Springer.

[6] Bluestein, H. B., M. M. French, I. PopStefanija, R. T. Bluth, and J. B. Knorr. 2010. A mobile, phased-array Doppler radar for the study of severe convective storms. Bulletin of the American Meteorological Society 91(5): 579-600.

[7] Bodine, D. J., R. D. Palmer, and G. Zhang. 2014. Dual-wavelength polarimetric radar analyses of tornadic debris signatures. Journal of Applied Meteorology and Climatology 53: 242-261.

[8] Bodine, D. J., R. D. Palmer, T. Maruyama, C. J. Fulton, Y. Zhu, and B. L. Cheong. 2016. Simulated frequency dependence of radar observations of tornadoes. Journal of Atmospheric and Oceanic Technology 33(9):

* 参考文献沿用原版书中内容，未改动

1825-1842.

[9] Bouniol, D. , et al. 2010. Using continuous ground-based radar and lidar measurements for evaluating the representation of clouds in four operational models. Journal of Applied Meteorology and Climatology 49: 1971-1991. https://doi.org/10.1175/2010JAMC2333.1.

[10] Bringi, V. N. , and V. Chandrasekar. 2001. Polarimetric Doppler Weather Radar: Principles and Applications. Cambridge: Cambridge University Press.

[11] Fabry, F. 2015. Radar Meteorology: Principles and Practice. Cambridge: Cambridge University Press.

[12] Federico, S. , E. Avolio, M. Petracca, G. Panegrossi, P. Sano, D. Casella, and S. Dietrich. 2014 Simulating lightning into the RAMS model: implementation and preliminary results. Natural Hazards and Earth System Sciences 14: 2933-2950. https://doi.org/10.5194/nhess-14-2933-2014.

[13] Formenton, M. , G. Panegrossi, D. Casella, S. Dietrich, A. Mugnai, P. Sano, F. Di Paola, H.-D. Betz, C. Price, and Y. Yair. 2013. Using a cloud electrification model to study relationships between lightning activity and cloud microphysical structure. Natural Hazards and Earth System Sciences 13: 1085-1104. https://doi.org/10.5194/nhess-13-1085-2013.

[14] Fukao, S. , K. Hamazu, and R. J. Doviak. 2014. Radar for Meteorological and Atmospheric Observations. Tokyo: Springer.

[15] Gettelman, A. , and S. C. Sherwood. 2016. Processes responsible for cloud feedback. Current Climate Change Reports 2: 179-189. https://doi.org/10.1007/s40641-016-0052-8.

[16] Grecu, M. , W. S. Olson, S. J. Munchak, S. Ringerud, L. Liao, Z. Haddad, B. L. Kelley, and S. F. McLaughlin. 2016. The GPM combined algorithm. Journal of Atmospheric and Oceanic Technology 33: 2225-2245. https://doi.org/10.1175/JTECH-D-16-0019.1.

[17] Haeffelin, M. , et al. 2016. Parallel developments and formal collaboration between European atmospheric profiling observatories and the U. S. ARM research program. In The Atmospheric Radiation Measurement (ARM) Program: The First 20 Years. Meteorological Monographs. Vol. 57. Boston: American Meteor Society. https://doi.org/10.1175/AMSMONOGRAPHS-D-15-0045.1.

[18] Hou, A. Y. , et al. 2014. The global precipitation measurement mission. Bulletin of the American Meteorological Society 95: 701-722. https://doi.org/10.1175/BAMS-D-13-00164.1.

[19] Huffman, G. J. , R. F. Adler, D. T. Bolvin, G. Gu, E. J. Nelkin, K. P. Bowman, Y. Hong, E. F. Stocker, and D. B. Wolff. 2007. The TRMM multi-satellite precipitation analysis: quasi-global, multi-year, combined-sensor precipitation estimates at fine scale. Journal of Hydrometeorology 8(1): 38-55.

[20] Iguchi, T. , T. Matsui, J. Shi, W. Tao, A. Khain, A. Hou, R. Cifelli, A. Heymsfield, and A. Tokay. 2012. Numerical analysis using WRF-SBM for the cloud microphysical structures in the C3VP field campaign: impacts of supercooled droplets and resultant riming on snow microphysics. Journal of Geophysical Research-Atmospheres 117. https://doi.org/10.1029/2012JD018101.

[21] Illingworth, A. J. , et al. 2007. CloudNet: continuous evaluations of cloud profiles in seven operational models using ground-based observations. Bulletin of the American Meteorological Society 88: 883-898.

[22] Khain, A. P. , K. D. Beheng, A. Heymsfield, A. Korolev, S. O. Krichak, Z. Levin, M. Pinsky, V. Phillips, T. Prabhakaran, and A. Teller. 2015. Representation of microphysical processes in cloud-resolving models: spectral (bin) microphysics versus bulk parameterization. Revista Geografica 53(2): 247-322.

[23] Kidd, C. , and V. Levizzani. 2011. Status of satellite precipitation retrievals. Hydrology and Earth System Sciences 15: 1109-1116. https://doi.org/10.5194/hess-15-1109-2011.

[24] Kirschbaum, D. B. , et al. 2017. NASA's remotely sensed precipitation: a reservoir for applications users. Bulletin of the American Meteorological Society 98:1169-1184. https://doi. org/10. 1175/BAMS-D-15-00296. 1.

[25] Kollias, P. , E. E. Clothiaux, M. A. Miller, B. A. Albrecht, G. L. Stephens, and T. P. Ackerman. 2007. Millimeter-wavelength radars: new frontier in atmospheric cloud and precipitation research. Bulletin of the American Meteorological Society 88:1608-1624. https://doi. org/10. 1175/BAMS-88-10-1608.

[26] Kucera, P. A. , E. E. Ebert, F. J. Turk, V. Levizzani, D. Kirschbaum, F. J. Tapiador, A. Loew, and M. Borsche. 2013. Precipitation from space: advancing earth system science. Bulletin of the American Meteorological Society 94:365-375. https://doi. org/10. 1175/BAMS-D-11-00171. 1.

[27] Kumjian, M. R. , and W. Deierling. 2015. Analysis of thundersnow storms over northern Colorado. Weather Forecast 30:1469-1490. https://doi. org/10. 1175/WAF-D-15-0007. 1.

[28] Kumjian, M. R. , and K. A. Lombardo. 2017. Insights into the evolving microphysical and kinematic structure of Northeastern U. S. winter storms from dual-polarization Doppler radar. Monthly Weather Review 145:1033-1061.

[29] Kummerow, C. , W. Barnes, T. Kozu, J. Shiue, and J. Simpson. 1998. The tropical rainfall measuring mission (TRMM) sensor package. Journal of Atmospheric and Oceanic Technology 15:809-817.

[30] Kummerow, C. , J. Simpson, O. Thiele, W. Barnes, A. T. C. Chang, E. Stocker, R. F. Adler, A. Hou, R. Kakar, F. Wentz, P. Ashcroft, T. Kozu, Y. Hong, K. Okamoto, T. Iguchi, K. Kuriowa, E. Im, Z. Haddad, G. Huffman, B. Ferrier, W. S. Olson, E. Zipser, E. A. Smith, T. T. Wilheit, G. North, T. Krishnamurti, and K. Nakamura. 2000. The status of the tropical rainfall measuring mission (TRMM) after two years in orbit. Journal of Applied Meteorology, Part 1 39(12):1965-1982.

[31] Kurdzo, J. M. , and R. D. Palmer. 2012. Objective optimization of weather radar networks for low-level coverage using a genetic algorithm. Journal of Atmospheric and Oceanic Technology 29(6):807-821.

[32] Kurdzo, J. M. , F. Nai, D. J. Bodine, T. A. Bonin, R. D. Palmer, B. L. Cheong, J. Lujan, A. Mahre, and A. D. Byrd. 2017. Observations of severe local storms and tornadoes with the atmospheric imaging radar. Bulletin of the American Meteorological Society 98(5):915-935.

[33] Levizzani, V. , P. Bauer, and F. J. Turk. 2007. Measuring Precipitation from Space: EURAINSAT and the Future. Vol. 28. Dordrecht: Springer Science and Business Media.

[34] Liu, Z. , D. Ostrenga, W. Teng, and S. Kempler. 2012. Tropical rainfall measuring mission (TRMM) precipitation data and services for research and applications. Bulletin of the American Meteorological Society 93:1317-1325. https://doi. org/10. 1175/BAMS-D-11-00152. 1.

[35] Maesaka, T. , K. Iwanami, S. -I. Suzuki, Y. Shusse, and N. Sakurai. 2015. Cloud radar network in Tokyo metropolitan area for early detection of cumulonimbus Generation. Paper Presented at 37th Conference on Radar Meteorology. Norman: American Meteorological Society. https://ams. confex. com/ams/37RADAR/webprogram/Paper275910. html.

[36] Marchand, R. 2016. ARM and satellite cloud validation. Meteorological Monographs 57:30. 1-30. 11. https://doi. org/10. 1175/AMSMONOGRAPHS-D-15-0038. 1.

[37] Mather, J. H. , D. D. Turner, and T. P. Ackerman. 2016. Scientific maturation of the ARM program. In The Atmospheric Radiation Measurement (ARM) Program: The First 20 Years. Meteorological Monographs. Vol. 57. Boston: American Meteorological Society. https://doi. org/10. 1175/AMSMONOGRAPHS-D-15-0053. 1.

[38] Meischner, P. 2003. Weather radar-principle and advanced applications. In Weather Radar-Principle and Advanced Applications Physics of Earth and Space Environments, 337. Berlin: Springer. ISBN: 3-540-000328-2.

[39] Michaelides, S. C. , ed. 2008. Precipitation: Advances in Measurement, Estimation and Prediction. Berlin: Springer Science and Business Media.

[40] Morrison, H. , J. A. Curry, and V. I. Khvorostyanov. 2005. A new double-moment microphysics parameterization for application in cloud and climate models. Part I: description. Journal of the Atmospheric Sciences 62:1665-1677. https://doi. org/10. 1175/JAS3446. 1.

[41] Mosier, R. M. , C. Schumacher, R. E. Orville, and L. D. Carey. 2011. Radar nowcasting ofcloudto-ground lightning over Houston, Texas. Weather Forecast 26: 199-212. https://doi. org/10. 1175/2010WAF 2222431. 1.

[42] Okamoto, H. , K. Sato, and Y. Hagihara. 2010. Global analysis of ice microphysics from CloudSat and CALIPSO: incorporation of specular reflection in lidar signals. Journal of Geophysical Research 115: D22209. https://doi. org/10. 1029/2009JD013383.

[43] Phillips, V. T. J. , P. J. DeMott, C. Andronache, K. Pratt, K. A. Prather, R. Subramanian, and C. Twohy. 2013. Improvements to an empirical parameterization of heterogeneous ice nucleation and its comparison with observations. Journal of the Atmospheric Sciences 70:378-409.

[44] Platnick, S. , K. G. Meyer, M. D. King, G. Wind, N. Amarasinghe, B. Marchant, G. T. Arnold, Z. Zhang, P. A. Hubanks, R. E. Holz, and P. Yang. 2017. The MODIS cloud optical and microphysical products: collection 6 updates and examples from Terra and Aqua. IEEE Transactions on Geoscience and Remote Sensing 55(1):502-525.

[45] Qu, J. J. , W. Gao, M. Kafatos, R. E. Murphy, and V. V. Salomonson, eds. 2006. Earth Science Satellite Remote Sensing: vol. 1: Science and Instruments. Beijing/Berlin, Heidelberg: Tsinghua University Press/ Springer.

[46] Qu, J. J. , W. Gao, M. Kafatos, R. E. Murphy, and V. V. Salomonson, eds. 2006. Earth Science Satellite Remote Sensing: vol. 2: Data, Computational Processing, and Tools. Beijing: Tsinghua University Press/Berlin, Heidelberg: Springer.

[47] Roberto, N. , E. Adirosi, L. Baldini, D. Casella, S. Dietrich, P. Gatlin, G. Panegrossi, M. Petracca, P. Sano, and A. Tokay. 2016. Multi-sensor analysis of convective activity in central Italy during the Hy-MeX SOP 1. 1. Atmospheric Measurement Techniques 9:535-552. https://doi. org/10. 5194/amt-9-535-2016.

[48] Seifert, A. , and K. D. Beheng. 2001. A double-moment parameterization for simulating autoconversion, accretion and selfcollection. Atmospheric Research 59:265-281.

[49] Seifert, A. , and K. D. Beheng. 2006. A two-moment cloud microphysics parameterization for mixed-phase clouds. Part 1: model description. Meteorology and Atmospheric Physics 92(1):45-66.

[50] Shupe, M. D. , J. M. Comstock, D. D. Turner, and G. G. Mace. 2016. Cloud property retrievals in the ARM program. In The Atmospheric Radiation Measurement (ARM) Program: The First 20Years. Meteorological Monographs. Vol. 57. Boston: American Meteor Society. https://doi. org/10. 1175/AMSMONOGRAPHS-D-15-0030. 1.

[51] Sisterson, D. , R. Peppler, T. S. Cress, P. Lamb, and D. D. Turner. 2016. The ARM southern great plains (SGP) site. In The Atmospheric Radiation Measurement (ARM) Program: The First 20 Years. Meteorological Monographs. Vol. 57. Boston: American Meteor Society. https://doi. org/10. 1175/AMSMONO-

GRAPHS-D-16-0004. 1.

[52] Skamarock, W. C. , J. B. Klemp, J. Dudhia, D. O. Gill, D. M. Barker, M. G. Duda, X. Y. Huang, W. Wang, and J. G. Powers. 2008. A description of the advanced research WRF version 3, NCAR technical note, Mesoscale and Microscale Meteorology Division, National Center for Atmospheric Research, Boulder, CO. Shupe, M. , et al. 2008. A focus on mixed-phase clouds: the status of ground-based observational methods. Bulletin of the American Meteorological Society 87:1549-1562.

[53] Skofronick-Jackson, G. , W. A. Petersen, W. Berg, C. Kidd, E. F. Stocker, D. B. Kirschbaum, R. Kakar, S. A. Braun, G. J. Huffman, T. Iguchi, P. E. Kirstetter, C. Kummerow, R. Meneghini, R. Oki, W. S. Olson, Y. N. Takayabu, K. Furukawa, and T. Wilheit. 2017. The global precipitation measurement (GPM) mission for science and society. Bulletin of the American Meteorological Society 98: 1679-1695. https://doi. org/10. 1175/BAMS-D-15-00306. 1.

[54] Stephens, G. L. , and C. D. Kummerow. 2007. The remote sensing of clouds and precipitation from space: a review. Journal of the Atmospheric Sciences 64:3742-3765.

[55] Stephens, G. L. , et al. 2002. The CLOUDSAT mission and the A-train-a new dimension of space-based observations of clouds and precipitation. Bulletin of the American Meteorological Society 83:1771-1790.

[56] Stokes, G. M. 2016. Original ARM concept and launch. In The Atmospheric Radiation Measurement (ARM) Program: The First 20 Years. Meteorological Monographs. Vol. 57. Boston: American Meteor Society. https://doi. org/10. 1175/AMSMONOGRAPHS-D-15-0054. 1.

[57] Tao , W. K. , ed. 2015. Cloud Systems, Hurricanes, and the Tropical Rainfall Measuring Mission (TRMM): A Tribute to Joanne Simpson. Berlin: Springer.

[58] Verlinde, J. , B. Zak, M. D. Shupe, M. Ivey, and K. Stamnes. 2016. The ARM North Slope of Alaska (NSA) sites. In The Atmospheric Radiation Measurement (ARM) Program: The First 20 Years. Meteorological Monographs. Vol. 57. Boston: American Meteor Society. https://doi. org/10. 1175/AMSMONOGRAPHS-D-15-0023. 1.

[59] Wakimoto, R. M. , and R. Srivastava, eds. 2003. Radar and Atmospheric Science: A Collection of Essays in Honor of David Atlas. Vol. 30(52). Boston: American Meteorological Society.

[60] Winker, D. M. , M. A. Vaughan, A. H. Omar, Y. Hu, K. A. Powell, Z. Liu, W. H. Hunt, and S. A. Young. 2009. Overview of the CALIPSO mission and CALIOP data processing algorithms. Journal of Atmospheric and Oceanic Technology 26:2310-2323. https://doi. org/10. 1175/2009JTECHA1281. 1.

[61] Yong, B. , D. Liu, J. J. Gourley, Y. Tian, G. J. Huffman, L. Ren, and Y. Hong. 2015. Global view of real-time TRMMmultisatellite precipitation analysis: implications for its successor global precipitation measurement mission. Bulletin of the American Meteorological Society 96:283-296. https://doi. org/10. 1175/BAMS-D-14-00017. 1.

第 2 章　天气雷达

Matthew R. Kumjian

2.1 引言

天气雷达是遥感云和降水最必不可少的工具。虽然雷达最初用于在军事上探测飞机和船只，但雷达操作员很快就注意到它探测降水的能力。自那时起，雷达工程、信号处理和气象学的发展提高了天气雷达的探测精度、分辨率和可用信息，使得预警更早、虚警更少，并提高了对灾害天气的认识，包括强对流和相关的冰雹、龙卷、洪水和破坏性强风，以及暴风雪和相关的大雪、相态转换和灾害性的混合相态降水。新天气雷达技术所增加的信息促进了降水分类方法的发展和细化，这些方法将雷达像素点分为各种降水类型、非降水信号或雷达回波。这使得业务雷达人员可以快速估计雷达扫描图像中的回波类型，这些改进通过提供非常有价值的定量估算降水信息也使水文气象学受益。近期，天气雷达数据正被同化到数值模式中，以改进模式分析和预报。这些内容将在本章和其他章节讨论。

"Radar"是"Radio Detection And Ranging"的首字母缩写，表示无线电探测和测距，由美国海军在 1940 年创造。雷达通过发射大功率无线电波工作，这是一种人眼不可见的电磁辐射。在空间或时间的任何一点上，这些电磁波可以用频率(以每秒或赫兹为周期的振荡速率)、振幅(振荡电场和磁场的强度)、相位(电场或磁场振荡周期中的点)和偏振(电场矢量振荡的方向，与波的传播方向正交)来表征。这些无线电波通过大气传播，可以与云和降水等水凝物粒子以及其他非气象粒子、生物和地面目标物发生相互作用。当被雷达信号照射时，这些粒子受电磁激励，吸取一部分雷达波的能量。散射过程是将一部分雷达波能量向各个方向辐射，吸收过程是消耗一部分雷达波能量并转化为热能，其中一些辐射可能被散射回雷达(后向散射)。无线电波在传播过程中由于散射和吸收而损失的能量称为衰减。

后向散射信号为表征云和降水提供了重要信息，例如，接收到信号的振幅可以反映有关降水强度的信息，因为电磁波以已知的速度(光速)在大气中传播，所以从发射信号到接收后向散射信号之间的这段时间反映了降水的距离或范围。测量到接收信号相位变化的时间速率是多普勒频移，它反映了粒子的运动是接近还是远离雷达；发射和/或接收不同偏振或频率的电磁辐射的雷达可以为水凝物的形状、取向和成分提供额外的信息。本章将从一个概念框架来概述这些内容和其他雷达基础知识，更深入的讨论参见雷达教科书[20,34,37,126]。

通常情况下，雷达的特性与它们发射电磁波的频率(或波长)相关。雷达的工作频率对其遥感探测云和降水的能力有多方面的影响，包括给定天线尺寸的有效波束宽度、可探测的最大多普勒速度和水凝物对入射电磁辐射的响应等。例如，对于许多业务天气雷达网所使用的抛物型反射面，辐射能量下降到其峰值功率一半时对应的有效波束宽度 θ(单位为度)与反射面直径 D_R 和雷达波长(λ)有关，$\theta \approx 70°\lambda/D_R$。因此，对于给定的天线反射面直径($D_R$)，波束宽度随波长($\lambda$)减小(频率增加)而减小；或者，对于给定的雷达波长(λ)，较大尺寸的反射面会造成较小的波束宽度。此外，不同 λ 的电磁辐射对云和降水粒子的响应也不同，例如，对于给定数量的水凝物，较小 λ 的衰减量比较长 λ 的衰减量大很多。在天气雷达系统设计中，天线反射面尺寸(如成本、便携性)、有效波束宽度(如分辨率)和探测能力(如衰减)之间的权衡需要重点考虑。

表 2.1　不同频段和波长的命名规范

名称	频率范围	波长范围
LF	30～300 kHz	10～1 km
MF	0.3～3 MHz	1000～100 m
HF	3～30 MHz	100～10 m
VHF	30～300 MHz	10～1 m
UHF	300～300 MHz	1～0.1 m
L	1～2 GHz	30～15 cm
S	2～4 GHz	15～8 cm
C	4～8 GHz	8～4 cm
X	8～12 GHz	4～2.5 cm
Ku	12～18 GHz	2.5～1.7 cm
K	18～27 GHz	1.7～1.2 cm
Ka	27～40 GHz	1.2～0.75 cm
W	75～110 GHz	4.0～2.73 mm
G	110～300 GHz	2.73～0.1 mm

按照惯例，不同的电磁辐射频段频带以字母命名。其中一些是在第二次世界大战期间开发的，目的是迷惑敌方间谍。这些波段及其名称已在表 1 中给出。通常，UHF 波段的低频雷达用于高层大气探测[19,54,57,58,145,167]，而高频雷达用于云和降水的遥感。传统天气雷达可分为降水雷达(S～X 波段)和云雷达(Ku～W 波段)，最近，G 波段雷达也被提出用于云的研究。在业务雷达网中，美国国家气象局的多普勒天气监测雷达－1988(WSR－88D)选择 S 波段运行，而加拿大、一些南美和欧洲国家则使用 C 波段雷达进行组网。最近，美国、亚洲和欧洲的一些业务雷达网已经补充了用于填补盲区的 X 波段雷达[9,105,108]，以改善低层数据覆盖[137]。X 波段雷达天线尺寸较小，雷达方便运输，常用作研究龙卷和强对流风暴[12,13,39,118,171,174]的移动雷达。高频云雷达主要用于科研[70,73,92－94,107]和星载应用[53]。

鉴于本书的重点是云和降水的遥感，本章内容仅限于天气回波。本章的其余部分是这样展开的：第 2.2 节介绍单个粒子的电磁散射；第 2.3 节将这一讨论扩展到单个非球形粒子在不同偏振条件下的散射；第 2.4 节介绍了粒子群的散射和等效反射率因子；第 2.5 节介绍了气象回波的探测和测距；第 2.6 节介绍了多普勒效应和多普勒速度估计的原理；第 2.7 节是对雷达系统的基本介绍；第 2.8 节介绍双偏振雷达变量及其物理意义；第 2.9 节简要概述了对很多新的天气雷达应用和研究路线；本章在第 2.10 节以一个简短的总结结束。

2.2 单个粒子的散射

水凝物对入射电磁辐射的散射与其的大小、形状、取向、组成(例如，冰、液体、混合相)、温度和辐射的频率有关。粒子的物理成分、温度和雷达波长的特征由介电常数或相对介电常数(ϵ_r)来表达，该变量是量化该粒子对电磁辐射照射响应的复数。对于天气雷达的工作频率，ϵ_r中较大的实数部分会导致来自特定尺度/形状的粒子返回更多后向散射能量，并且波在传播过程中穿过这种粒子群时产生较大的相位偏移。虚数部分越大，表示粒子的吸收能力越强。例如，在 S 波段(美国 WSR-88D 雷达网的波长)，液态水ϵ_r的实数部分相比冰要大得多。因此，液态雨滴的后向散射能力比同等大小的冰粒子大得多。对于降水雷达，ϵ_r与复折射指数 m 有关，$\epsilon_r=m^2$。复折射指数的实部可以认为是波在真空中的相位速度与在介质中的相位速度之比。同样地，介质中电磁辐射的波长比在真空中短。因此，雷达波在电介质中传播时，其相位速度(或波长)比在周围空气中小。

考虑一个由某种电介质材料组成的任意形状的粒子。如图 2.1a 中的球体所示，可以将该粒子分解为微小的有限散射元[16]。当一个入射电场作用于粒子(图 2.1b)，小球体内电子和质子受静电力作用而产生净偶极矩，即粒子内所有小球体各自的电荷大小乘以电荷间距的矢量和。这可以认为是在每个小球体中感应的偶极子，均按电场矢量方向排列(图 2.1b)。如果入射电场是振荡的(就像雷达波一样)，那么这些感应偶极子以与入射辐射相同的频率振荡。因此，每个微小球体都表现为一个偶极子振荡器，向各个方向发射电磁子波，这些子波在远离粒子的某一点的总和就是总散射波。

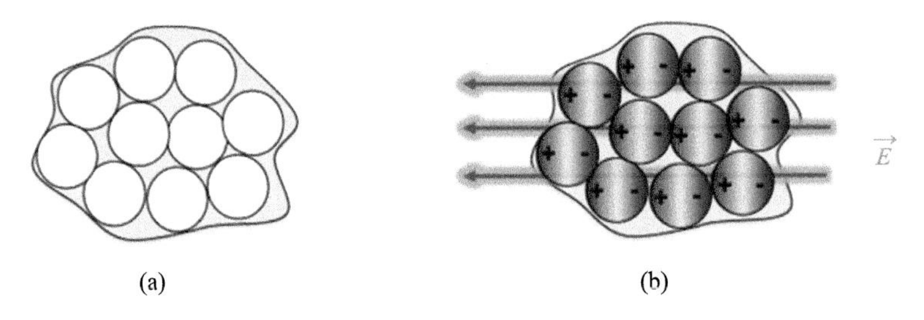

图 2.1 (a)图中所示为由微小球形有限散射单元(白色圆圈)组成的任意粒子；(b)当电场(绿色矢量)作用于粒子时，感应产生偶极矩(红色阴影和加号表示净正电荷；蓝色阴影和负号表示净负电荷)，而偶极子则沿着电场矢量的方向排列

粒子尺度与辐射波波长之比对计算总散射很重要。通常，尺度参数被用来评估粒子尺度和波长之间的关系：$x=2\pi a/\lambda$，其中 a 是粒子特征长度(例如，粒子等效球体半径)。这个尺度参数可以认为是单个粒子特征长度对应辐射波的周期数。当粒子远小于辐射波长时(即 $x\ll1$)，那么在某一时刻，粒子内所有微小球体的电场几乎没有变化(图 2.2a)。因此，粒子内所有的感应偶极子的振荡相位相同。对于小的球形粒子，由此产生的散射辐射模式类似于偶极子天线。

非球形粒子将在下一节讨论。相比之下,对于一个比波长大的粒子,在给定的时刻,穿过粒子的电场可能存在较大的变化(图 2.2b)。因此,感应偶极子的振荡相位不同,会导致散射子波之间的相长和相消干涉。当共振参数 K 接近 1 时[131],干涉或“共振”效应的可能性增加,K 的表达式如下:

$$K=\frac{D\sqrt{|\epsilon_r|}}{\lambda} \tag{2.1}$$

式中 D 为粒子的等体积球体直径。因为式(2.1)中包含了相对介电常数ϵ_r,所以ϵ_r大的粒子具有更强的散射效果(也就是说,更有可能出现强共振效应)。

图 2.2 (a)粒子粒径小于贯穿它的均匀电场(绿色矢量)的波长(轨迹为绿色线条);(b)粒子粒径大于贯穿它的非均匀电场的波长。(a)中感应偶极子彼此同步振荡;(b)中感应偶极子之间的振荡相位不同

从目标物散射回雷达的辐射量与目标物的后向散射截面 σ_b 或雷达截面(习惯上,这两个术语在雷达气象学中可以互换使用)有关。由于无法了解特定范围内精确的目标物散射特性,所以假设雷达接收到的能量来自一个假想的散射辐射各向同性的目标。后向散射截面是这个假想的各向同性散射目标的截面面积。注意,后向散射截面通常不对应于实际粒子的几何面积。例如,在 S 波段,由于上述相对介电常数的差异,1 mm 雨滴的后向散射截面比 1 mm 冰球大。

Mie[110] 给出了介电球体的 σ_b 的精确解。如果球形粒子粒径远小于波长,可以用瑞利散射近似处理介电球体粒子后向散射截面的 Mie 散射解:

$$\sigma_b \approx \frac{\pi^5}{\lambda^4} D^6 |K|^2 \tag{2.2}$$

式中 λ 为雷达波长,介电因子 K 为复折射指数 m 的函数:

$$|K|^2 \equiv \left|\frac{m^2-1}{m^2+2}\right|^2 \tag{2.3}$$

对于在 S 和 C 波段工作的天气雷达,许多云和降水粒子的粒径远小于雷达波长。例外情况包括大雨滴和正在融化的大冰晶,如冰雹和一些雪花聚集体。此外,许多水凝物不是球形的,这一重要事实将在第 2.3 节双偏振雷达中讨论。非球形粒子后向散射截面的近似公式可以通过修改瑞利散射近似得到,可在文献中查阅到[16]。

2.3 偏振

回想一下,电磁辐射的偏振描述了其电场矢量的方向。由于电磁波是横波,偏振方向总是垂直于波的传播方向。为了理解偏振和非球形粒子的重要性,再次假设一个由大量微小球体组成的电磁小粒子。这次,考虑一个六边形的平板晶体,它与它的主轴大致在水平方向上对齐

(图 2.3),该晶体是非球形的,因此它对入射辐射的响应取决于入射辐射的偏振。考虑一个水平偏振的入射波。它通过在每个小球体中感应一个按电场方向(在本例中,是水平方向)排列的偶极子来激励小球体。每个小球体的辐射情况如图 2.4 顶部所示。

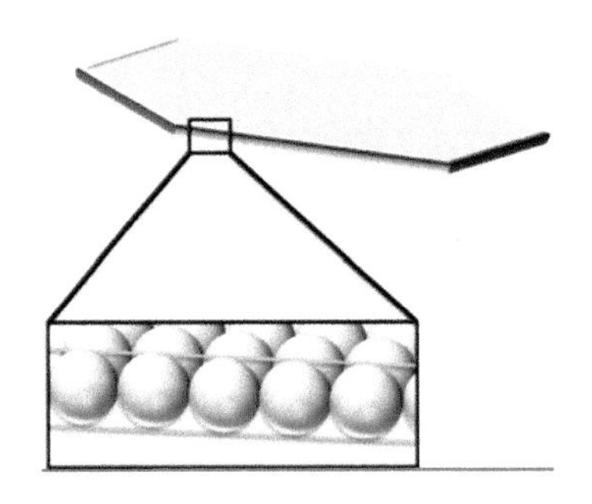

图 2.3 非球形粒子散射的概念模型。考虑一个六边形平板状冰晶,由大量的微小球体(如放大的部分所示)组成,当被电磁辐射照射时,它们表现为偶极子振荡器

虽然感应偶极子之间在相位上互相振荡,但每个小球体的激励电场也会影响其邻近的小球体(图 2.4 底部)。对于一个给定的小球体,其左邻和右邻区域内的电场(绿色)与其内部的感应电场(黑色)方向相同。因此,这些场正向叠加,一个球体的总内部场(紫色)就通过左右邻近球体被增强了。由于球体的内部电场增强了,从它散射出来的辐射也比孤立球体更强。因此,这些所谓的近场相互作用[16,101]也会增强冰晶在水平偏振条件下的总散射辐射。

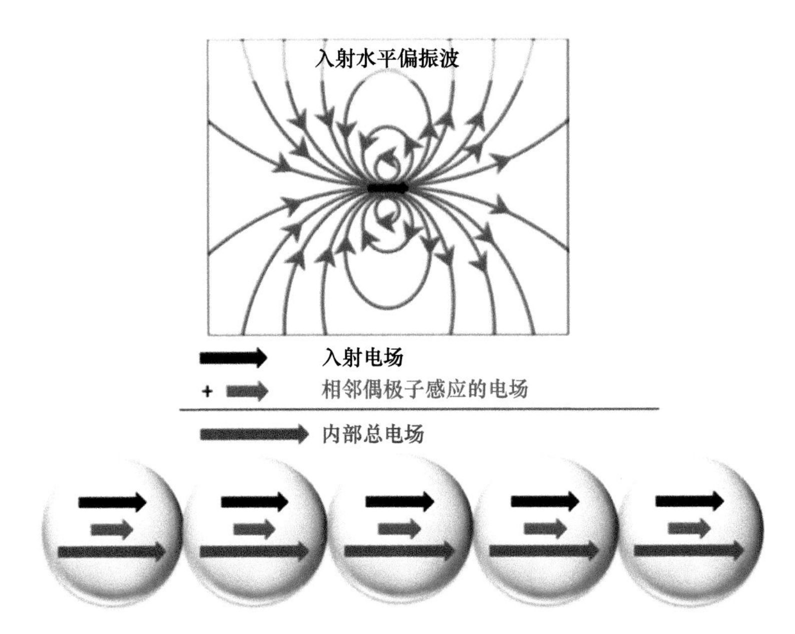

图 2.4 上图:由入射水平偏振电场(黑色矢量)激发的水平排列偶极子的电场线(绿色);下图:水平方向粒子的概念模型,该粒子由许多小球体组成(见图 2.3),被入射的水平偏振波辐照(黑色箭头表示入射电场)。每个小球体的辐射形式都如图 2.4 上部分所示。每个球内由其相邻球所感应的电场用绿色矢量表示。每个球激发的内部总电场矢量和是入射电场和相邻电场(黑色+绿色)的总和,用紫色表示。相邻球的电场增强了入射电场,激发出更强的内部总电场。改编自 Lu 等[101]

现在考虑同样的六边形晶体被垂直偏振波照射。同样,每个小球体都有一个感应偶极子,这次是垂直指向的(图 2.5 上部)。但是,现在一个给定球体左右相邻的电场(图 2.5 底部,绿色)与入射场(黑色)的方向相反。这就导致了相消干涉,使得存在相邻球体的情况下总内部场(紫色)相对于孤立球体减小。因此,这种球体散射的辐射就减少了,所以近场相互作用减少了垂直偏振时冰晶的总散射辐射。

从图 2.4 和图 2.5 可以看出,入射的水平偏振波激发的冰晶内电场比垂直偏振波强,因此来自六边形冰晶的散射波在水平偏振时的振幅比垂直偏振时的振幅大。这是雷达偏振背后的原理:非球形粒子将在不同的偏振类型下以不同的方式散射辐射,这一信息可以被用来更好地分析云和降水的特征。

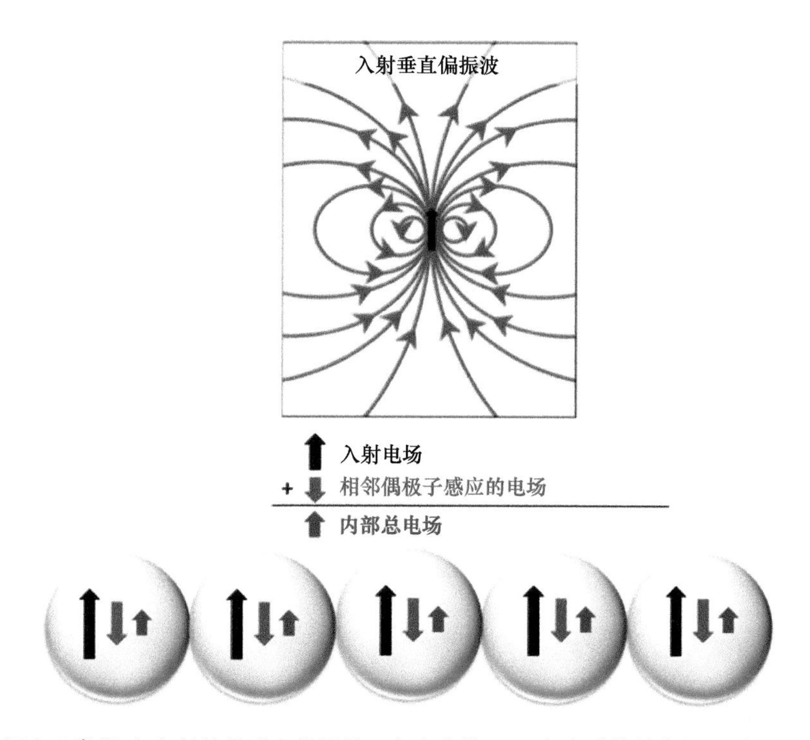

图 2.5　与图 2.4 类似,但入射的是垂直偏振波。在这种情况下,相邻球体的电场(绿色)作用方向相反,从而对入射场(黑色)产生相消干涉,由此产生的内部场(紫色)的振幅更小

考虑一个水凝物,它的长轴不再与入射偏振方向完全对齐(图 2.6)。对于入射的水平偏振波,每个小球体都有一个水平偏振的感应偶极子辐射模式,与前述一样。然而这一次,球体的指向使得它们经受到来自其相邻球体的具有垂直分量的电场。因此,每个球体激发的内部总电场在水平方向和垂直方向都有分量(图 2.6 中的紫色虚线)。因此,尽管入射波只有水平偏振,现在散射波既有水平偏振的分量,也有垂直偏振的分量。粒子在散射时改变辐射偏振的过程称为去偏振。一些天气雷达以一种偏振通道发射辐射,以两个正交偏振通道接收辐射,以测量云和降水导致的去偏振。去偏振现象也可以用来分析云和降水粒子的物理特性,这一点将在第 2.8 节中讨论。

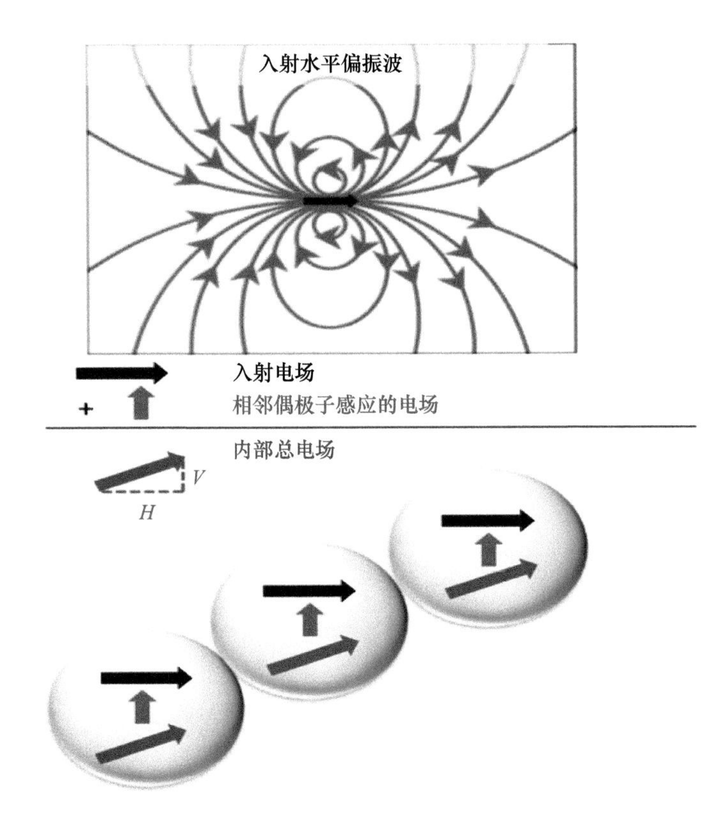

图 2.6　与图 2.4 和图 2.5 类似，但入射的是水平偏振波且粒子排列方向与入射偏振方向成一定角度，在这种情况下，来自邻近球体的电场有一个正交于入射波偏振方向的分量。由此产生的内部总场在水平偏振和垂直偏振方向都有分量。因此，尽管入射波只有水平偏振，但后向散射波在 H 和 V 偏振都有分量

2.4　弥散目标物

在大气中，水凝物通常不是孤立、单一的散射体。相反，它们大量分布在云内和周围。因此，当雷达波遇到这样一个群体时，它同时照射了许多粒子。在某一给定时刻被雷达探测的大气区域称为雷达采样体积。采样体积的尺寸取决于波束宽度(方位和仰角两个方向)。对于脉冲体制的雷达，采样体积的径向尺寸则由脉冲宽度决定。采样体积中的每个粒子都对电磁辐射进行散射，而总的后向散射信号是每个粒子产生散射波的总和。单位体积的后向散射截面之和称为雷达反射率 η：

$$\eta \equiv \int_0^{\infty} \sigma_b(D) N(D) \mathrm{d}D \tag{2.4}$$

式中 $N(D)\mathrm{d}D$ 为采样体积中等效球体直径 $D \sim D+\mathrm{d}D$ 范围内的粒子数浓度。因此，瑞利近似条件满足时式(2.4)可以写成：

$$\eta \approx \frac{\pi^5}{\lambda^4} |K|^2 Z \tag{2.5}$$

这里的雷达反射率因子(Z)定义为：

$$Z \equiv \int_0^\infty N(D) D^6 \mathrm{d}D \tag{2.6}$$

当然，只有提前知道目标物的特性，这些表达式才有实际作用。一般来说，采样体积中目标物的介电特性(K)是未知的。此外，瑞利近似是否有效通常是未知的。由于这些不确定性，定义了另一个量：等效雷达反射率因子(Z_e)，它假设散射体由小的球形液滴组成，其后向散射截面可以用瑞利近似描述。因此，

$$\eta \approx \frac{\pi^5}{\lambda^4} |K_w|^2 Z_e \tag{2.7}$$

其中 K_w 为液态水的介电因子。对于典型的天气雷达工作波长，$|K_w|^2 = 0.93$。注意，$Z = Z_e$ 只在目标是小液滴的情况下。天气雷达显示器上显示的 Z_e，通常以 10 的对数表示，单位为分贝(dB)或 dBZ：

$$Z_e[\mathrm{dBZ}] = 10\log_{10}\left[\frac{Z_e[\mathrm{mm^6\,m^{-3}}]}{1\mathrm{mm^6\,m^{-3}}}\right] \tag{2.8}$$

使用这个对数表达式是因为在云和降水中，Z_e 值可以跨越多个数量级，从非降水云的 $<10^{-2}$ mm^6 m^{-3}(<-20 dBZ)到带雹强对流风暴的 $>10^7$ mm^6 m^{-3}(>70 dBZ)。

每个粒子的散射子波能够对采样体积中其他粒子的散射子波进行相长或相消干涉。当粒子随着气流下降或移动时，它们在采样体积内的相对位置“重新排列”，这导致接收到的信号功率波动。因此，需要多个样本来估算平均 Z_e，且每个样本之间有足够的时间间隔，在统计上是相互独立的。对于一般大气条件下的业务天气雷达，充分重新采样和保证样本独立所需的时间大约为 $\leqslant 0.01$ s。通常情况下，使用 10 或 10^2 个样本获得一个较好的 Z_e 估计。足够数量的样本以及样本通过重新调整达到相互独立所需的时间共同决定了雷达对大气特定区域进行采样所需的驻留时间，从而限制了扫描的时间分辨率。但是，一些科研雷达采用先进的技术来克服这个问题，实现更快地采样。这包括通过一种有时被称为调频的技术[13,34,44,55,56,118]在较短的时间内发射频率稍有不同的脉冲，以在短时间内提供独立的样本。一些雷达使用多个频率以不同的固定仰角[172]几乎同时发射多个波束，使雷达能够一次对大气进行更大范围的扫描。还有一些雷达使用先进的信号处理技术，如数字波束赋形[64]，快速对大气的大片区域成像。这些技术进步极大地提高了天气雷达对灾害性风暴进行快速采样的能力。

尽管 Z_e 是最常用和最熟悉的量之一，但要将其校准到定量使用所需的精度仍是一项挑战。例如，为了定量估计降水的误差可接受，大多数业务天气雷达网要求 Z_e 的定标误差在 1 dB 以内[47,130]。Atlas[4]总结了许多用于雷达绝对定标的技术，包括使用已知后向散射截面的目标，如角反射器或由气球或飞机系留的金属球，以及与地面雨滴谱仪的测量结果进行比较。此外，雨中双偏振雷达变量之间的自洽性已被用于绝对雷达定标[45-47,63,130]。在实践中，对最优的定标方法尚无共识。

2.5 天气回波探测及测距

对于雷达来说，要在给定的范围内探测到云或降水回波，接收到的信号必须足够强，能够

从噪声中分辨出来。雷达接收到的从目标返回的功率取决于电磁波在往返目标物之间传播路径的特性、雷达系统(如天线)的特性以及目标物本身的特性。这些都体现在雷达方程中;标准雷达说明文档[7,20,34,37,126]中可以找到不同形式的雷达方程及其推导过程。雷达方程很重要,因为它使得等效反射率因子 Z_e 能够根据雷达接收的返回功率进行估算。例如,可以通过增加雷达的发射功率、增加发射脉冲的持续时间和/或增加天线反射面的尺寸(这有助于聚焦发射波束并收集接收到的信号)来增加返回信号的强度。另外,如果目标物将更多的辐射散射回雷达和/或它们距离雷达的位置更近,更大的功率会被雷达接收。因为辐射强度随 $1/r^2$(其中 r 是雷达到辐射源的距离)成比例减小。

许多现代天气雷达是脉冲多普勒雷达系统,通常每秒发射 $10^2\sim10^3$ 个脉冲,脉冲持续时间就是脉冲长度,通常是微秒级,这就产生了几百米量级的脉冲宽度。这些脉冲的传输速率称为脉冲重复频率(PRF)。反之,由 PRF 可以得到脉冲重复时间(PRT)或连续脉冲时间间隔。发射的电磁波以光速 c 在大气中传播,因此,到目标物的距离,或雷达探测距离(r),可由脉冲发出并返回雷达所需的时间 Δt 决定:

$$r=\frac{c\Delta t}{2} \tag{2.9}$$

上式中,系数 2 是因为脉冲必须到达目标物然后返回。雷达每个 PRT 发射一次脉冲,这就限制了下一个脉冲发射之前当前脉冲可以往返的最大距离:$r_{max}=c\times PRT/2$。如果目标位于 r_{max}之外,则雷达接收到该目标散射回的信号是在下一个脉冲发射后。在这种情况下,雷达会认为回波来自散射第二个脉冲的目标物,结果致使目标物的真实距离被混淆或折叠,并被错误地指出位于 0 到 r_{max}之间(图 2.7)。现代天气雷达使用更先进的技术来减轻这种所谓的距离模糊[10,159,160]。

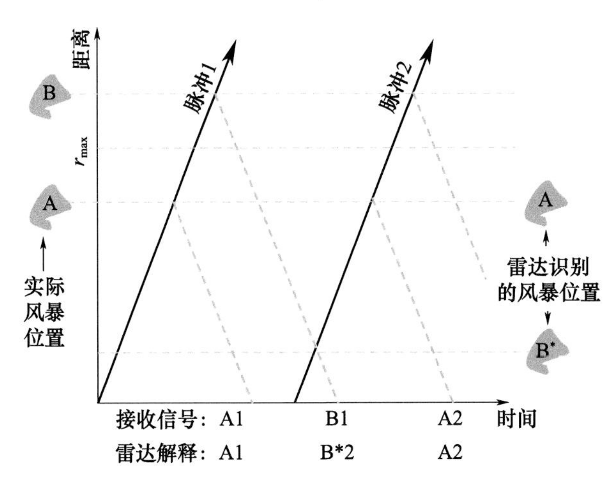

图 2.7 距离模糊的概念模型。雷达在初始时间发射脉冲 1(倾斜的黑色箭头)。有两个风暴,A 和 B,位于图的左边,最大不模糊距离(r_{max})显示为黑色虚线,来自风暴 A 和 B 的后向散射信号显示为灰色虚线。图的最下面一行显示了实际接收到这些后向散射信号的时间,如其中"A1"代表来自风暴 A 的脉冲 1 的信号。然而,由于风暴 B 位于 r_{max}之外,雷达将其回波解释为来自脉冲 2 的一个更近距离(相当于风暴 B 的实际距离减去 r_{max})的风暴 B* 的回波。因此,风暴 B 的距离被称为"混淆"或"折叠"

许多天气雷达在监测扫描模式下工作，在这种模式下，雷达天线被设置成固定仰角并按方位角旋转。通常，数据从“鸟瞰”视角显示，以往称为平面位置显示或 PPI。与 PPI 同义的更现代的术语包括“sweep”(扫描)和“surveillance scan”(监测扫描)。在这种监测扫描模式下，雷达像素位置由距离和方位坐标指定，这些坐标有时被插值到笛卡尔网格上(图 2.8a)。业务扫描方式通常包括几个在不同的仰角上的监测扫描，称为体积覆盖模式或 VCP。例如，美国国家气象局用于降水监测的典型 VCP 在 0.5°和 19.5°之间有 10～15 个仰角。VCP 的数据可以插值到某一高度，即为等高平面位置显示(CAPPI)。

如果雷达天线以方位角保持不变，而仰角变化的方式扫描，雷达可以采集穿过云和降水的垂直截面数据，称为距离-高度显示(RHI)(图 2.8b)，其中雷达像素位置由距离和高度指定。一些雷达垂直指向，当云和降水移动到雷达所在位置时采集样本，结果显示在时间-高度图上(图 2.8c)，这种天顶指向雷达一般有一根固定的天线或者它可能在垂直指向时旋转方位角，后者通常用于校准目的。因为指向天顶的抛物面天线的反射面像一个碗或水盆(bird－bath)，这种扫描被通俗地称为“水盆”扫描。雷达气象学家可以通过将一种或多种这些扫描结合到一个扫描模式中，以从不同视角观测云和降水。

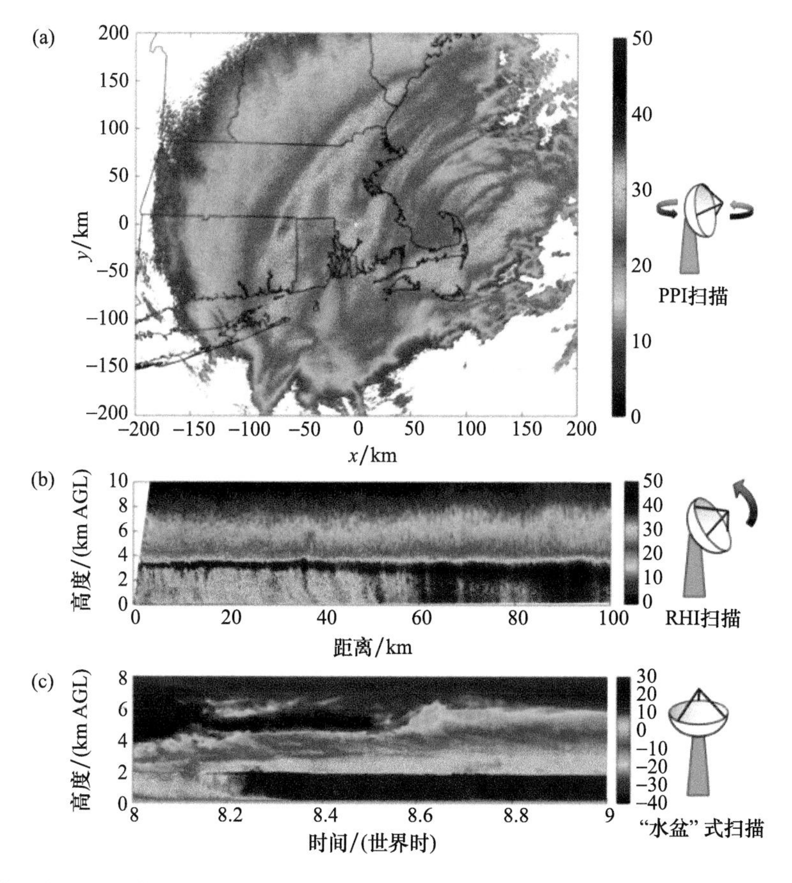

图 2.8 各种类型的雷达扫描显示。(a)雷达天线处于固定仰角和变化方位角的水平偏振等效反射率因子(Z_H，单位为 dBZ，根据色标着色)的 PPI 或监视扫描，雷达位于坐标系的原点；(b)雷达天线固定方位角和变化仰角的 Z_H(dBZ，根据色标着色)的 RHI 扫描；(c)雷达天线垂直指向的 Z_e(dBZ，根据色标着色)天顶指向或“水盆”式扫描，有时，天线在垂直指向时旋转方位角(右侧天线扫描方式的卡通图)

2.6 多普勒效应

任何一个听到列车鸣笛或救护车警笛的人都体会过多普勒效应：当鸣笛或警笛的源头向人移动的时候，音调会发生变化。这种音调的变化表明声波频率的变化：声源移向观察者时频率增加，远离观察者时频率下降。

为了从电磁波散射的角度理解多普勒效应，这里引入一个概念模型[34]：考虑一只静止的昆虫，它将在 t_0 时刻被一个发射的脉冲照射（图 2.9a 左）；在某个短暂的瞬间之后（$t_0+\delta t$；图 2.9 右），昆虫已经被照射了几个电磁波周期（图中是 2 个）；这种时变的电磁场迫使昆虫中的极性分子以相同的频率振动，以相同的频率向雷达散射辐射。在这种情况下，没有多普勒频移。

现在考虑一只飞向雷达的昆虫（图 2.9b 左）。相对于静止的昆虫，飞向雷达的昆虫“经历”了更快速的电磁场振荡：小虫在同一时间段 δt 内经历了 3 个完整的周期（图 2.9b 右），在给定的时间内，更多的周期意味着更高的频率，因为虫子每秒被照射的周期更多，它的分子被迫以更高的频率振动，从而将更高频率的辐射散射回雷达。这个频移决定于目标物在给定时间内移动的距离或其径向速度 v_r，反比于雷达波完成一个周期的距离，即雷达波长 λ。因此多普勒频移为 $f_D=2v_r/\lambda$，这里的系数 2 来自于：(1)昆虫以 v_r 接近雷达时每秒经受更多的发射波周期；(2)以移动的昆虫为参考系，雷达以速度 v_r 接近昆虫，因此每秒经历了更多的散射波周期。所以，散射信号相对于发射波有频率偏移，这就是多普勒频移。目标移向雷达导致频率增加，而远离雷达导致频率下降。

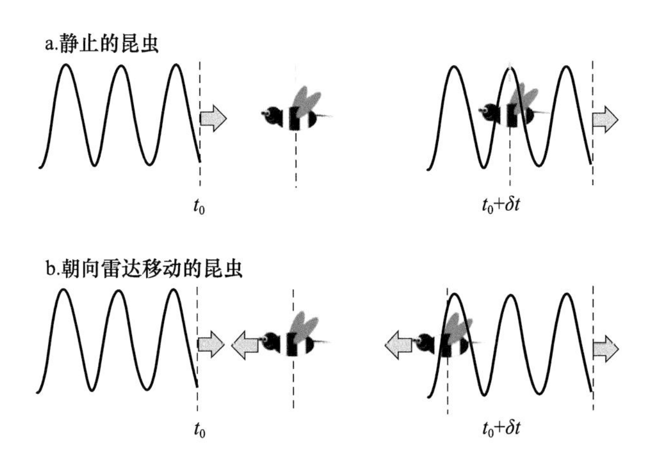

图 2.9 在电磁散射的背景下，多普勒频移背后的物理机制示意图（图中包含两个时间：左列 t_0 和稍后的一个右列 $t_0+\delta t$）(a)中的波向昆虫传播并照射它时，昆虫是静止的；(b)中虫子正在向波移动

由于这种频移，频移后的散射波将在某一点不再与发射波同步。因此，多普勒频移后的波将以每秒 $2\pi f_D$ 弧度的速率积累相位偏移（相对于发射波）。通过跟踪相位偏移随时间的变化，雷达原则上可以估计出多普勒频移。然而，气象应用中的多普勒频移太小，无法在单个脉冲的持续时间内测量。例如，假设一个气象目标以 $v_r=10$ m/s 的速度移动。对于一个 10 cm

波长(S 波段)的雷达来说，多普勒频移是 $f_D=200$ Hz。在常用的 1 μs 脉冲持续时间内，产生的相移只有 0.072°。所以，相移的测量是从一个脉冲到下一个脉冲，或者每一 PRT 时间间隔。许多现代天气雷达的 PRT 大约是 10^{-3} s。这个时间要相对长得多，所以目标物能够移动一定的距离，并产生可测量的相移。

但是，就像雷达测量云和降水的最大范围是有限的一样，脉冲之间也有一个可探测的最大径向速度，称为 Nyquist 速度。在脉冲间隔时间内，如果目标物移向雷达的距离跨越了半个波长，那么雷达就不能将这种情况和远离雷达半个波长的情况进行区分。换句话说，连续发射脉冲时雷达所能探测到的最大速度与产生 ±180°相移的速度相等，这个就是最大不模糊速度 $v_{\max}=\pm\lambda/(4\times \mathrm{PRT})$，正负两个符号都是可能的，因为多普勒速度可以是正的或负的。对于天气雷达，习惯上认为目标物接近雷达时为负速度，远离雷达时为正速度。如果目标的真实径向速度在量级上超过 $v_{\max}$，则显示出的速度是经过折叠的(图 2.10)。

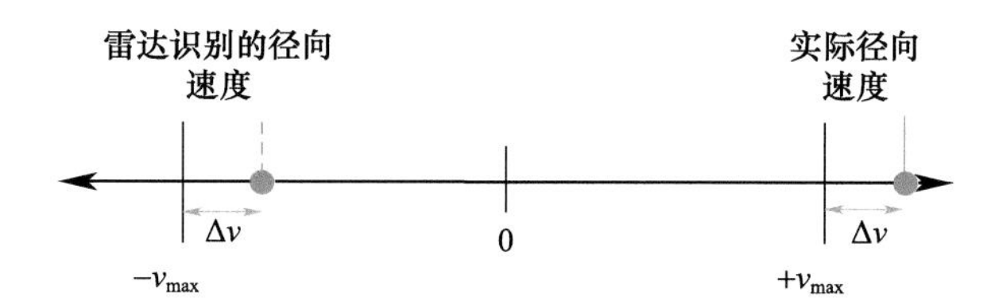

图 2.10 多普勒速度模糊的图解。如果真实径向速度(灰线)超过 Nyquist 速度的大小(± $v_{\max}$)，则雷达识别出的径向速度如灰色虚线所示。注意，在这种情况下，折叠后的多普勒速度与实际径向速度的符号相反

图 2.11 给出了由移动式 X 波段多普勒雷达观测的一个龙卷中多普勒速度模糊的示例。这张图片展现了龙卷中心一个清晰的低 Z_H 眼，它由正在经历离心运动的碎片和降水粒子产生。因为龙卷中的粒子在接近或远离雷达时的速度超过了 Nyquist 速度(在这种情况下约为 ±55 m/s)，所以产生了速度折叠。由图可见，龙卷东北侧有一个从大的正速度(远离雷达)到大的负速度(移向雷达)的突变，以及在西南侧有一个较小范围的从大的负速度(移向雷达)到大的正速度(远离雷达)的突变。

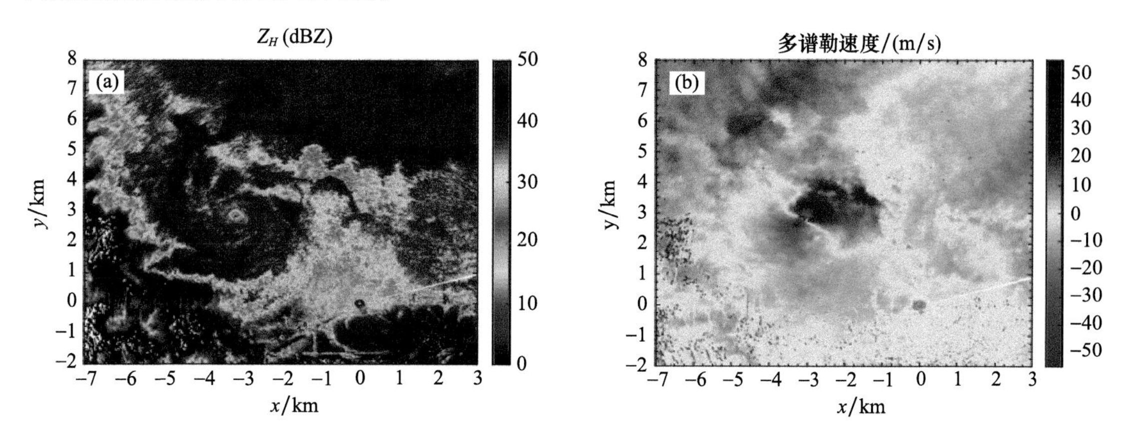

图 2.11 龙卷中多普勒速度模糊的示例(a) Z_H(dBZ，根据色标着色)和(b)多普勒速度(根据色标着色)的监测扫描图。由 X 波段移动式多普勒雷达于 2013 年 5 月 19 日 0045UTC 收集的数据，由 Karen Kosiba 博士(灾害天气研究中心)提供。正速度(红色)表示向外，负速度(蓝色)表示向内。雷达位于 $x=0$ km，$y=0$ km，龙卷位于 $x=-3$ km，$y=3$ km

需要注意的是 PRT 同时存在于 r_{max} 和 v_{max} 的表达式中：较大的 PRT（表示脉冲之间的时间间隔较长）导致较大的 r_{max} 而较小的 v_{max}。这就是所谓的多普勒两难，可以在数学上表示为 $|v_{max}| \times r_{max} = c\lambda/8 =$ 常数。现代雷达系统使用各种技术来减少这些模糊，包括使用多个或交错的 PRT 对云和降水进行采样[10,159,160]。

回忆前面讲到的：雷达接收的信号是采样体积内所有散射点的贡献之和。因为采样体积通常比雷达波长大得多（几十到几百米对比降水雷达几厘米的波长），所以在给定的位置或时间测量所有信号的相位对气象目标来说是没有意义的。接收信号相位的脉冲间变化与采样体积内粒子的平均径向速度有关。请注意，静止地物杂波目标出现的较长时间尺度的相位变化可能与不断变化的大气条件有关，因为大气条件的变化影响大气折射指数或折射率[14,38]。一般来说，雷达采样体积中的水凝物可能会以不同的径向速度移动，因为它们的下落速度、采样体积内的风切变、湍流等不同。总的后向散射信号是每个粒子的散射子波之和，因此它可能包含一个多普勒频移谱。通过对接收信号的时间序列进行傅里叶分析，可以恢复这个频移谱，称为多普勒谱。

多普勒谱表示随多普勒速度的变化，具有相同多普勒速度的散射体对总信号功率贡献的多少。对整个谱的积分就是总功率，与 Z_e 相关。这个谱的功率加权平均值是平均多普勒速度：在雷达图形显示中，它被分配给一个像素。采样体积内多普勒速度的变化由多普勒谱宽表征。例如，宽多普勒谱可能与风暴中的强湍流和/或增强的风切变区域相关联，并可用于识别边界层辐合带和龙卷[150]，也有研究使用多普勒谱信息来估计云中的湍流和夹卷[1,74,109]。最近一些对垂直指向云雷达的研究也探索了其他谱参量，如偏度和峰度及其与微物理过程的关系[43,75,103,161]。

图 2.12 给出了垂直指向雷达在层状降水期间的观测个例，时间-高度图显示 Z_e 在 2000 m 以下有明显增强（图 2.12a），并伴随着平均多普勒速度的增加（图 2.12b）。这标志着从散射较弱、下落较慢的雪融化为散射较强、下落较快的雨，这种转变也可以从多普勒谱宽的增加中得到体现（图 2.12c），且表明雨滴的下落速度相比融化层上方的雪和冰粒子有更大的变化。图中红色点标记处的多普勒谱见图 2.12d。由此图可知，红色点处返回雷达的大部分能量来自于多普勒速度在 −2～0 m/s 之间的粒子。灰色虚线表示功率加权平均多普勒速度（−1.2 m/s），灰色箭头表示两倍的多普勒谱宽（0.5 m/s）。总之，多普勒谱可以提供非常详细的关于云和降水的信息。

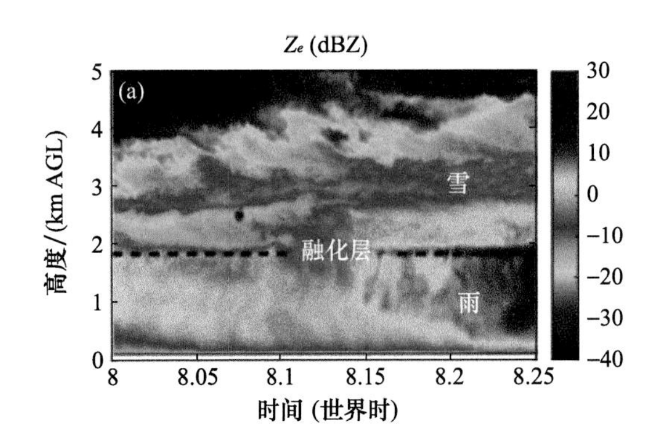

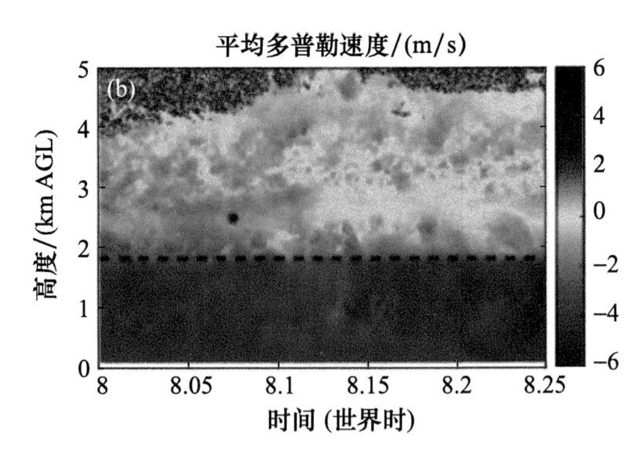

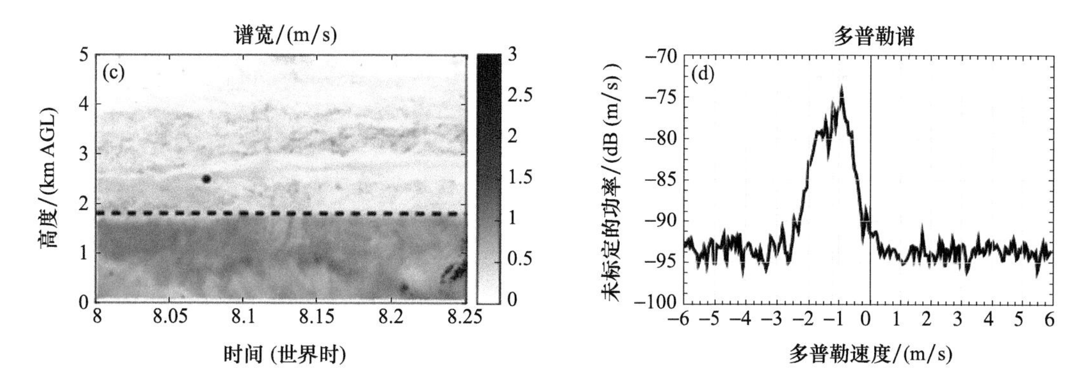

图 2.12 大气辐射测量计划(ARM)位于俄克拉何马州中北部的 Ka 波段垂直指向雷达于 2011 年 4 月 27 日观测的一次层状降水个例。时间－高度图分别显示(a)等效反射率因子 Z_e(dBZ)；(b)平均多普勒速度 v_r(m/s)；(c)谱宽(m/s)。图(d)为在(a)－(c)上洋红和黑色圆圈所示的点上的多普勒谱。垂直的灰色虚线表示平均多普勒速度(1.2 m/s)，双向箭头表示多普勒谱宽(0.5 m/s)的两倍，负多普勒速度表明水凝物正在下落

2.7 雷达系统

虽然雷达技术在不断改进，但现代天气雷达的许多组成部分在几十年里基本保持不变。图 2.13 是一个脉冲多普勒雷达系统的简化框图；更多的技术处理方法参见标准的雷达工程文献[20,34,147]。雷达信号的产生始于稳定本机振荡器(STALO)和相干振荡器(COHO)，前者产生频率为 f_S 的纯正弦信号，后者产生频率为 f_C 的纯正弦信号并与 STALO 信号锁相。COHO 的频率 f_C 通常比雷达的工作频率低得多，例如，对于许多 S 波段雷达(2～4 GHz)，f_C 约为 30～60 MHz，这个较低的频率称为中频(IF)。雷达工作频率 f_0(有时称为载波频率)通常取 COHO 和 STALO 频率之和($f_0=f_S+f_C$)，由 COHO 和 STALO 信号混合得到。图 2.13 中，由圆圈包围着×的符号表示混频器。

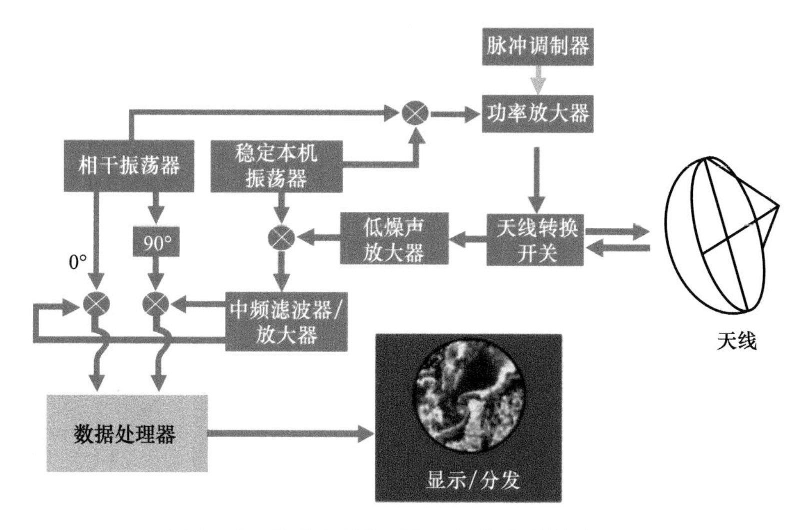

图 2.13 脉冲多普勒雷达主要部件的简化框图

对这个连续波信号的调整在脉冲调制器中进行，它控制信号的时间和宽度以产生一系列脉冲。随后，功率放大器用于增加这些脉冲的强度。在许多现代天气雷达中，脉冲被速调管放大。速调管本质上是一种电子枪，通过施加振荡电压将电子“聚集”在一起。有时候，另一种用于天气雷达的功率放大器是行波管(TWT)。在过去，许多雷达使用磁控管功率振荡器。对于给定的输出功率，磁控管更小、更便宜，但产生的信号是随机相位的。这意味着信号相位必须进行逐个脉冲追踪，以便在估计多普勒速度时有一个参考系。最近，成本更低的固态发射机越来越受欢迎，对发射机类型的选择涉及到功率、成本和性能之间的权衡。

脉冲被放大后，被送到天线转换开关。天线转换开关的重要工作在于发射时向天线发送高功率信号，而其他时间让天线接收的低功率信号进入接收链路。如果高功率发射信号泄漏到接收链路中的敏感电子元件中，它们可能会受到严重损坏或破坏。当发射时，脉冲信号被从天线转换开关传送到雷达天线。雷达天线用于传输、聚焦、发射和接收辐射。许多现代天气雷达使用抛物面反射器，它将辐射聚焦成一个狭窄的波束(通常约1°或更少是理想的)。辐射通过位于抛物面反射器焦点处的馈源喇叭送入抛物面反射器。这个馈源喇叭通常由一个或多个支柱固定。其他几何形状的天线用于云和降水雷达，包括CSU－CHILL雷达的双偏置格里高利天线[69]。最近的研究集中发展相控阵雷达技术，使用平面或圆柱形结构的小型微波发射/接收元件阵列[176]。这种相控阵雷达的优点是可通过控制每个发射元的相位差以电子方式引导或控制波束，而不需要移动反射器来引导电磁辐射[178]。然而，与使用传统反射面天线的雷达相比，这种雷达相当昂贵。

对于单站雷达，向大气发射辐射的天线通常也用于接收来自云、降水或其他目标物的后向散射信号，接收到的信号通过天线转换开关发送回来，进入接收链路。接收链路的一个主要作用是放大所需的信号并滤除不需要的噪声。请注意，接收信号的振幅往往比发射信号弱许多数量级，因此需要一个低噪声放大器。接收信号的频率 f_R 等于发射信号的频率与任意多普勒频移之和：$f_R = f_0 \pm f_D = f_S + f_C \pm f_D$。对于接收频率低于发射频率(即低于 f_0)的信号，天气雷达中电子器件的工作效率更高。因此，放大的接收信号与来自STALO的基准信号混合。这一过程称为差拍变频，由此产生的信号被“降频变换”至频率 $f_C \pm f_D$，也就是中频加上任意多普勒频移。一个中频滤波器(有时称为匹配滤波器)被用于隔离这种被进一步放大的降频变换信号。

为了分离多普勒频率，中频信号(频率 $f_C \pm f_D$)再次混合在相位检波器，这时有两个来自COHO频率为 f_C 的信号：一个相对发射信号0°相移，另一个90°相移。回顾上文，可以利用脉冲与脉冲之间信号相位的差异反演多普勒速度。需要同相(0°)和正交(90°)信号来确定目标是否移向或远离雷达。然后，这些信号被发送到数据处理器。在这里得到气象、水文在业务和科研中需要的传统雷达参量。最后，这些参量被输出用于分发和/或显示。

2.8 双偏振雷达变量

世界各地的业务天气雷达网已经或将要升级至双偏振，即天气雷达在两个相互正交的偏振方向发射和/或接收电磁波。这一部分归因于具备双偏振能力的雷达能够获得丰富的信息。关于双偏振雷达原理和应用的更详细的综述可以在文献中找到[177,77-79,26]。双偏振(dual－polarization)或测定偏振(polarimetric)雷达(注意：“双测定偏振的(dual-polarimetric)”是冗余的，因此是不正确的术语）发射水平（H）和垂直（V）偏振波，可以同时或交替。在 H 偏振波和 V 偏振波同时发射和接收模式（STAR）中，电场矢量一般在垂直于波传播方向的平面上沿椭圆运动。假设 H 偏振和 V 偏振信号的振幅相等，如果雷达发射的 H 和 V 偏振波之间的相位差为 $\pm 90°$，电场矢量的轨迹就形成一个圆，而这个波就称为圆偏振波；如果相位差是 $0°$或$\pm 180°$，电场矢量在 H 和 V 偏振方向之间（即 45°或 135°）沿一条倾斜的线运动，有时这被称为倾斜45°线偏振。在交替发射和接收模式中，一次发射一个偏振波，同时接收两个偏振波，这种模式存在多种配置：只发射 H 偏振波、只发射 V 偏振波，或者 H 和 V 偏振波交替发射（尽管仍然是一次一个）。大多数现有的业务天气雷达网使用 STAR 模式，这两种运行模式的优缺点在文献中进行了讨论[35,60-62,179]。

通过比较两个偏振方向的后向散射辐射，可以获得水凝物的形状和类型的信息。这一信息包含在偏振雷达变量中，将在下面的小节中单独讨论，更全面的介绍参见文献[20,34,77]。

2.8.1 差分反射率 Z_{DR}

第 2.3 节介绍了非球形粒子如何因照射辐射的偏振方向不同而产生不同的散射辐射。回想一下，Z_e与雷达采样体积内水凝物的后向散射功率成正比，通过比较不同偏振方向的等效雷达反射率因子，可以获得关于粒子形状的信息。差分反射率 Z_{DR}被 Seliga 和 Bringi[142]首次引入，定义为 H 与 V 偏振方向的等效雷达反射率因子（Z_H 和 Z_V）之比，Z_H 和 Z_V 单位为 mm^6/m^3。当 Z_H 和 Z_V 单位为 dBZ 时，$Z_{DR}=Z_H-Z_V$，因此两个偏振方向散射辐射相同的粒子 $Z_{DR}=0$ dB。对于具备电磁性质的小粒子，水平方向上质量较大的粒子 $Z_{DR}>0$ dB，垂直方向上质量较大的粒子 $Z_{DR}<0$ dB。如第 2.3 节所述，这是因为粒子中微小的有限散射元之间的近场相互作用相长干涉增强了一个偏振方向的内部总电场，而相消干涉减小了另一个偏振方向的总内电场。例如，以枝状和平板状冰晶为代表的高度非球形水凝物的最大尺寸方向在下落时倾向于水平，能够产生非常大的 Z_{DR}值（极端纵横比情况下>6 dB）。相反，锥形霰粒子的最大尺寸方向在下落时倾向于垂直，则产生负的值 Z_{DR}。随着雨滴尺寸的增大，雨滴的形变在空气阻力作用下越来越严重，从而导致正值 Z_{DR}。对于球形粒子或那些随机翻转的粒子，H 和 V 偏振方向的散射辐射相等，导致 $Z_{DR}=0$ dB。在 S 波段，不同类型降水的典型取值范围可参考文献[77]。

对于给定的粒子形状，增加相对介电常数会增大粒子内部的激励电场，从而增强微小散射元的近场相互作用，致使 Z_{DR} 增大。例如，非球形雨滴的 Z_{DR} 比相同大小和形状的冰粒子的 Z_{DR}大。以松散的雪聚集体（在科学文献中有时以相对介电常数非常低的球形或球状粒子表征）为代表的质量高度分散的粒子，具有非常弱的相长/相消近场相互作用，因此 Z_{DR} 接近于 0 dB。事实就是如此，虽然雪聚集体往往具有高度非球形和不规则的形状[67]。雪聚集体在下落过程中的倾角也有很大的变化，这也导致 Z_{DR}接近于 0。

如果具备电磁性质的粒子较大，Z_{DR}的解译会因为共振散射效应而变复杂，因为它不再直接与粒子形状有关。对于降水雷达，出现大冰雹时尤其需要注意，例如，在 S 波段进行的计算[86]显示最大尺寸>5 cm 的一定取向的湿椭球冰雹的 Z_{DR}值为负。然而，考虑到大冰雹往往会翻滚，形状效应往往会在采样体积中被平均掉，导致观测到的 Z_{DR}值通常接近于 0 dB。Z_{DR}对雷达采样体积内的水凝物信号进行了以 Z_e为权重的加权处理，因此，观测到的 Z_{DR}偏向于代表

那些对整体 Z_e 贡献最大的粒子。对于采样体积内不同粒子类型的粒子群，观测到的 Z_{DR} 可能介于每个粒子群的“本征”Z_{DR} 之间，这取决于每个粒子群对总 Z_e 的贡献。

2.8.2 差分相移 Φ_{DP}

相对于在真空中传播，电磁波在相同的距离内穿过介质如液体或冰晶时，它的波会产生相移。这是因为当波通过液体或冰晶(折射率大于 1.0)时，相速度会降低；等效地，电磁辐射的波长在这样的介质中减小。因此，相对于在介质外传播，介质内传播的波在同样距离上会经历更多的以 360°为周期的相位震荡，这就导致了介质内传播的波会相对于真空中传播产生相移。在概念上，这个问题通常被抽象为一种介电材料平板，该平板在与波传播方向正交的方向上无限延伸，而在波的传播方向上是有限厚度。在这种“无限平板”模型中，穿过材料的波相对于真空中传播，在相同距离内出现了相移。

对于粒径比波长小的粒子，可以用粒子充塞的平板代替介电材料平板达到相同的效果(图 2.14；也可参见文献[16])。然而，这在物理机制上是不同的。考虑用雷达脉冲对这个粒子平板进行采样，假设雷达采样体积沿径向比雷达波长延伸得更远(注意图 2.14 中没有显示这一点)。因此，根据粒子在采样体积内位置的不同，粒子中微小散射元的偶极子振荡将被入射波相位周期的不同部分激发。Bohren 和 Huffman[16] 表明，对于均匀充塞水凝物的采样体积，平板外一定距离处前向散射波的预期平均相移相对于自由传播的波为 $90°+\theta$，其中 θ 是入射波与微小散射元自身振荡之间的相移。对于天气雷达，θ 很小，此处忽略不计。在图 2.14 中，前向散射波用彩色细线表示，真空中传播的入射波用黑色粗线表示。合成的波是采样体积中被照射的每个粒子产生的前向相移散射波与原始发射波的总和(图 2.14 中的彩色虚线)。因此，相对于在自由空间传播的发射波，总传播相移在 0°和 90°之间，这取决于前向散射子波的贡献。反过来，这些小波的贡献取决于采样体积中粒子的大小和浓度。

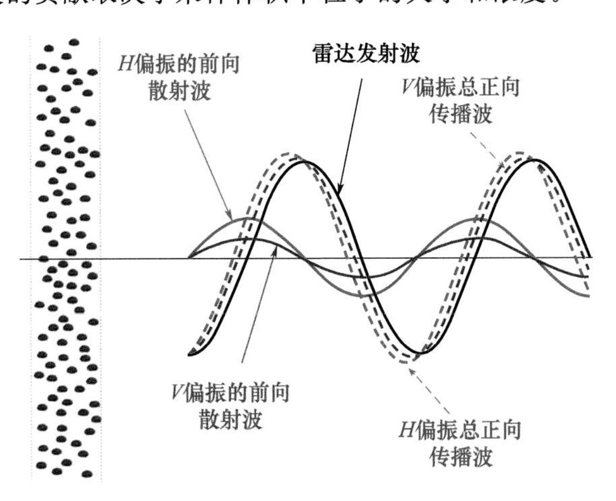

图 2.14 概念图显示波通过一片由相同雨滴(每一个都假设为比波长小的具备电磁性质的粒子)组成的无限长平板。实黑线是雷达发射的原始波。蓝色和橙色的细线分别代表了 H 偏振和 V 偏振的前向散射波，它们产生于这片雨滴，相移值为 90°。注意，H 偏振波的振幅更大，因为雨滴是椭球且按其最大尺寸水平排列。总正向传播波是这两种波的和，如虚线所示。需要注意的是，合成的 H 偏振波(蓝色虚线)比合成的 V 偏振波(橙色虚线)有更大的相位滞后；这一差异为 Φ_{DP} 的 1/2。该图源自 Bohren 和 Huffman[16]，进行了一些修改以表现偏振关系

对于具有电磁性质的小粒子，预期 H 和 V 偏振方向的总传播相移之间没有差异，除非 H 和 V 偏振方向的这些粒子前向散射的振幅存在差异，会导致各自方向总传播波相移的不同。这只会发生在非球形粒子的情况下，是因为第 3 节所述的相长/相消近场相互作用而产生的。注意，对于具有电磁性质的小粒子，前向和后向散射的振幅是相同的，因此上面用于后向散射的结论也适用于前向散射。因此，对于水平排列的椭球粒子（如雨滴），相比于 V 偏振的情况，H 偏振前向散射波对总前向传播波的贡献更大。于是，比较两个偏振方向前向散射波对总传播相移的影响，H 偏振的更大，导致 H 偏振相移比 V 偏振更大（如图 2.14 中的蓝色和橙色点状曲线）。该相位差为传播差分相移 Φ_{DP} 的一半；因为所测得的 Φ_{DP} 来自于采样体积范围内往返的双向传播。

Seliga 和 Bringi[143]、Jameson[65]、Sachidananda 和 Zrnic[136] 主张使用 Φ_{DP} 来估算降雨。对于水平排列的具有电磁性质的小粒子（如较长波长下的雨滴），Φ_{DP} 随距离呈单调增加趋势。如前所述，Φ_{DP} 的大小随着非球形粒子的浓度、尺度和相对介电常数的增加而增加。然而，与 Z_H 或 Z_V 不同，Φ_{DP} 不受球形粒子或其他在 H 和 V 偏振方向散射性质相同的粒子的影响。这使得 Φ_{DP} 对于识别混合的降水粒子类型非常有用，例如雨与冰雹混合或小冰晶粒子与雪聚集体/霰混合。后一种情况下，Z_{DR} 可能由较大的雪聚集体或霰主导，导致 Z_{DR} 接近 0 dB。但是，这些较大的各向同性散射粒子对 Φ_{DP} 贡献不大，而较小的非球形冰晶对 Φ_{DP} 贡献很大，它们导致了 Z_{DR} 在接近 0 dB 时[81,140] Φ_{DP} 增加。

气象学中经常使用 Φ_{DP} 距离微分的一半：差分传播相移 K_{DP}，它表示单位径向距离内的相移，可以更容易地识别含有非球形粒子（例如雨）的强降水区域；可以看作是采样体积中非球形粒子数量和/或大小的度量。由于 Φ_{DP} 和 K_{DP} 的值是基于相位测量而不是功率，所以不受衰减或差分衰减的影响，除非雷达信号完全消失，因此这些变量经常被用于估计和校正衰减[18,22,49,133,134,148,154]。

在实际操作中，雷达测量的总差分相移 Ψ_{DP} 是由传播差分相移 Φ_{DP}、由雷达硬件系统引起的发射波的差分相位（有时称为“系统差分相位”）和后向散射造成的差分相移（称为后向散射差分相位 δ）组合而成。这种后向散射差分相位通常来自于非球形的电磁大粒子。近期有研究关注在雨和融雪中 δ 包含的微物理信息[162,163]。

2.8.3　同偏振相关系数 ρ_{HV}

同偏振相关系数 ρ_{HV} 是在 H 和 V 偏振通道接收到的相同偏振方向的信号（即发射 H 且接收 H；发射 V 且接收 V）之间的相关性。它可以被认为是雷达采样体积内粒子变异性的一个度量，值为 1.0 表示性质完全相同的降水，低于 1.0 表示粒子存在多样性。具体来说，ρ_{HV} 随着采样体积内粒子形状、倾角和/或相对介电常数的不均一性增加而降低。考虑 ρ_{HV} 最简单的方法是将其作为雷达采样体积内粒子本征 Z_{DR} 变化的度量：如果采样体积内粒子本征 Z_{DR} 值的分布很窄，那么 ρ_{HV} 接近 1.0。当粒子在采样体积内的 Z_{DR} 变化剧烈时，ρ_{HV} 降低。此外，当采样体积内的差分相移 Ψ_{DP} 显著变化时，ρ_{HV} 减小。这可能是由于采样体积中存在产生 δ 的粒子，或采样体积内存在 Φ_{DP} 梯度（例如，当波束不是均匀填充时可参考文献[79,128]）。这种不均匀的波束填充在对流风暴中是常见的，呈现出放射条纹状的 ρ_{HV} 低值区由强降水内核向外延伸。

在 S 波段，ρ_{HV}的值在均匀降水中常高（>0.98），例如纯雨或纯雪聚集体。对于小冰晶与大雪花聚集体混合的情况，粒子形状多样性的增加导致 ρ_{HV}略减小（>0.95）[77]。由于 5～6 mm 雨滴在 C 波段的共振散射效应对 δ 有贡献，C 波段降雨中的 ρ_{HV}可降至 0.93。相比之下，在融化的雪和冰雹中，ρ_{HV}可能低得多（<0.85）。对于昆虫、鸟类、烟雾和火山灰、军用箔片和龙卷碎片等非气象散射物，ρ_{HV}可能极低（<0.7）。因此，ρ_{HV}经常用于业务天气雷达数据中区分降水和非降水回波，并用于探测融化层[41,42]。此外，它还有助于识别正在发生的龙卷[15,82,129,141,164]，这将在下一节中讨论。

2.8.4 线性退偏振比 LDR

双偏振天气雷达通过交替发射和接收 H 和 V 偏振信号（如发射 H，接收 H 和 V；发射 V，接收 H 和 V 等）可以测量降水中发生的退偏振量。如果雷达在 H 偏振通道发射辐射，则 V 偏振方向上接收信号的功率称为交叉偏振功率，反映到等效雷达反射率因子上，一般写作 Z_{VH}（标准表达：第二个下标表示发射信号偏振，第一个下标表示接收信号偏振）。Z_{VH} 与 Z_H 的比值（如果以对数单位表示，则为它们的差 $Z_{VH}-Z_H$）被称为线性退偏振比或 L_{DR}。由于后向散射信号的交叉偏振分量通常非常弱，因此大多数水凝物的 L_{DR} 值（dB）为负。雷达系统噪声限制最低可探测的 L_{DR} 约为−40 dB。L_{DR} 仅适用于 H 和 V 偏振交替发射和接收的雷达：如果只发射 H 方向偏振波，则接收到的 V 偏振信号来自目标物的退偏振。在 STAR 模式下，不能区分接收的偏振信号里分别来自共偏振或交叉偏振的贡献。因此，由于美国 WSR-88D 雷达网络以 STAR 模式运行，他们没有 L_{DR} 观测数据。

当非球形粒子具有一个相对于偏振面的平均倾角时，L_{DR} 增强，这个角简称倾角。在大多数情况下，水凝物的平均倾角大约为零，使得 L_{DR} 在降雨、雪聚合物等情况下相当低。然而，对于摇摆的具有电磁性质的非球形大粒子，特别是形状不规则的湿冰雹[72]，L_{DR} 会增大。也有证据表明，对流风暴上升气流中摇摆、冻结的雨滴可以产生明显更大的 L_{DR} 值[59,87]，融化层"亮带"通常也具有局部增强的 L_{DR} 值（图 2.15）。最近的研究表明，用天顶指向雷达探测柱状冰晶时，L_{DR} 也很有用[116]。

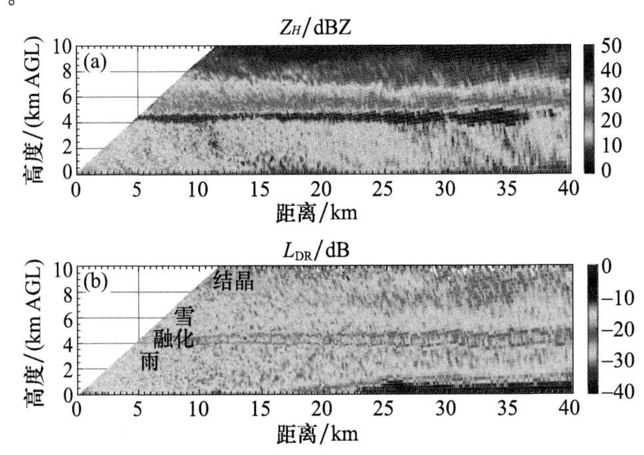

图 2.15 层状降水中(a)Z_H和(b)L_{DR}的垂直截面图，在印度洋的一次外场试验中使用 SPolKa 雷达采样。在 Z_H 和 L_{DR} 图中，融化层亮带清晰。在 L_{DR} 图中标注了水凝物类型。L_{DR} 在融化层增强，在高空小冰晶区域略有增强；在雪花聚集体和雨中，L_{DR} 值非常低

2.9 天气雷达的应用

双偏振和多普勒天气雷达提供的丰富信息深化了对云和降水的认识并催生出很多新的应用。本节提供了其中一些应用的简单概述;其余的将在本卷的其他部分讨论。

强对流风暴中灾害的探测已成为人们关注的焦点。多普勒速度信息使业务雷达人员能够识别对流风暴中的强风和下击暴流[91,104,125],以及识别雷达临近方位角之间的与中气旋和龙卷环流相关的强切变区域[23,24]。此外,基于天气雷达的动力场反演大大提高了对灾害性风暴的理解,包括强对流和热带气旋。当两个或多个多普勒雷达几乎同时从不同角度探测同一场风暴时,它们测量粒子运动在不同方向上的分量,尤其需要指出是,它们测量粒子的三维运动矢量在雷达径向上的投影。通常假设粒子的水平运动矢量等于水平风矢量,尽管该假设不宜用于强烈的垂直风切变或在离心力明显的强环流中[36,114,169]。结合将水平和垂直运动耦合起来的质量连续性方程和对粒子下落速度的估计(对于扫描雷达,通常利用 Z_e 粗略估计下落速度)可以估计出三维风场。在文献中,有被称为双多普勒或多多普勒的三维多普勒速度合成技术[3,29,40,76,144]。三维风的合成在以下条件时会更加精确:多个雷达波束以采样点为圆心形成的角度>30°且<150°;两个波束都处于较低的仰角;雷达同时采样风暴内的相同区域。利用多普勒速度合成技术估计得到的三维风场可以计算出有用的运动参量,用于以认识灾害性风暴。图2.16显示了一个龙卷超级单体在100 m(AGL)时的垂直速度和垂直涡度。由垂直涡度最大值的变化可知,上升气流和下沉气流的区域在龙卷周围呈螺旋式运动。

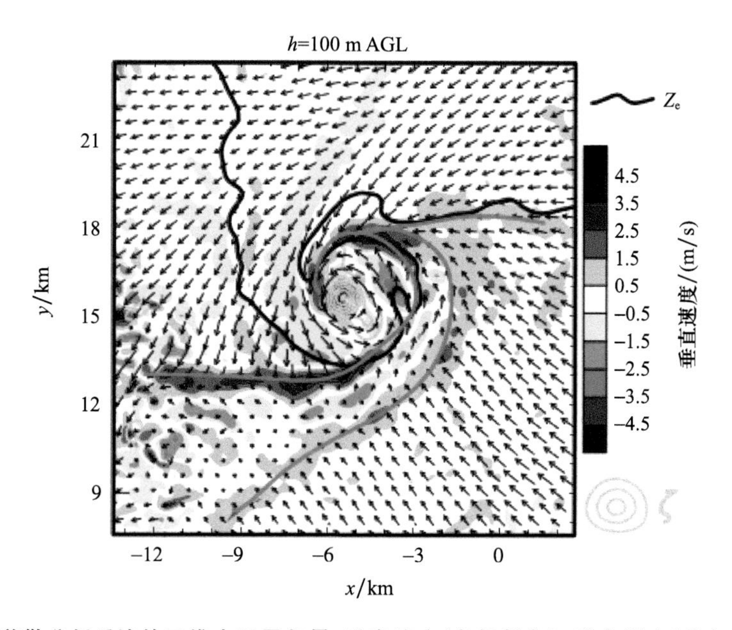

图2.16 双多普勒分析反演的二维水平风矢量,垂直速度(色标颜色),垂直涡度(黄色等高线),以及100 m(AGL)处的 Z_e 回波结构(黑色)。灰色线表示与初级和次级阵风锋相关的辐合线。图片由 Karen Kosiba 博士提供(灾害性天气研究中心)

双偏振雷达的出现促进了帮助业务雷达人员进行水凝物分类的业务算法的快速发展[97,98,117,151,157,168]，通过这些算法，每个雷达像素上的不同的雷达观测量被划分为一组水凝物或非气象回波类型的最佳分类(图 2.17)。此外，双偏振提供的附加滴谱信息与回波分类的结合大大改善了定量降水估计(QPE)，尽管这些 QPE 算法正在持续改进中[28,41,48,134]，但算法上的进步已经使洪水监测得到提升[17,27]。

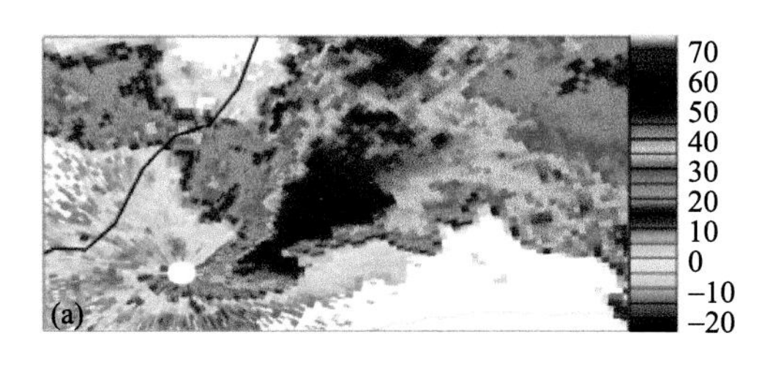

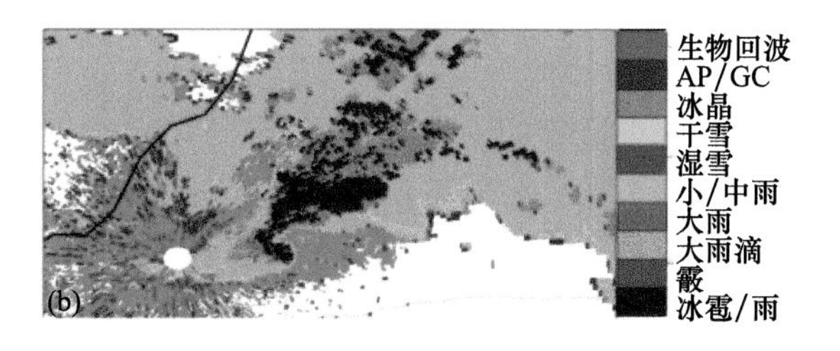

图 2.17　2017 年 3 月 1 日在美国印第安纳州西南部的龙卷风暴的低空雷达观测图(a)Z_H(根据色标着色，单位为 dBZ)和(b)业务水凝物分类算法的产品(根据色标着色)。雷达数据由埃文斯维尔附近的双偏振 WSR-88D 雷达(KVWX)于 0432 UTC 采样。在水凝物分类中，品红色表示“未知分类”

多普勒速度信息和偏振观测量的结合可以改进龙卷的探测：多普勒速度中显示的强旋转，加上 ρ_{HV} 显著降低，表明龙卷正在掀起碎片，这是 Ryzhkov 等首次发现的[129]。碎片不规则形状和指向一起导致了 ρ_{HV} 值远低于降水情况下的预期值(图 2.18)，这一龙卷碎片特征一直是许多近期研究的焦点[15,82,141,164]。在一些强龙卷个例中，碎片可能被抛到很高的地方，并落在风暴的左侧，如图 2.18 所示。当前的研究仍在探索利用龙卷碎片特征来实时获取反映龙卷剧烈程度的信息[15]。

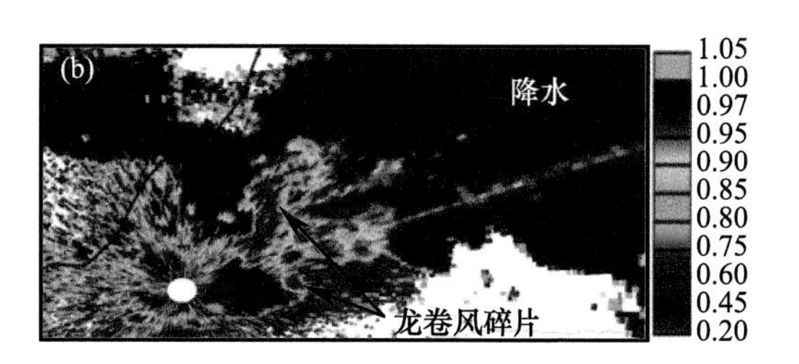

图 2.18　与图 2.17 类似，但显示(a)多普勒速度(m/s，根据色标着色)和(b)ρ_{HV}(根据色标着色)。(a)中箭头表示多普勒速度指示的旋转区域，(b)中箭头表示龙卷碎屑/碎片的 ρ_{HV} 极低，因为较高 ρ_{HV} 值表示有降水

对致灾性冰雹的探测非常受业务雷达人员的关注。由于雨滴和冰雹的散射特征不同[5,21]，双偏振雷达可以用于探测冰雹。特别是，大冰雹下落时的指向比雨滴更混乱，导致 Z_{DR} 值在 0 dB 附近。这种特征通常与较大的 Z_H(＞55 dBZ)和较小的 ρ_{HV} 相伴随，而跟形状无关。最近的研究工作集中在提高冰雹大小的估计精度[115,132,133]。用 S 波段双偏振雷达[52,133]能够相对简单地从大冰雹中区分小的、典型的非致灾性冰雹(直径＜2.5 cm)。这是因为小冰雹在融化过程中会覆盖一层水膜[124]，导致其散射特征类似于具有高 Z_{DR} 和 K_{DP} 值的大雨滴。然而，事实证明，区分大冰雹(＞2.5 cm)和超大冰雹(＞5 cm)仍具挑战[11,115]。一些研究建议通过识别 0℃层之上的上升气流中 ρ_{HV} 的大幅降低来识别超大冰雹[86,119]。该想法的依据是湿增长导致的不规则形状和雹瓣，和(或)显著的共振效应和后向散射差分相位。最近的研究还发现，在小冰雹浓度极高以至于可累积几厘米厚的情况下，S 波段的 K_{DP} 值异常高(>10 °/km)[80]。这些都是正在开展研究的领域。

因为天气雷达可以为观测风暴的最新演变及其相关灾害提供线索，对流风暴特征的雷达监测也受到了关注。20 世纪 80 年代，双偏振雷达第一次发现了对流风暴存在一个延伸到 0℃层之上[50]，Z_{DR} 呈正值的柱状区域，被称为 Z_{DR} 柱。图 2.19 是一次对流风暴的 RHI 扫描，图中出现了两个 Z_{DR} 柱。虽然其周围以霰和雪为主的区域 Z_{DR} 接近于零，Z_{DR} 柱中 Z_{DR} 的正值表明上升气流将液态粒子(雨滴或液态水含量较大的湿雪)抬升至 0℃层以上。对于 WSR-88D 雷达网使用的监测扫描模式，Z_{DR} 柱可被识别为 0℃层以上局部增强的 Z_{DR} 区域(图 2.20)。Z_{DR} 柱是对流风暴中上升气流的标志，此外，Z_{DR} 柱的特征与上升气流的强度和演变有关[89]，特别是，在 0℃层以上 Z_{DR} 柱的升高与上升气流和近地面降水的增强有关。目前正在开发一种业务算法来监测 Z_{DR} 柱高度的这些变化[149]，Z_{DR} 柱的宽度可能是衡量风暴产生大冰雹能力的重要决定因素[33,78]。与 Z_{DR} 柱类似，增强的 K_{DP} 柱[59,82,166]也提供了风暴特征和短期演变的诊断信息。

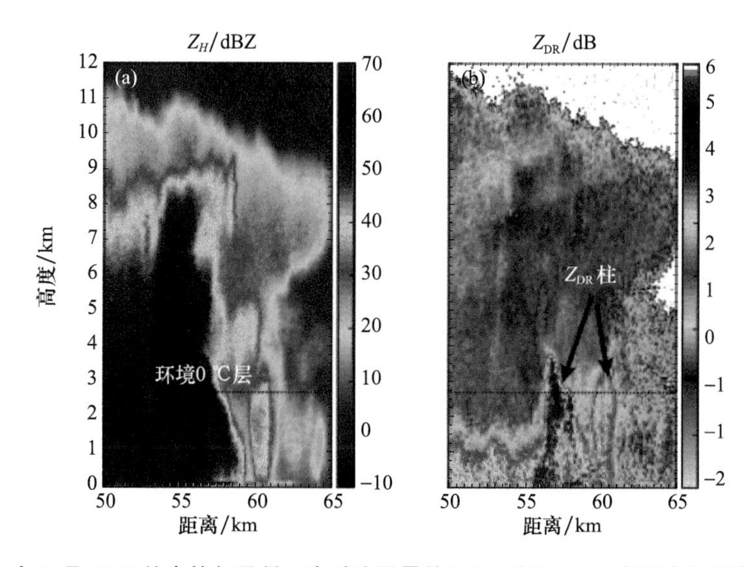

图 2.19 2009 年 2 月 10 日俄克拉何马州一次对流风暴的(a) Z_H 和(b) Z_{DR}(根据色标着色)垂直剖面图。存在两个明显的 Z_{DR} 柱，0℃层(在两图上用虚线表示)之上 Z_{DR} 值为正。数据来自俄克拉何马州诺曼的 KOUN 雷达，由 Valery Melnikov 博士提供(国家强风暴实验室)

最近，另外一种探测对流风暴中发生水凝物粒子分选区域的算法正在开发。这种粒子分选机制可能与对流上升气流的发展有关，因为小云滴被抬升，而更大、下落更快的水滴和小的融化冰粒能够在上升气流中下落，导致 Z_{DR} 较大而 Z_H 相对较低[84]。正在开发的基于粒子分选的雷暴风险评估和临近预报(TRENDSS)算法[121]就是用以识别这些区域(图 2.21)。首先，该算法在每次 PPI 扫描中在给定 Z_H 值范围内识别正的 Z_{DR} 异常值，然后将这些 Z_{DR} 异常数据进行合成(图 2.21a 中的蓝色阴影)，并用作短期预测风暴趋势的产品。在图 2.21a 中，最左边单元格的南侧和北侧都有粒子分选区域；20 min 后，该风暴分裂(图 2.21b)。

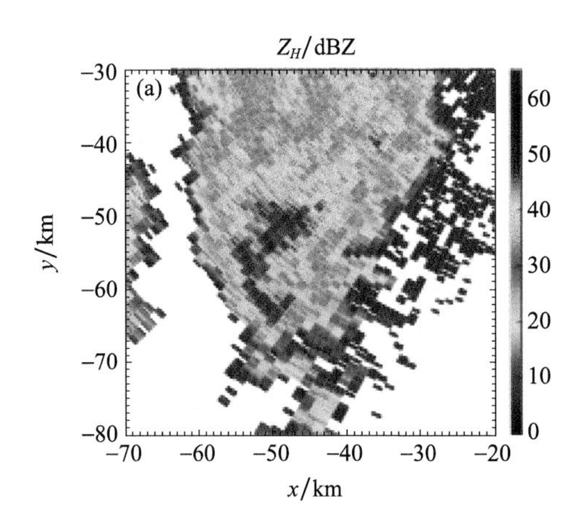

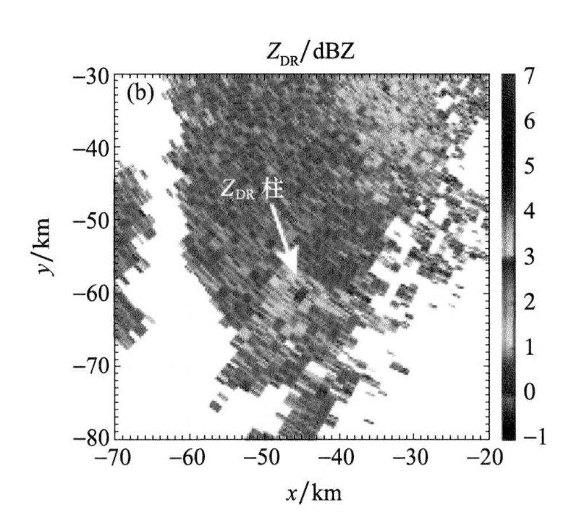

图 2.20　以 3.2°仰角的监测扫描为例，美国南达科他州西南部一场严重冰雹的(a)Z_H和(b)Z_{DR}(根据色标着色)场。数据来自 2013 年 6 月 21 日南达科他州拉皮德城附近的双偏振 WSR-88D 雷达(KUDX)。Z_{DR}柱处的雷达采样空间位于雷达高度之上约 4.5 km 处

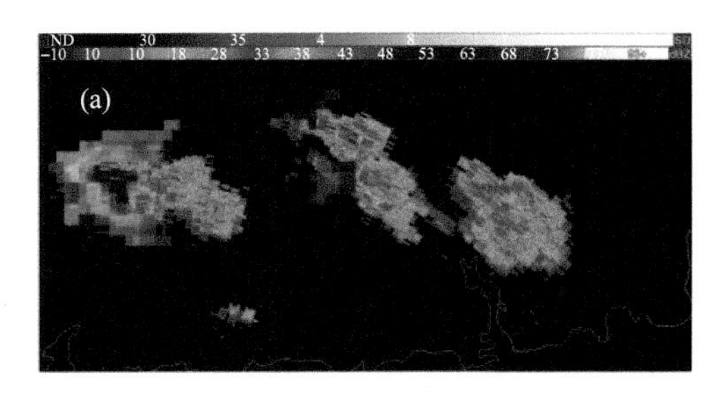

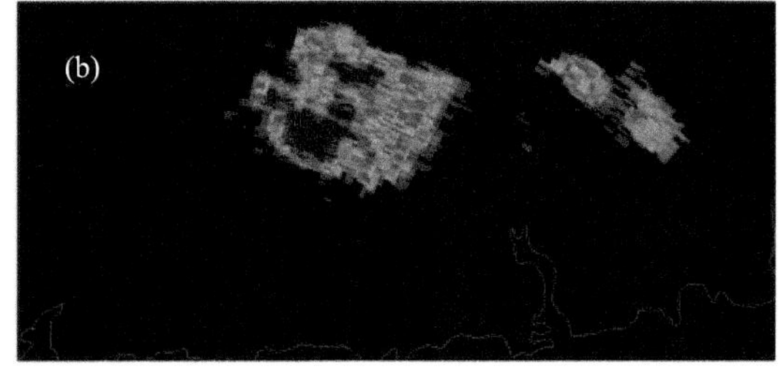

图 2.21　间隔 20 min 的两个不同 Z_H(单位：dBZ，彩色标示)监测扫描示例。顶部图(a)还包括 TRENDSS 算法的输出(蓝色阴影)，表明最左边单体两侧的粒子分选区和潜在增长区。实际上，风暴在 20 min 后分裂如图(b)。2015 年 6 月 23 日，该数据由多部 WSR-88D 雷达于康涅狄格州南部上空记录。图片由 Joey Picca 提供

预测风暴产生龙卷的潜势是业务雷达人员关注的一个重点问题，因为它可以提高预警时间以挽救生命。一个有前景的活跃研究领域是探索低层双偏振雷达特征和风暴附近环境之间的关系。在超级单体和一些非超级单体龙卷风暴中，研究人员发现了具有特殊诊断价值的低层增强的 Z_{DR}和 K_{DP}区域，分别称为 Z_{DR}弧和 K_{DP}足[81,2]。Z_{DR}增强区一般位于以 Z_H梯度为中心且距 K_{DP}增强区有一定的偏移，K_{DP}增强区一般位于靠近风暴 Z_H强降水区的核心。这种低层偏振特征的分布是由于对流层最低几千米处存在相对于风暴的非静止气流导致的粒子分选，而这往往与垂直风切变有关[31,32,82,83]。当 Z_{DR}增强区的长轴与风暴运动方向平行时，通常意味着低层显著的风暴相对螺旋度(图 2.22)，这是业务雷达人员用来评估风暴中龙卷潜势的一个重要参数[30,106,123,155,156]。定量研究风暴运动的 Z_{DR}和 K_{DP}增强区的分离距离和方位角的工作正在进行中[99,100]。

偏振雷达也为冬季的灾害性降水提供了重要的信息，包括降水类型的转化和强降雪。图 2.23 显示了纽约长岛区域在降水类型转变期间 ρ_{HV}的一系列监测扫描结果。如前文所述，水

凝物形状和成分的多样性会导致 ρ_{HV} 降低，因此在混合相降水中，水凝物处于融化或冻结状态时 ρ_{HV} 降低。如图 2.23 中个例所示，艾斯利普机场附近雷达报告降水类型从雪过渡到雪/雨/冰丸的混合物，此时减小的 ρ_{HV} 带经过其上空（图 2.23a、b）。当减小的 ρ_{HV} 带移动到雷达的北部时，降水类型就完全变为雨（图 2.23c、d）。业务雷达人员利用双偏振雷达的这些重要特征来实时评估降水的相态转换[120]。此外，最近的研究还发现了与冰丸形成和从冰粒转变为冻雨有关的 Z_{DR} 特征[85,88,165]。尽管这种所谓的重新冻结特征在很多个例中都被观察到且是冰丸存在的一个有力证据，导致其出现的确切微物理过程正在被研究且仍然不清楚。

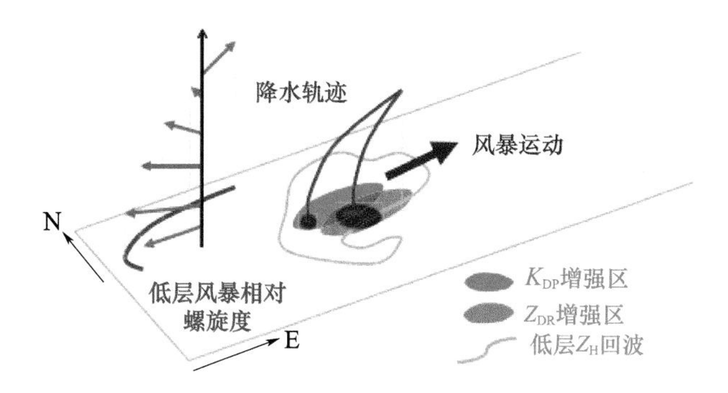

图 2.22　低层风暴相对螺旋度与低层 Z_{DR} 和 K_{DP} 增强区偏移特征的示意图。低层 Z_H 回波显示为灰色，Z_{DR} 和 K_{DP} 增强区分别显示为橙色和蓝色。垂直风的廓线由不同高度上的绿色矢量表示，低层风暴的相对速度曲线由左侧的蓝色粗曲线表示。红色矢量表示风暴运动，紫红色表示降水轨迹即大雨滴和小雨滴的下落位置

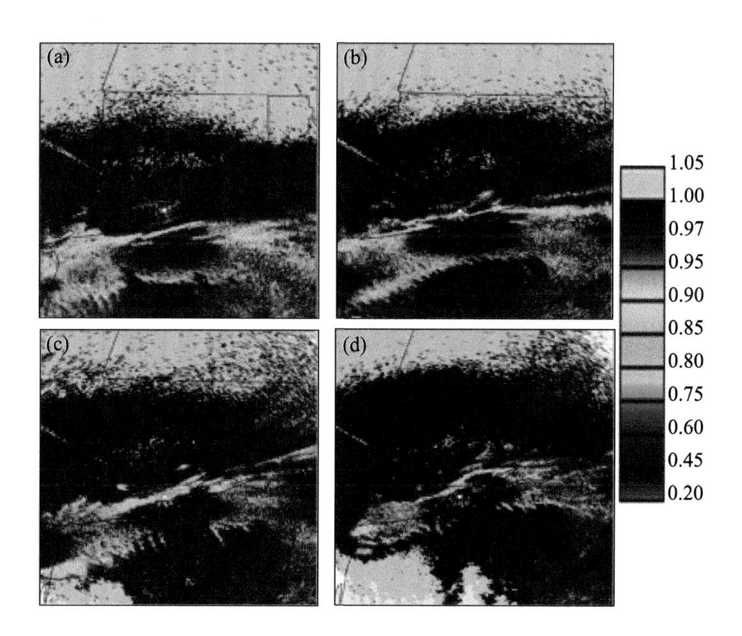

图 2.23　2014 年 2 月 13 日，美国纽约长岛上空一次冬季风暴降水过渡的 ρ_{HV}（阴影代表取值范围）0.5°仰角监测扫描的时间序列图：(a) 1426 UTC；(b) 1456 UTC；(c) 1526 UTC；(d) 1556 UTC。数据来自附近的纽约厄普顿双偏振 WSR-88D 雷达（KOKX）

识别大雪区域可帮助预报员监测积雪量大的地点。近期，一系列研究发现－15℃附近高度上 K_{DP} 的增强与近地面降水的增强有关[2,8,71,111,138－140]。K_{DP}（有时是 Z_{DR}）增大意味着旺盛的平板状晶体生长和随后的高效聚合，它们共同导致下方更强的降雪，该区域有时被称为枝状生长区。用时间-高度图来描述这种特征的方式被称为准垂直剖面（QVPs）[135]，其显示了增强的空中 K_{DP} 与下方强降雪（大 Z_H）之间的明显关联（图 2.24）。最近，QVP 技术被 Kumjian 和 Lombardo[81] 进行了扩展，包括多普勒速度和对雷达上方中尺度辐散/辐合的估计；这些工作也显示了增强的中尺度上升和－15℃附近旺盛的平板状晶体生长特征之间的明显联系（图 2.24）。在图 2.24 所示的个例中，K_{DP} 增大区下方的降雪率有时超过 15 cm/h。

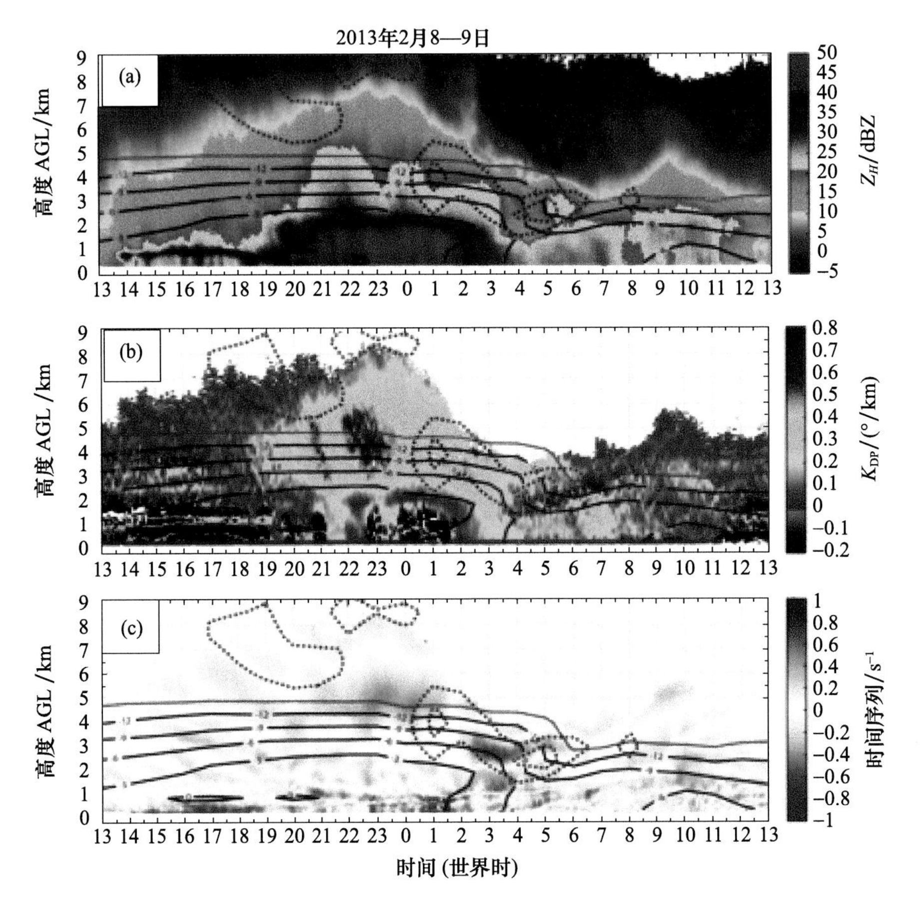

图 2.24　2013 年 2 月 8－9 日暴雪期间(a) Z_H、(b) K_{DP} 和(c)大尺度辐散的准垂直剖面(QVPs)时间序列图(彩色标示)。数据来自纽约附近的双偏振 WSR-88D 雷达(KOKX)。叠加的是来自快速更新模型的温度等值线(黑色曲线，以 3℃的增量从－12 到 0℃；品红线表示－15℃)和中尺度上升气流(黑色点虚线，从－1 Pa/s 开始，以 1 Pa/s 为增量)。改编自 Kumjian 和 Lombardo[81]，并做出了一些调整

冷云中的过冷水会对飞机造成严重的结冰危险。然而，由于云滴尺度小，在有其他大得多的降水粒子存在时，低频降水雷达无法直接探测到云滴，因此探测过冷水仍然是天气雷达的一个重大挑战。研究人员正在寻找存在过冷水的间接证据，比如正在发生的凇附和二次冰晶的产生。在双偏振天气雷达观测中一个很可能的特征是温度在－3～－8℃时 K_{DP} 的升高与接近于零的 Z_{DR}。这个特征被认为可以表明存在近球形的凇附粒子（霰）和/或大量针状冰晶与雪

花聚集体混合物[43,81,90,146]。为了验证这些特征和过冷水之间的联系，需要原位测量和偏振雷达数据的结合。

天气雷达面临的最大挑战之一是将先进的天气雷达观测结果与数值模型相结合，特别需要指出的是，双偏振雷达数据同化是一个正在进行的研究领域[68,122,175]。这是很困难的，因为许多业务使用的数值天气预报(NWP)模型采用了简化的微物理方案，没有提供关于粒子形状、下落特征或混合相水凝物中水含量的信息。至关重要的是，这些粒子特性对观测到的雷达参量影响最大，而雷达观测值受采样体积中最大粒子的影响较大，与常用的模式诊断量(如总粒子浓度、总质量)之间的关系具有高度的不确定性。这些诊断量并不如粒子谱分布中位于尾端的大粒子对雷达观测量影响大。目前，许多前向算子(模型预测变量和雷达观测变量之间的桥梁)都建立了关于许多重要的粒子特性假设[68,131]，但是许多用于雷达数据同化的前向算子过于简单[95,96,158]，导致较大的不确定性[25]。然而，数值模拟的最新进展表明，对粒子特性的预测[51,66,112,113]正在向前推进，这可能有助于 NWP 模型和天气雷达数据的融合[152,153]。未来改进复杂粒子形状(如雪晶)电磁散射计算的工作将会改进前向算子[102]。

2.10 总结和结论

本章重点介绍了云和降水的雷达遥感基础以及天气雷达在水文气象应用方面的最新进展。雷达一直是探测云和降水最重要的工具，特别是双偏振多普勒天气雷达观测带来了对云和降水物理的新认识，并为业务上探测和监测灾害性天气带来重要进展，包括强对流风暴(冰雹、龙卷、破坏性大风)和冬季风暴(大雪、降水类型转化)。近几十年来，雷达回波分类和降水定量估计的算法也有了很大的改进，毫无疑问，未来的突破将包括提高对云和降水的基本物理过程的理解，以及通过数据同化将数值天气预报模型和偏振多普勒天气雷达观测相结合。本卷的其他部分涵盖了雷达遥感的许多重要应用。

致谢 作者很荣幸能为本卷做出贡献，并感谢 Constantin Andronache 博士促成此事。此外，还要感谢 Joseph Picca(美国国家气象局风暴预测中心)、Israel Silber(宾夕法尼亚州立大学)、Marcus van Lier-Walqui(哥伦比亚大学和美国国家航空航天局戈达德空间研究所)、Hughbert Morrison(美国国家大气研究中心)和 Olivier Prat(北卡罗来纳州立大学北卡罗来纳气候研究所)。特别感谢宾夕法尼亚州立大学的同事们在过去几年里进行的无数次讨论：Kültegin Aydin、Craig Bohren、Anthony dilake、Eugene Clothiaux、Steven Greybush、Jerry Harrington、Paul Markowski、James Marquis、Yvette Richardson 和 Johannes Verlinde。最后，作者要感谢他在宾州州立大学雷达小组的优秀学生：Robert Schrom、Dana Tobin、Kyle Elliott、Scott Loeffler、Steven Naegele、John Banghoff、Charlotte Martinkus 和 Zhiyuan Jiang。本章中使用的雷达数据个例由美国国家海洋和大气管理局和美国能源部大气系统研究项目提供。作者要感谢 Scott Collis(阿尔贡国家实验室)、Karen Kosiba(灾害天气研究中心)和 Angela Rowe(华盛顿大学)对这项工作很有裨益的评论。

参考文献*

[1] Albrecht, B. , M. Fang, and V. Ghate. 2016. Exploring stratocumulus cloud-top entrainment processes and parameterizations by using Doppler cloud radar observations. Journal of the Atmospheric Sciences 73:729-742.

[2] Andric, J. , M. R. Kumjian, D. S. Zrnic, J. M. Straka, and V. M. Melnikov. 2013. Polarimetric signatures above the melting layer in winter storms: an observational and modeling study. Journal of Applied Meteorology and Climatology 52:682-700.

[3] Armijo, L. 1969. A theory for the determination of wind and precipitation velocities with Doppler radars. Journal of the Atmospheric Sciences 26:570-573.

[4] Atlas, D. 2002. Radar calibration: Some simple approaches. Bulletin of the American Meteorological Society 83:1313-1316.

[5] Aydin, K. , T. A. Seliga, and V. Balaji. 1986. Remote sensing of hail with a dual linear polarized radar. Journal of Climate and Applied Meteorology, 25:1475-1484.

[6] Battaglia, A. , C. D. Westbrook, S. Kneifel, P. Kollias, N. Humpage, U. Löhnert, J. Tyynelä, and G. W. Petty. 2014. G band atmospheric radars: a new frontier in cloud physics. Atmospheric Measurement Techniques 7:1527-1546.

[7] Battan, L. J. 1973. Radar Observation of the Atmosphere, 324. Chicago: University of Chicago Press.

[8] Bechini, R. , L. Baldini, and V. Chandrasekar. 2013. Polarimetric radar observations of the ice region of precipitation clouds at C-band and X-band radar frequencies. Journal of Applied Meteorology and Climatology 52:1147-1169.

[9] Beck, J. , and O. Bousquet. 2013. Using gap-filling radars in mountainous regions to complement a national radar network improvements in multiple-Doppler wind syntheses. Journal of Applied Meteorology and Climatology 52:1836-1850.

[10] Bharadwaj, N. , and V. Chandrasekar. 2007. Phase coding for range ambiguity mitigation in dual-polarized Doppler weather radars. Journal of Atmospheric and Oceanic Technology 24:1351-1363.

[11] Blair, S. F. , D. R. Deroche, J. M. Boustead, J. W. Leighton, B. L. Barjenbruch, and W. P. Gargan. 2011. A radar-based assessment of the detectability of giant hail. Electronic Journal of Severe Storms Meteorology 6(7):1-30.

[12] Bluestein, H. B. , W. P. Unruh, D. C. Dowell, T. A. Hutchinson, T. M. Crawford, A. C. Wood, and H. Stein. 1997. Doppler radar analysis of the Northfield, Texas, tornado of 25 May 1994. Monthly Weather Review 125:212-230.

[13] Bluestein, H. B. , M. M. French, I. PopStefanija, R. T. Bluth, and J. B. Knorr. 2010. A mobile, phased-array Doppler radar for the study of severe convective storms The MWR-05XP. Bulletin of the American Meteorological Society 91:579-600.

[14] Bodine, D. J. , D. Michaud, R. D. Palmer, P. L. Heinselman, J. Brotzge, N. Gasperoni, B. L. Cheong, M. Xue, and J. Gao. 2011. Understanding radar refractivity sources of uncertainty. Journal of Applied Meteorology and Climatology 50:2543-2560.

[15] Bodine, D. J. , M. R. Kumjian, R. D. Palmer, P. L. Heinselman, and A. V. Ryzhkov. 2012. Tornado damage

* 参考文献沿用原版书中内容,未改动

estimation using polarimetric radar. Weather and Forecasting 28:139-158.

[16] Bohren, C., and D. Huffman. 1983. Absorption and Scattering of Light by Small Particles, 530. Wiley, New York.

[17] Boodoo, S., D. Hudak, A. Ryzhkov, P. Zhang, N. Donaldson, D. Sills, and J. Reid. 2015. Quantitative precipitation estimation from a C-band dual-polarized radar for the 8 July 2013 flood in Toronto, Canada. Journal of Hydrometeorology 16:2027-2044.

[18] Borowska, L., A. V. Ryzhkov, D. S. Zrnic, C. Simmer, and R. D. Palmer. 2011. Attenuation and differential attenuation of 5-cm-wavelength radiation in melting hail. Journal of Applied Meteorology and Climatology 50:59-76.

[19] Bremer, J., and U. Berger. 2002. Mesospheric temperature trends derived from ground-based LF phase-height observations at mid-latitudes comparison with model simulations. Journal of Atmospheric and Solar-Terrestrial Physics 64:805-816.

[20] Bringi, V., and V. Chandrasekar. 2001. Polarimetric Doppler Weather Radar Principles and Applications, 636. Cambridge: Cambridge University Press.

[21] Bringi, V. N., J. Vivekanandan, and J. D. Tuttle. 1986. Multiparameter radar measurements in Colorado convective storms. Part II: Hail detection studies. Journal of the Atmospheric Sciences 43:2564-2577.

[22] Bringi, V. N., V. Chandrasekar, N. Balakrishnan, and D. S. Zrni′c. 1990. An examination of propagation effects on radar measurements at microwave frequencies. Journal of Atmospheric and Oceanic Technology 7:829-840.

[23] Brown, R. A., L. R. Lemon, and D. W. Burgess. 1978. Tornado detection by pulsed Doppler radar. Monthly Weather Review 106:29-38.

[24] Burgess, D. W., L. D. Hennington, R. J. Doviak, and P. S. Ray. 1976. Multimoment Doppler display for severe storm identification. Journal of Applied Meteorology 15:1302-1306.

[25] Carlin, J. T., A. V. Ryzhkov, J. C. Snyder, and A. P. Khain. 2016. Hydrometeor mixing ratio retrievals for storm-scale radar data assimilation Utility of current relations and potential benefits of polarimetry. Monthly Weather Review 144:2981-3001.

[26] Chandrasekar, V., R. Keranen, S. Lim, and D. Moisseev. 2013. Recent advances in classification of observations from dual polarization weather radars. Atmospheric Research 119:97-111.

[27] Chang, W.-Y., J. Vivekanandan, K. Ikeda, and P.-L. Lin. 2016. Quantitative precipitation estimation of the epic 2013 Colorado flood event: Polarization radar-based variational scheme. Journal of Applied Meteorology and Climatology 55:1477-1495.

[28] Cifelli, R., V. Chandrasekar, S. Lim, P. C. Kennedy, Y. Wang, and S. A. Rutledge. 2011. A new dual-polarization radar rainfall algorithm application in Colorado precipitation events. Journal of Atmospheric and Oceanic Technology 28:352-364.

[29] Collis, S., A. Protat, P. T. May, and C. Williams. 2013. Statistics of storm updraft velocities from TWP-ICE including verification with profiling measurements. Journal of Applied Meteorology and Climatology 52:1909-1922.

[30] Davies-Jones, R., D. W. Burgess, and M. Foster. 1990. Test of helicity as a forecast parameter. In 16th Conference on Severe Local Storms, Kananaskis Park, AB, Canada. 588-592. Boston: American Meteor Society(Preprints).

[31] Dawson, D. T., E. R. Mansell, Y. Jung, L. J. Wicker, M. R. Kumjian, and M. Xue. 2014. Lowlevel ZDR sig-

natures in supercell forward flanks:The role of size sorting and melting of hail. Journal of the Atmospheric Sciences 71:276-299.

[32] Dawson,D. T. ,E. R. Mansell,and M. R. Kumjian. 2015. Does wind shear cause hydrometeor size sorting? Journal of the Atmospheric Sciences 72:340-348.

[33] Dennis,E. J. ,and M. R. Kumjian. 2017. The impact of vertical wind shear on hail growth in simulated supercells. Journal of the Atmospheric Sciences 74:641-663.

[34] Doviak,R. J. ,and D. S. Zrni'c. 1993. Doppler Radar and Weather Observations,562. San Diego:Academic Press.

[35] Doviak,R. J. ,V. Bringi,A. Ryzhkov,A. Zahrai,and D. S. Zrni'c. 2000. Considerations for polarimetric upgrades to operational WSR-88D radars. Journal of Atmospheric and Oceanic Technology 17:257-278.

[36] Dowell,D. C. ,C. R. Alexander,J. Wurman,and L. J. Wicker. 2005. Centrifuging of hydrometeors and debris in tornadoes:radar-reflectivity patterns and wind-measurement errors. Monthly Weather Review 133:1501-1524.

[37] Fabry,F. 2015. Radar Meteorology Principles and Practice,256. Cambridge:Cambridge University Press.

[38] Fabry,F. ,C. Frush,I. Zawadzki,and A. Kilambi . 1997. On the extraction of near-surface index of refraction using radar phase measurements from ground targets. Journal of Atmospheric and Oceanic Technology 14:978-987.

[39] French,M. M. ,D. W. Burgess,E. R. Mansell,and L. J. Wicker. 2015. Bulk hook echo raindrop sizes retrieved using mobile,polarimetric Doppler radar observations. Journal of Applied Meteorology and Climatology 54:423-450.

[40] Gal-Chen,T. 1978. A method for the initialization of the anelastic equations:implications for matching models with observations. Monthly Weather Review 106:587-606

[41] Giangrande,S. E. ,and A. V. Ryzhkov. 2008. Estimation of rainfall based on the results of polarimetric echo classification. Journal of Applied Meteorology and Climatology 47:2445-2462.

[42] Giangrande,S. E. ,J. M. Krause,and A. V. Ryzhkov. 2008. Automatic designation of the melting layer with a polarimetric prototype of the WSR-88D radar. Journal of Applied Meteorology and Climatology 47:1354-1364.

[43] Giangrande,S. E. ,T. Toto,A. Bansemer,M. R. Kumjian,S. Mishra,and A. V. Ryzhkov. 2016. Insights into riming and aggregation processes as revealed by aircraft,radar,and disdrometer observations for a 27 April 2011 widespread precipitation event. Journal of Geophysical Research-Atmospheres 121:5846-5863.

[44] Girardin-Gondeau,J. ,F. Baudin,and J. Testud. 1991. Comparison of coded waveforms for an airborne meteorological Doppler radar. Journal of Atmospheric and Oceanic Technology 8:234-246.

[45] Goddard,J. W. F. ,J. Tan,and M. Thurai. 1994. Technique for calibration of meteorological radar using differential phase. Electronics Letters 30:166-167.

[46] Gorgucci,E. ,G. Scarchilli,and V. Chandrasekar. 1992. Calibration of radars using polarimetric techniques. IEEE Transactions on Geoscience and Remote Sensing 30:853-858.

[47] Gourley,J. J. ,A. J. Illingworth,and P. Tabary. 2009. Absolute calibration of radar reflectivity using redundancy of the polarization observations and implied constraints on drop shapes. Journal of Atmospheric and Oceanic Technology 26:689-703.

[48] Gourley,J. J. ,Y. Hong,Z. L. Flamig,J. Wang,H. Vergara,and E. N. Anagnostou. 2011. Hydrologic evaluation of rainfall estimates from radar,satellite,gauge,and combinations on Ft. Cobb Basin,Oklaho-

ma. Journal of Hydrometeorology 12:973-988.

[49] Gu,J.-Y.,A. V. Ryzhkov,P. Zhang,P. Neilley,M. Knight,B. Wolf,and D.-I. Lee. 2011. Polarimetric attenuation correction in heavy rain at C band. Journal of Applied Meteorology and Climatology 50:39-58.

[50] Hall,M. P. M.,S. M. Cherry,J. W. F. Goddard,and G. R. Kennedy. 1980. Rain drop sizes and rainfall rate measured by dual-polarization radar. Nature 285:195-198.

[51] Harrington,J. Y.,K. Sulia,and H. Morrison. 2013. A method for adaptive habit prediction in bulk microphysical models. Part I:Theoretical development. Journal of the Atmospheric Sciences 70:349-364.

[52] Heinselman,P. L.,and A. V. Ryzhkov. 2006. Validation of polarimetric hail detection. Weather and Forecasting 21:839-850.

[53] Heymsfield,G. M.,L. Tian,L. Li,M. McLinden,and J. Cervantes. 2013. Airborne radar observations of severe hailstorms Implications for future spaceborne radar. Journal of Applied Meteorology and Climatology 52:1851-1867.

[54] Hibbins,R. E.,J. D. Shanklin,P. J. Epsy,M. J. Jarvis,D. M. Riggin,D. C. Fritts,and F.-J. Lübken. 2005. Seasonal variations in the horizontal wind structure from 0-100 km above Rothera station,Antarctica (67S,68W). Atmospheric Chemistry and Physics 5:2973-2980.

[55] Hildebrand,P. H.,and R. K. Moore. 1990. Meteorological radar observations from mobile platforms. In Radar in Meteorology,ed. D. Atlas,287-314. Boston:American Meteorological Society.

[56] Hildebrand,P. H.,W.-C. Lee,C. A. Walther,C. Frush,M. Randall,E. Loew,R. Neitzel,R. Parsons,J. Testud,F. Baudin,and A. LeCornec. 1996. The ELDORA/ASTRAIA Airborne Doppler weather radar:high-resolution observations from TOGA COARE. Bulletin of the American Meteorological Society 77:213-232.

[57] Hocking,W. K. 1999. Temperatures using radar-meteor decay times. Geophysical Research Letters 26:3297-3300. https://doi.org/10.1029/1999GL003618.

[58] Hocking,W. K.,B. Fuller,and B. Vandepeer. 1999. Real-time determination of meteorrelated parameters utilizing digital technology. Journal of Atmospheric and Solar—Terrestrial Physics 63:155-169.

[59] Hubbert,J. C.,V. N. Bringi,L. D. Carey,and S. Bolen. 1998. CSU-CHILL polarimetric radar measurements from a severe hail storm in eastern Colorado. Journal of Applied Meteorology 37:749-775.

[60] Hubbert,J. C.,S. M. Ellis,M. Dixon,and G. Meymaris. 2010. Modeling,error analysis,and evaluation of dual-polarization variables obtained from simultaneous horizontal and vertical polarization transmit radar. Part I:Modeling and antenna errors. Journal of Atmospheric and Oceanic Technology 27:1583-1598.

[61] Hubbert,J. C.,S. M. Ellis,M. Dixon,and G. Meymaris. 2010. Modeling,error analysis,and evaluation of dual-polarization variables obtained from simultaneous horizontal and vertical polarization transmit radar. Part II:Experimental data. Journal of Atmospheric and Oceanic Technology 27:1599-1607.

[62] Hubbert,J. C.,S. M. Ellis,W.-Y. Chang,S. Rutledge,and M. Dixon. 2014. Modeling and interpretation of S-band ice crystal depolarization signatures from data obtained by simultaneously transmitting horizontally and vertically polarized fields. Journal of Applied Meteorology and Climatology 53:1659-1677.

[63] Illingworth,A.,and T. Blackman. 2002. The need to represent raindrop size spectra as normalized gamma distributions for the interpretation of polarization radar observations. Journal of Applied Meteorology 41:286-297.

[64] Isom,B.,R. D. Palmer,R. Kelley,J. Meier,D. Bodine,M. Yeary,B. L. Cheong,Y. Zhang,T.-Y. Yu,and M. I. Biggerstaff. 2013. The atmospheric imaging radar:Simultaneous volumetric observations using a

phased array weather radar. Journal of Atmospheric and Oceanic Technology 30:655-675.

[65] Jameson, A. R. 1985. Microphysical interpretation of multiparameter radar measurements in rain. Part Ⅲ: Interpretation and measurement of propagation differential phase shift between orthogonal linear polarizations. Journal of the Atmospheric Sciences 42:607-614.

[66] Jensen, A. A., and J. Y. Harrington. 2015. Modeling ice crystal aspect ratio evolution during riming: a single-particle growth model. Journal of the Atmospheric Sciences 72:2569-2590.

[67] Jiang, Z., M. Oue, J. Verlinde, E. E. Clothiaux, K. Aydin, G. Botta, and Y. Lu. 2017. What can we conclude about the real aspect ratios of ice particle aggregates from two-dimensional images? Journal of Applied Meteorology and Climatology 56:725-734.

[68] Jung, Y., G. Zhang, and M. Xue. 2008. Assimilation of simulated polarimetric radar data for a convective storm using the ensemble Kalman filter. Part I: Observation operators for reflectivity and polarimetric variables. Monthly Weather Review 136:2228-2245.

[69] Junyent, F., V. Chandrasekar, V. N. Bringi, S. A. Rutledge, P. C. Kennedy, D. Brunkow, J. George, and R. Bowie. 2015. Transformation of the CSU-CHILL radar facility to a dualfrequency, dual-polarization Doppler system. Bulletin of the American Meteorological Society 96:975-996.

[70] Kalesse, H., W. Szyrmer, S. Kneifel, P. Kollias, and E. Luke. 2016. Fingerprints of a riming event on cloud radar: Doppler spectra observations and modeling. Atmospheric Chemistry and Physics 16: 2997-3012. https://doi. org/10. 5194/acp-16-2997-2016.

[71] Kennedy, P. C., and S. A. Rutledge. 2011. S-band dual-polarization radar observations of winter storms. Journal of Applied Meteorology and Climatology 50:844-858.

[72] Kennedy, P. C., S. A. Rutledge, W. A. Petersen, and V. N. Bringi. 2001. Polarimetric radar observations of hail formation. Journal of Applied Meteorology 40:1347-1366.

[73] Kneifel, S., A. von Lerber, J. Tiira, D. Moisseev, P. Kollias, and J. Leinonen. 2015. Observed relations between snowfall microphysics and triple-frequency radar measurements. Journal of Geophysical Research-Atmospheres 120:6034-6055. https://doi. org/10. 1002/2015JD023156.

[74] Kollias, P., B. A. Albrecht, R. Lhermitte, and A. Savtchenko. 2001. Radar observations of updrafts, downdrafts, and turbulence in fair-weather cumuli. Journal of the Atmospheric Sciences 58:1750-1766.

[75] Kollias, P., J. Remillard, E. Luke, and W. Szyrmer. 2011. Cloud radar Doppler spectra in drizzling stratiform clouds 1. Forward modeling and remote sensing applications. Journal of Geophysical Research 116:D13201.

[76] Kosiba, K. A., J. Wurman, P. Markowski, Y. Richardson, P. Robinson, and J. Marquis. 2013. Genesis of the Goshen County, WY tornado on 05 June 2009 during VORTEX2. Monthly Weather Review 141:1157-1181.

[77] Kumjian, M. R. 2013. Principles and applications of dual-polarization weather radar. Part 1: Description of the polarimetric radar variables. Journal of Operational Meteorology 1(19):226-242.

[78] Kumjian, M. R. 2013. Principles and applications of dual-polarization weather radar. Part 2: Warm and cold season applications. Journal of Operational Meteorology 1(20):243-264.

[79] Kumjian, M. R. 2013. Principles and applications of dual-polarization weather radar. Part 3: Artifacts. Journal of Operational Meteorology 1(21):265-274.

[80] Kumjian, M. R., and Z. J. Lebo. 2016. Large accumulations of small hail. In 28th Conference on Severe Local Storms. Portland, OR: American Meteorological Society. 8A. 4.

[81] Kumjian, M. R., and K. A. Lombardo. 2017. Insights into the evolving microphysical and kinematic structure of northeastern U. S. winter storms from dual-polarization Doppler radar. Monthly Weather Review

145:1033-1061.

[82] Kumjian,M. R. ,and A. V. Ryzhkov. 2008. Polarimetric signatures in supercell thunderstorms. Journal of Applied Meteorology and Climatology 47:1940-1961.

[83] Kumjian,M. R. ,and A. V. Ryzhkov. 2009. Storm-relative helicity revealed from polarimetric radar measurements. Journal of the Atmospheric Sciences 66:667-685.

[84] Kumjian, M. R. , and A. V. Ryzhkov. 2012. The impact of size sorting on the polarimetric radar variables. Journal of the Atmospheric Sciences 69:2042-2060.

[85] Kumjian, M. R. , and A. D. Schenkman. 2014. The curious case of ice pellets in Middle Tennessee on 1 March 2014. Journal of Operational Meteorology—Image of Note 2(17):209-213.

[86] Kumjian, M. R. , J. C. Picca, S. M. Ganson, A. V. Ryzhkov, J. Krause, D. S. Zrni′ c, and A. P. Khain. 2010. Polarimetric radar characteristics of large hail. In 25th Conference on Severe Local Storms,Denver, CO. 11. 2. Boston:American Meteorological Society(Preprints).

[87] Kumjian,M. R. ,S. Ganson,and A. V. Ryzhkov. 2012. Raindrop freezing in deep convective updrafts:a microphysical and polarimetric model. Journal of the Atmospheric Sciences 69:3471-3490.

[88] Kumjian,M. R. ,A. V. Ryzhkov,H. D. Reeves,and T. J. Schuur. 2013. A dual-polarization radar signature of hydrometeor refreezing in winter storms. Journal of Applied Meteorology and Climatology 52: 2549-2566.

[89] Kumjian,M. R. ,A. P. Khain,N. BenMoshe,E. Ilotoviz,A. V. Ryzhkov,and V. T. J. Phillips. 2014. The anatomy and physics of ZDR columns Investigating a polarimetric radar signature with a spectral bin microphysical model. Journal of Applied Meteorology and Climatology 53:1820-1843.

[90] Kumjian,M. R. ,S. Mishra,S. E. Giangrande,T. Toto,A. V. Ryzhkov,and A. R. Bansemer. 2016. Polarimetric radar and aircraft observations of saggy bright bands during MC3E. Journal of Geophysical Research-Atmospheres 121:3584-3607.

[91] Kuster,C. M. ,P. L. Heinselman,and T. J. Schuur. 2016. Rapid-update radar observations of downbursts occurring within an intense multicell thunderstorm on 14 June 2011. Weather and Forecasting 31: 827-851.

[92] Lhermitte,R. M. 1987. A 94 GHz Doppler radar for cloud observations. Journal of Atmospheric and Oceanic Technology 4:36-48.

[93] Lhermitte,R. M. 1988. Observation of rain at vertical incidence with a 94 GHz Doppler radar:an insight on Mie scattering. Geophysical Research Letters 15:1125-1128.

[94] Lhermitte,R. M. 1990. Attenuation and scattering of millimeter wavelength radiation by clouds and precipitation. Journal of Atmospheric and Oceanic Technology 7:464-479.

[95] Li,X. ,and J. R. Mecikalski. 2012. Impact of the dual-polarization Doppler radar data on two convective storms with a warm-rain radar forward operator. Monthly Weather Review 140:2147-2167.

[96] Li,X. ,J. R. Mecikalski,and D. Posselt. 2017. An ice-phase microphysics forward model and preliminary results of polarimetric radar data assimilation. Monthly Weather Review 145:683-708.

[97] Lim,S. ,V. Chandrasekar,and V. N. Bringi. 2005. Hydrometeor classification system using dual-polarization radar measurements:model improvements and in situ verification. IEEE Transactions on Geoscience and Remote Sensing 43:792-801.

[98] Liu,H. ,and V. Chandrasekar. 2000. Classification of hydrometeors based on polarimetric radar measurements:development of fuzzy logic and neuro-fuzzy systems,and in situ verification. Journal of Atmospher-

ic and Oceanic Technology 17:140-164.

[99] Loeffler,S. M. ,andM. R. Kumjian. 2016. Analysis of polarimetric radar signatures in tornadic non-supercellular storms. In 28th Conference on Severe Local Storms,12A. 4. Boston:American Meteorological Society.

[100] Loeffler, S. M. , and M. R. Kumjian. 2016. Quantifying ZDR _ KDP separation in severe convective storms to assess tornadic potential. In 28th Conference on Severe Local Storms,Poster 169. Boston:American Meteorological Society.

[101] Lu, Y. , E. E. Clothiaux, K. Aydin, G. Botta, and J. Verlinde. 2013. Modeling variability in dendritic ice crystal backscattering cross sections at millimeter wavelengths using a modified Rayleigh-Gans theory. Journal of Quantitative Spectroscopy and Radiation Transfer 131:95-104.

[102] Lu,Y. ,Z. Jiang,K. Aydin,J. Verlinde,E. E. Clothiaux,and G. Botta. 2016. A polarimetric scattering database for non-spherical ice particles at microwave wavelengths. Atmospheric Measurement Techniques 9:5119-5134.

[103] Luke,E. ,and P. Kollias. 2013. Separating cloud and drizzle radar moments during precipitation onset using Doppler spectra. Journal of Atmospheric and Oceanic Technology 30:1656-1671.

[104] Mahale, V. N. , G. Zhang, and M. Xue. 2016. Characterization of the 14 June 2011 Norman, Oklahoma downburst through dual-polarization radar observations and hydrometeor classification. Journal of Applied Meteorology and Climatology 55:2635-2655.

[105] Maki, M. , et al. 2012. Tokyo metropolitan area convection study for extreme weather resilient cities (TOMACS). Extended Abstracts. In 7th European Conference on Radar in Meteorology and Hydrology,Toulouse,France.

[106] Markowski,P. M. ,and Y. P. Richardson. 2010. Mesoscale Meteorology in Midlatitudes. 1st ed. ,407 pp. Oxford:Wiley-Blackwell.

[107] Matrosov,S. Y. 2017. Characteristic raindrop size retrievals from measurements of differences in vertical Doppler velocities at Ka-and W-band radar frequencies. Journal of Atmospheric and Oceanic Technology 34:65-71.

[108] McLaughlin,D. ,et al. 2009. Short-wavelength technology and the potential for distributed networks of small radar systems. Bulletin of the American Meteorological Society 90:1797-1817.

[109] Melnikov,V. M. ,and R. J. Doviak. 2009. Turbulence and wind shear in layers of large Doppler spectrum width in stratiform precipitation. Journal of Atmospheric and Oceanic Technology 26:430-443.

[110] Mie,G. 1908. Beiträge zur Optik trüber Medien, speziell kolloidaler Metallösungen. Annals of Physics 330:377-445.

[111] Moisseev,D. N. ,S. Lautaportti,J. Tyynela,and S. Lim. 2015. Dual-polarization radar signatures in snowstorms:role of snowflake aggregation. Journal of Geophysical Research-Atmospheres 120:12644-12655.

[112] Morrison,H. ,and J. A. Milbrandt. 2015. Parameterization of cloud microphysics based on the prediction of bulk ice particle properties. Part I:Scheme description and idealized tests. Journal of the Atmospheric Sciences 72:287-311.

[113] Morrison,H. ,J. A. Milbrandt,G. H. Bryan,K. Ikeda,S. A. Tessendorf,and G. Thompson. 2015. Parameterization of cloud microphysics based on the prediction of bulk ice particle properties. Part II:Case study and comparisons with observations and other schemes. Journal of the Atmospheric Sciences 72:312-339.

[114] Nolan,D. S. 2013. On the use of Doppler radar-derived wind fields to diagnose the secondary circulations

of tornadoes. Journal of the Atmospheric Sciences 70:1160-1171

[115] Ortega,K. L. ,J. M. Krause,and A. V. Ryzhkov. 2016. Polarimetric radar characteristics of melting hail. Part III:Validation of the algorithm for hail size discrimination. Journal of Applied Meteorology and Climatology 55:829-848.

[116] Oue,M. ,M. R. Kumjian,Y. Lu,J. Verlinde,K. Aydin,and E. Clothiaux. 2015. Linear depolarization ratios of columnar ice crystals in a deep precipitation system over the Arctic observed by zenith-pointing Ka-band Doppler radar. Journal of Applied Meteorology and Climatology 54:1060-1068.

[117] Park,H. S. ,A. V. Ryzhkov,D. S. Zrni′c,and K. -E. Kim. 2009. The hydrometeor classification algorithm for the polarimetric WSR-88D: description and application to an MCS. Weather and Forecasting 24: 730-748.

[118] Pazmany,A. L. ,J. B. Mead,H. B. Bluestein,J. C. Snyder,and J. B. Houser. 2013. A mobile,rapid-scanning X-band polarimetric(RaXPol)Doppler radar system. Journal of Atmospheric and Oceanic Technology 30:1398-1413.

[119] Picca,J. ,and A. V. Ryzhkov. 2012. A dual-wavelength polarimetric analysis of the 16 May 2010 Oklahoma City extreme hailstorm. Monthly Weather Review 140:1385-1403.

[120] Picca,J. C. ,D. M. Schultz,B. A. Colle,S. Ganetis,D. R. Novak,and M. J. Sienkiewicz. 2014. The value of dual-polarization radar in diagnosing the complex microphysical evolution of an intense snowband. Bulletin of the American Meteorological Society 95:1825-1834.

[121] Picca,J. C. ,D. M. Kingfield,and A. V. Ryzhkov. 2017. Utilizing a polarimetric size sorting signature to develop a convective nowcasting algorithm. In 18th Conference on Aviation,Range,and Aerospace Meteorology,Seattle,WA. Boston:American Meteorological Society,14. 5.

[122] Posselt,D. J. ,X. Li,S. A. Tushaus,and J. R. Mecikalski. 2015. Assimilation of dualpolarization radar observations in mixed-and ice-phase regions of convective storms:information content and forward model errors. Monthly Weather Review 143:2611-2636.

[123] Rasmussen,E. N. ,and D. O. Blanchard. 1998. A baseline climatology of sounding-derived supercell and tornado forecast parameters. Weather and Forecasting 13:1146-1164.

[124] Rasmussen,R. M. ,and A. J. Heymsfield. 1987. Melting and shedding of graupel and hail. Part I:Model physics. Journal of the Atmospheric Sciences 44:2754-2763.

[125] Richter,H. ,J. Peter,and S. Collis. 2014. Analysis of a destructive wind storm on 16 November 2008 in Brisbane,Australia. Monthly Weather Review 142:3038-3060.

[126] Rinehart,R. E. 2004. Radar for Meteorologists,482 pp. Columbia,MO:Rinehart Publications.

[127] Romine,G. S. ,D. W. Burgess,and R. B. Wilhelmson. 2008. A dual-polarization-radar-based assessment of the 8 May 2003 Oklahoma City area tornadic supercell. Monthly Weather Review 136:2849-2870.

[128] Ryzhkov,A. V. 2007. The impact of beam broadening on the quality of radar polarimetric data. Journal of Atmospheric and Oceanic Technology 24:729-744.

[129] Ryzhkov,A. V. ,T. J. Schuur,D. W. Burgess,and D. S. Zrni′c. 2005. Polarimetric tornado detection. Journal of Applied Meteorology 44:557-570.

[130] Ryzhkov,A. V. ,S. E. Giangrande,V. M. Melnikov,and T. J. Schuur. 2005. Calibration issues of dual-polarization radar measurements. Journal of Atmospheric and Oceanic Technology 22:1138-1155.

[131] Ryzhkov,A. V. ,M. Pinsky,A. Pokrovsky,and A. P. Khain. 2011. Polarimetric radar observation operator for a cloud model with spectral microphysics. Journal of Applied Meteorology and Climatology 50:

873-894.

[132] Ryzhkov, A. V. , M. R. Kumjian, S. M. Ganson, and A. P. Khain. 2013. Polarimetric radar characteristics of melting hail. Part I: Theoretical simulations using spectral microphysical modeling. Journal of Applied Meteorology and Climatology 52:2849-2870.

[133] Ryzhkov, A. V. , M. R. Kumjian, S. M. Ganson, and P. Zhang. 2013. Polarimetric radar characteristics of melting hail. Part II: Practical implications. Journal of Applied Meteorology and Climatology 52: 2871-2886.

[134] Ryzhkov, A. V. , M. Diederich, P. Zhang, and C. Simmer. 2014. Potential utilization of specific attenuation for rainfall estimation, mitigation of partial beam blockage, and radar networking. Journal of Atmospheric and Oceanic Technology 31:599-619.

[135] Ryzhkov, A. V. , P. Zhang, H. Reeves, M. Kumjian, T. Tschallener, S. Trömel, and C. Simmer. 2016. Quasi-vertical profiles-a new way to look at polarimetric radar data. Journal of Atmospheric and Oceanic Technology 33:551-562.

[136] Sachidananda, M. , and D. S. Zrni′c. 1986. Differential propagation phase shift and rainfall rate estimation. Radio Science 21:235-247.

[137] Schenkman, A. D. , M. Xue, A. Shapiro, K. Brewster, and J. Gao. 2011. Impact of CASA radar and Oklahoma Mesonet data assimilation on the analysis and prediction of tornadic mesovortices in an MCS. Monthly Weather Review 139:3422-3445.

[138] Schneebeli, M. , N. Dawes, M. Lehning, and A. Berne. 2013. High-resolution vertical profiles of X-band polarimetric radar observables during snowfall in the Swiss Alps. Journal of Applied Meteorology and Climatology 52:378-394.

[139] Schrom, R. S. , and M. R. Kumjian. 2016. Connecting microphysical processes in Colorado winter storms with vertical profiles of radar observations. Journal of Applied Meteorology and Climatology 55: 1771-1787.

[140] Schrom, R. S. , M. R. Kumjian, and Y. Lu. 2015. Polarimetric radar signatures of dendritic growth zones within Colorado winter storms. Journal of Applied Meteorology and Climatology 54:2365-2388.

[141] Schultz, C. J. , et al. 2012. Dual-polarization tornadic debris signatures. Part I: Examples and utility in an operational setting. Electronic Journal of Operational Meteorology 13(9):120-137.

[142] Seliga, T. A. , and V. N. Bringi. 1976. Potential use of radar differential reflectivity measurements at orthogonal polarizations for measuring precipitation. Journal of Applied Meteorology 15:69-76.

[143] Seliga, T. A. , and V. N. Bringi. 1978. Differential reflectivity and differential phase shift: applications in radar meteorology. Radio Science 13:271-275.

[144] Shapiro, A. M. , C. K. Potvin, and J. Gao. 2009. Use of a vertical vorticity equation in variational dual-Doppler wind analysis. Journal of Atmospheric and Oceanic Technology 26:2089-2106.

[145] Silber, I. , and C. Price. 2016. On the use of VLF narrowband measurements to study the lower ionosphere and the mesosphere-lower thermosphere. Surveys in Geophysics. https://doi. org/10. 1007/s10712-016-9396-9.

[146] Sinclair, V. A. , D. Moisseev, and A. von Lerber. 2016. How dual-polarization radar observations can be used to verify model representation of secondary ice. Journal of Geophysical Research-Atmospheres 121: 10954-10970.

[147] Skolnik, M. I. 2001. Introduction to Radar Systems, 772 pp. New York: McGraw Hill.

[148] Snyder,J. C. ,H. B. Bluestein,G. Zhang,and S. J. Frasier. 2010. Attenuation correction and hydrometeor classification of high-resolution,X-band,dual-polarized mobile radar measurements in severe convective storms. Journal of Atmospheric and Oceanic Technology27:1979-2001.

[149] Snyder,J. C. ,A. V. Ryzhkov,M. R. Kumjian,A. P. Khain,and J. C. Picca. 2015. A ZDR column detection algorithm to examine convective storm updrafts. Weather and Forecasting 30:1819-1844.

[150] Spoden,P. J. ,R. A. Wolf,and L. R. Lemon. 2012. Operational uses of spectrum width. Electronic Journal of Severe Storms Meteorology 7(2):1-28.

[151] Straka,J. M. ,and D. S. Zrni'c. 1993. An algorithm to deduce hydrometeor types and contents from multiparameter radar data. In 26th Conference on Radar Meteorology,Norman,OK,513-516. Boston:American Meteorological Society(Preprints).

[152] Sulia,K. J. ,and M. R. Kumjian. 2017. Simulated polarimetric fields of ice vapor growth using the adaptive habit model. Part I:Large-eddy simulations. Monthly Weather Review 145:2281-2302.

[153] Sulia,K. J. ,and M. R. Kumjian. 2017. Simulated polarimetric fields of ice vapor growth using the adaptive habit model. Part II:A case study from the FROST experiment. Monthly Weather Review 145:2303-2323.

[154] Testud,J. ,E. Le Bouar,E. Obligis,and M. Ali-Mehenni . 2000. The rain profiling algorithm applied to polarimetric weather radar. Journal of Atmospheric and Oceanic Technology 17:332-356. 155. Thompson,R. L. ,R. Edwards,J. A. Hart,K. L. Elmore,and P. M. Markowski. 2003. Close proximity soundings within supercell environments obtained from the Rapid Update Cycle. Weather and Forecasting 18:1243-1261.

[156] Thompson,R. L. ,C. M. Mead,and R. Edwards. 2007. Effective storm-relative helicity and bulk shear in supercell thunderstorm environments. Weather and Forecasting 22:102-115.

[157] Thompson,E. J. ,S. A. Rutledge,B. Dolan,V. Chandrasekar,and B. L. Cheong. 2014. A dualpolarization radar hydrometeor classification algorithm for winter precipitation. Journal of Atmospheric and Oceanic Technology 31:1457-1481.

[158] Tong,M. ,and M. Xue. 2005. Ensemble Kalman filter assimilation of Doppler radar data with a compressible nonhydrostatic model OSS experiments. Monthly Weather Review 133:1789-1807.

[159] Torres,S. ,Y. F. Dubel,and D. S. Zrni'c. 2004. Design,implementation,and demonstration of a staggered PRT algorithm for the WSR-88D. Journal of Atmospheric and Oceanic Technology 21:1389-1399.

[160] Torres, S. , R. Passarelli Jr. , A. Siggia, and P. Karhunen. 2010. Alternating dual-pulse, dualfrequency techniques for range and velocity ambiguity mitigation on weather radars. Journal of Atmospheric and Oceanic Technology 27:1461-1475.

[161] Tridon,F. ,and A. Battaglia. 2015. Dual-frequency radar Doppler spectral retrieval of rain drop size distributions and entangled dynamics variables. Journal of Geophysical Research-Atmospheres 120:5585-5601.

[162] Trömel, S. , M. R. Kumjian, A. V. Ryzhkov, C. Simmer, and M. Diederich. 2013. Backscatter differential phase—estimation and variability. Journal of Applied Meteorology and Climatology 52:2529-2548.

[163] Trömel,S. ,A. V. Ryzhkov,P. Zhang,and C. Simmer. 2014. Investigations of backscatter differential phase in the melting layer. Journal of Applied Meteorology and Climatology 53:2344-2359.

[164] Van Den Broeke,M. S. ,and S. T. Jauernic. 2014. Spatial and temporal characteristics of tornadic debris signatures. Journal of Applied Meteorology and Climatology 53:2217-2231.

[165] Van Den Broeke, M. S. , D. M. Tobin, andM. R. Kumjian. 2016. Polarimetric radar observations of precipitation type and rate from the 2-3 March 2014 winter storm in Oklahoma and Arkansas. Weather and Forecasting 31:1179-1196.

[166] Van Lier-Walqui, M. , A. M. Fridlind, A. S. Ackerman, S. Collis, J. Helmus, D. R. MacGorman, K. North, P. Kollias, and D. J. Posselt. 2016. On polarimetric radar signatures of deep convection for model evaluation: columns of specific differential phase observed during MC3E. Monthly Weather Review 144: 737-758.

[167] Vincent, R. A. , and I. M. Reid. 1983. HF Doppler measurements of mesospheric gravity waves momentum fluxes. Journal of the Atmospheric Sciences 40:1321-1333.

[168] Vivekanandan, J. , D. S. Zrni′c, S. Ellis, D. Oye, A. V. Ryzhkov, and J. M. Straka. 1999. Cloud microphysics retrieval using S-band dual-polarization radar measurements. Bulletin of the American Meteorological Society 80:381-388.

[169] Wakimoto, R. M. , P. Stauffer, W. -C. Lee, N. T. Atkins, and J. Wurman. 2012. Finescale structure of the LaGrange, Wyoming, tornado during VORTEX2: GBVTD and photogrammetric analyses. Monthly Weather Review 140:3397-3418.

[170] Wurman, J. 1994. Vector winds from a single-transmitter bistatic dual-Doppler radar network. Bulletin of the American Meteorological Society 75:983-994.

[171] Wurman, J. M. , and K. A. Kosiba. 2013. Finescale radar observations of tornado and mesocyclone structures. Weather and Forecasting 28:1157-1174.

[172] Wurman, J. , and M. Randall. 2001. An inexpensive, mobile rapid-scan radar. In 30th Conference on Radar Meteorology, Munich, Germany, P3. 4. Boston: American Meteorological Society(Preprints).

[173] Wurman, J. , S. Heckman, and D. Boccippio. 1993. A bistatic multiple-Doppler radar network. Journal of Applied Meteorology 32:1802-1814.

[174] Wurman, J. M. , J. Straka, E. Rasmussen, M. Randall, and A. Zahrai. 1997. Design and deployment of a portable, pencil-beam, pulsed, 3-cm Doppler radar. Journal of Atmosphericand Oceanic Technology 14: 1502-1512.

[175] Yokota, S. , H. Sako, M. Kunii, H. Yamauchi, and H. Niino. 2016. The tornadic supercell on the Kanto Plain on 6 May 2012: polarimetric radar and surface data assimilation with EnKF and ensemble-based sensitivity analysis. Monthly Weather Review 144:3133-3157.

[176] Zhang, G. , R. J. Doviak, D. S. Zrni′c, R. D. Palmer, L. Lei, and Y. Al-Rashid. 2011. Polarimetric phased-array radar for weather measurement: a planar or cylindrical configuration? Journalof Atmospheric and Oceanic Technology 28:63 73.

[177] Zrni′c, D. S. , and A. V. Ryzhkov. 1999. Polarimetry for weather surveillance radars. Bulletin ofthe American Meteorological Society 80:389-406.

[178] Zrni′c, D. S. , J. F. Kimpel, D. E. Forsyth, A. Shapiro, G. Crain, R. Ferek, J. Heimmer, W. Benner, T. J. McNellis, and R. J. Vogt. 2007. Agile-beam phased array radar for weather observations. Bulletin of the American Meteorological Society 88:1753-1766.

[179] Zrni′c, D. S. , R. J. Doviak, G. Zhang, and A. V. Ryzhkov. 2010. Bias in differential reflectivitydue to cross coupling through the radiation patterns of polarimetric weather radars. Journalof Atmospheric and Oceanic Technology 27:1624-1637.

第3章　龙卷观测的地基雷达技术

David J. Bodine,James M. Kurdzo

3.1 雷达在龙卷研究中的作用

观测数据显示,地球上最强的风由龙卷产生,最高可超过 135 m/s[174,203]。在美国,平均每年有 70 人因龙卷而死亡,而强龙卷(如 2011 年 4 月 27 日)可导致数百人死亡,单个龙卷的爆发也可造成数十亿美元的损失。美国平均每年被报道的龙卷事件有 1250 起,包括欧洲、中国、澳洲阿根廷和孟加拉国在内的世界其他地区也有大量的龙卷事件发生[72,75]。

对局地强风暴和龙卷的观测受制于相对有限的技术手段,包括原位探测、遥感以及实地灾情调查。由于原位探测站点的密度相对较低[84],雷达为人们认识这些现象提供了大量的精细观测[118]。虽然重要气象参量的直接测量可由原位探测完成,但原位探测主要是在地表进行,而且只可能在一个龙卷超级单体内和附近的少数地点进行。对龙卷的原位探测非常难,而且相关数据稀少,因为测量仪器必须放置在龙卷路径上。在另一方面,雷达测量可穿透母体风暴和龙卷,实现风场和降水场的三维高时空分辨率成像,并且可以在距离龙卷相对安全的距离上完成采样。实地灾情调查提供了关于龙卷风速和直径的重要信息,并可分析得到雷达无法识别的更小尺度的龙卷结构,如吸管涡旋。灾情调查信息结合龙卷照片判读技术可以实现龙卷动力结构的间接测量,而多普勒天气雷达更直接地测量风速。因此,雷达观测提供了风速测量,形成了关于龙卷的一般风速和直径典型范围的知识基础。

为了改进龙卷的预报并减少龙卷的风险和危害,龙卷研究人员已经在几个重要的方向开展工作。为了更好地预测龙卷,科学家必须首先了解龙卷是如何形成的,然后将这种理解应用到通过气象探测(包括雷达)或数值天气预报(其中一些同化了雷达数据)来识别龙卷形成的技术上。多普勒雷达提供了龙卷生成期间风和降水场变化情况的重要信息,因此,多普勒雷达可在许多情况下用于探测龙卷的发生过程。龙卷研究的另一个重点是研究风场和龙卷导致的民居和其他建筑物或树木碎片的三维分布。对风场和龙卷碎片三维特征的认识可帮助科学家和工程师评估龙卷如何影响不同的建筑结构,然后通过改进建筑的结构设计来减轻龙卷对生命和财产的威胁。

几十年来,地基雷达技术的新发展大大提高了人们对龙卷和龙卷风暴的科学认识。这些研究成果提高了龙卷预警的能力,使预警时间从 NEXRAD(新一代天气雷达)网中部署多普勒雷达之前的 5 min 延长到现在的 14 min 左右[6,169]。自NEXRAD 网部署以来,龙卷造成的伤亡人数减少了近 50%[169]。尽管这些改进令人印象深刻,但在缓解龙卷对社会的危害方面仍存在许多重大挑战，而地基雷达系统的持续科学进步是应对这些挑战的关键。

提高龙卷的预报水平和加深对龙卷的社会风险的认识面临许多挑战，而新的雷达技术有潜力应对这些挑战。随着天气雷达技术的成熟，多普勒天气雷达完成一次风暴体积扫描所需的时间(称为更新时间)已从约 5 min 提高到 5～10 s。本章将讨论实现这些快速扫描的新雷达技术，包括其在未来业务雷达网中应用的可行性。新的雷达系统还纳入了双偏振雷达技术，即雷达发射和接收正交偏振的两种电磁波。双偏振技术使科学家和预报员能够更好地诊断龙卷风暴中的降水粒子类型(冰晶霰、干冰雹、冰雹伴随降雨、雨滴)分布，并增加了远程探测龙卷碎片的能力[22,97,158]。

在本章中，将讨论用于研究龙卷的地基雷达系统，并探讨科学家如何使用这些雷达系统来推进有关龙卷的科学认识和提高龙卷预报水平。还将介绍地基雷达系统的历史概况，包括雷达技术的概述，讨论如何利用这些雷达开展外场科学试验，并详细探讨使用不同的雷达系统获得的关于龙卷成因和龙卷动力学的科学认知。本章的总结部分讨论了未来的地基雷达系统，包括未来的雷达系统将如何帮助科学家认识尚未解决的关于龙卷的重要科学问题。

3.2 龙卷的理论和模拟

在讨论龙卷的地基雷达观测之前，这里将龙卷的理论和模拟作为背景介绍，可以在有关龙卷动力学的综述论文中找到关于这些课题更深入的讨论[53,151]。龙卷理论、涡室实验和数值模拟的开展使龙卷流场结构的概念模型得以建立。图 3.1 展示了由大涡模拟(LES)[24,126]得到的龙卷内部流场。龙卷外流区的特征是角动量恒定($\Gamma=vr$)，因此随着龙卷中心距离的增加切向速度成比例下降到 $1/r$。在龙卷核心区域内，类似一个固态核心，切线速度 v 与距离 r 成正比，角动量沿径向向外流区增加。流场表现为旋转平衡，即向内的径向气压梯度力被向外的离心力所平衡。

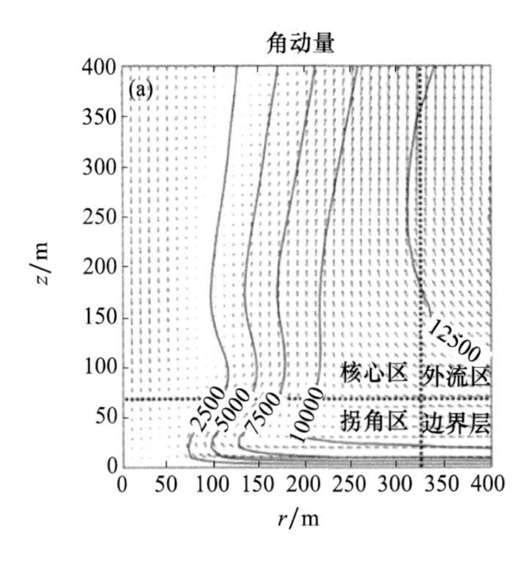

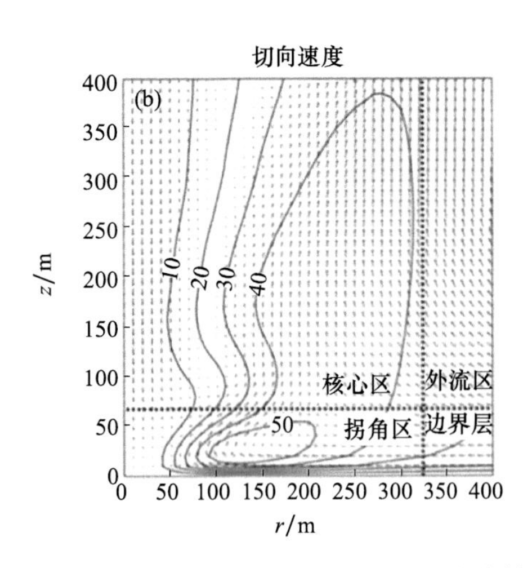

图 3.1 大涡模拟所得龙卷模型示例。图中，r-z 平面的风矢量叠加了(a)角动量和(b)切向速度的等值线。拐角区、边界层(B. L.)、核心区和外流区在每个图上都有标注

对于拐角流和边界层区域，摩擦会导致旋转平衡的破坏。摩擦导致了切向速度的降低和离心力的减少，而龙卷中心巨大的气压差基本不受影响。因此，在较大的径向压力梯度的驱动下，近地表气流朝内流向龙卷（辐合）。虽然地表附近的摩擦效应最大，但最大的切向速度通常在拐角流区内观测到。这是因为剧烈的沿径向向内急流将角动量输送到相比高空更小的半径内，从而产生更强烈的龙卷[61,111,113]。龙卷的拐角气流和边界层区域较浅，最大高度在地面以上 10～100 m。在非常靠近地面处（例如<10 m AGL），由于摩擦耗散，角动量会减少，这个区域称为摩擦层（图 3.1b）。用多普勒雷达观测这些区域很困难，因为雷达的波束高度随着距离的增加而升高，并且除了非常近的近邻区（例如，小于几千米），波束都在这些区域之上。

早期的实验室研究发现，随着旋流比的变化，龙卷的环流类型发生显著的变化[46,51,150]，特别是旋流比的变化往往会影响由径向风和垂直风组成的次级环流。旋流比是一个无量纲参数，通过取切向速度（或环流）与垂直速度（或质量通量）之比来计算。在低旋流比时，径向气流向中心轴汇聚，在龙卷中心形成向上旋转的抛射气流，形成单胞涡旋。龙卷结构由低旋流比到高旋流比的演变如图 3.2 所示。单胞涡旋的大涡模拟（图 3.2a）显示了一个核心直径窄的龙卷，伴随着贯穿整个龙卷的高架的高切向速度和上升气流区域。

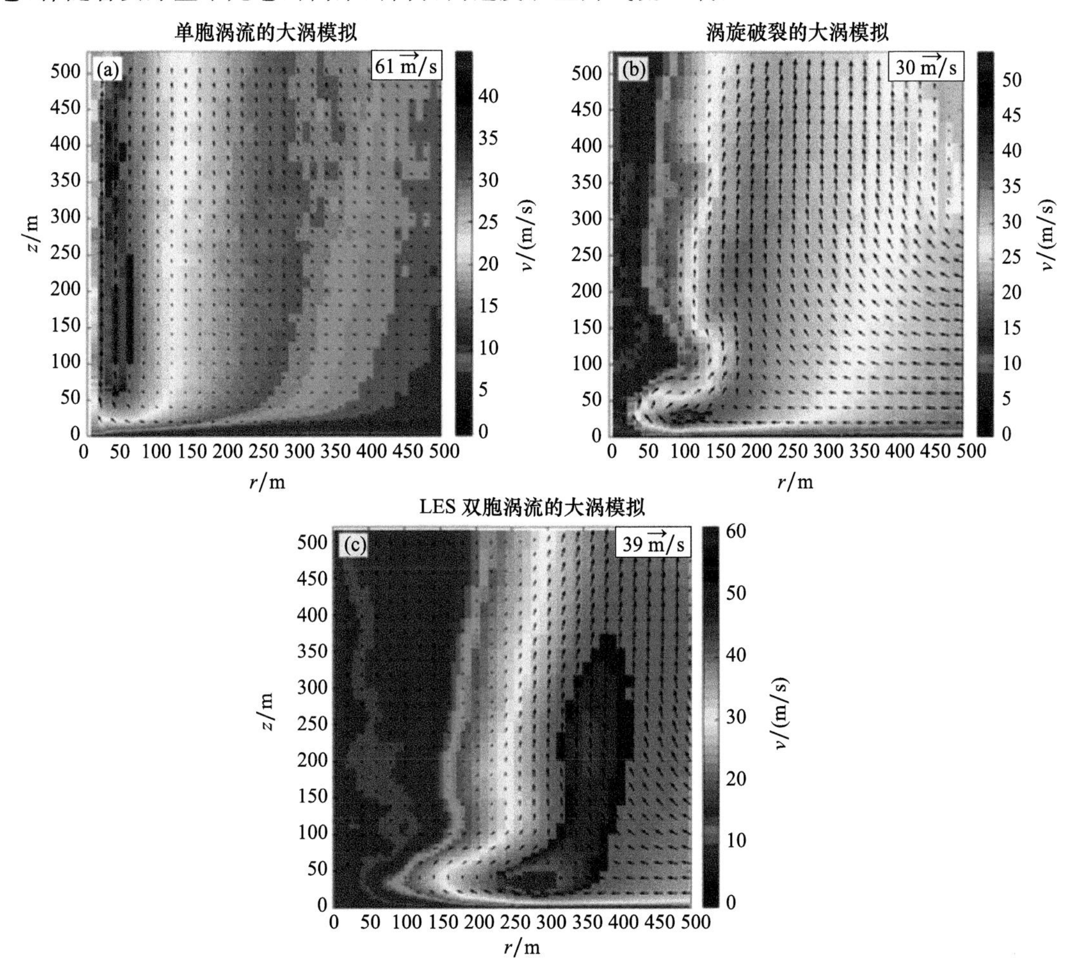

图 3.2　(a)单胞涡流；(b)涡旋破裂；(c)双胞涡流的大涡模拟。该数据由每 20 m 计算 1 次风的轴对称平均值得到。风矢量显示的气流分量在 *r-z* 平面，切向速度按色标着色

随着旋流比的增加，龙卷中心形成下沉气流，并逐渐向下渗透到地表（图 3.2b），这种类型的流场称为涡旋破裂[5]，涡旋破裂发生在沿中轴上升气流和下沉气流交汇的地方。在涡旋破裂时，除了垂直运动发生反转外，涡旋破裂上方也发生了由层流向湍流的过渡。图 3.2b 所示为 100 m AGL 以下上升气流和 100 m AGL 以上下沉气流形成的涡旋破裂的大涡模拟。涡旋破裂导致近地面的切向速度的增强明显比高空显著。处于临界旋流比时，涡旋破裂恰好发生在地表之上，地面附近龙卷相对于高空表现出强度的最大增长，这通常被称为最佳配置[111,151]。随着旋流比的进一步增大，龙卷流场呈现出中心下沉气流被上升气流环绕的双胞涡旋结构。图 3.2c 给出了一个中心下沉气流延伸到地表的双胞涡旋的大涡模拟。

当旋流比非常高的时候，龙卷会形成多个强涡旋围绕着母涡旋打转，称为吸管旋涡或子旋涡。在某个时刻，子旋涡的数量可以在 2～6，子旋涡可以在多个尺度上存在，也可以具有与上述流场类型类似的流场结构，例如，子涡旋可以包含更小的子涡旋或具有单/双胞涡旋。图 3.3 显示了一个用大涡模拟得到的多涡龙卷的例子：该模拟是利用多普勒雷达在 2010 年 5 月 10 日俄克拉何马州穆尔市观测到的 EF-4 龙卷的三维轴对称风进行初始化的，该模拟产生了四个大的子涡，这些子涡在模拟区域内产生了最强的水平和垂直速度以及最高的湍流动能（TKE）。在更大的尺度上，龙卷的特征是围绕着广阔的中心下沉气流的狭窄强上升气流带。在上升-下沉气流交界处，龙卷子涡沿着垂直速度的巨大水平梯度形成，子涡旋的产生可导致水平风速比龙卷内平均气流的速度高 2～3 倍[60,113,136]，并常常加剧龙卷的灾害性[67,68]。

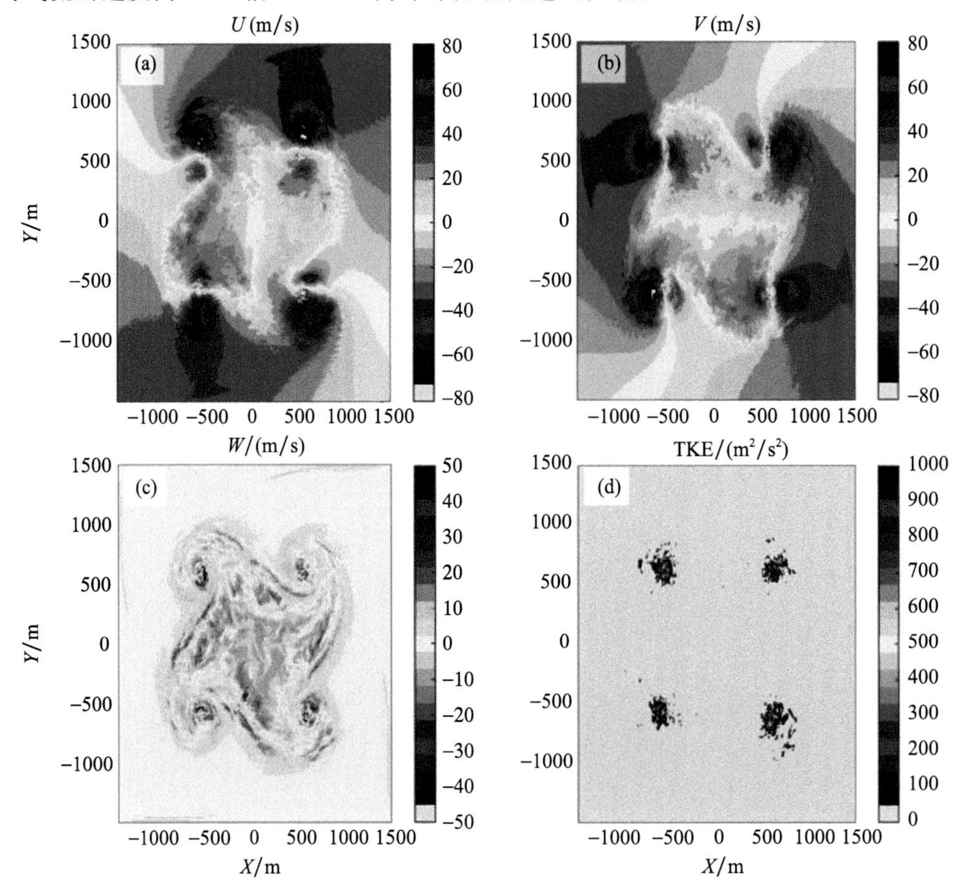

图 3.3 利用多普勒雷达数据初始化侧边界和上边界条件对 2010 年 5 月 10 日俄克拉何马州穆尔市龙卷进行的大涡模拟。在 64 m(AGL)处的瞬时(a)纬向风(U)，(b)经向风(V)，(c)垂直速度(W)，(d)1.2 s 平均湍流动能。模拟的大子涡旋与文献中的单和双多普勒分析一致[74]

3.3 地基龙卷观测雷达的技术发展史

使用雷达观测龙卷的历史可以追溯到20世纪40年代，形成了一个不断扩充的龙卷数据集和相关知识的数据库。雷达系统和能力在这段时间内得到了显著的发展，产生了一系列具有不同空间和时间分辨率、灵敏度、基于多普勒效应的测速能力、偏振类型、天线设计、发射机种类和波长的广泛而多样的实例。早期龙卷观测主要由固定地点的雷达收集，包括军队系统、国家强风暴实验室（NSSL）的原型雷达和一些天气监测雷达（WSR-57、74和88D系统）。然而，固定地点雷达的缺点是空间分辨率有限，如果目标风暴没有碰巧靠近雷达所在地点，则缺乏低层波束覆盖。移动雷达历来为观测和分析对流风暴及龙卷提供了许多最佳机会，因为它们能够相对较近地观测风暴[10,207,209]。

3.3.1 早期雷达观测

对龙卷预警的需求推动了许多早期使用雷达对龙卷的观测和研究。在天气雷达出现之前，军事雷达经常被用来为即将到来的风暴进行预警。1948年3月25日，在俄克拉何马州俄克拉何马市的Tinker空军基地，Ernest J. Fawbush少校和Robert C. Miller上尉第一次对龙卷进行了详细预报，Bradford描述了这一过程[28]。Fawbush和Miller根据当天天气流型与5天前袭击基地的龙卷的相似性，并结合军用雷达认识到雷暴在俄克拉何马城西北60英里①到西南100英里处形成，发布了他们的龙卷临近预报[28]。尽管人们普遍认为5天后龙卷再次袭击Tinker空军基地的可能性不大，这个成功龙卷预警具有很幸运的成分，但这一预警还是给了人们希望，认为龙卷预报和预警确实是可能的。

5年后的1953年4月9日，当伊利诺伊州水资源调查局的研究人员在伊利诺伊州香槟市附近一部重新组装的天气雷达上第一次观测到钩状回波时，这个想法向前迈进了一大步[11]。在发现雷达上这个奇怪的形状附近有龙卷后，钩状回波第一次与龙卷联系在一起。仅仅一个月后，1953年5月11日，得克萨斯农工大学的研究人员利用改进的S波段SO-7N雷达在一个事后确认的龙卷雷暴中观察到了类似的特征[28]，得克萨斯州韦科市在这场风暴中遭受了巨大的破坏，114人丧生。研究人员将钩状回波与由此产生的龙卷联系起来，从而实现了使用包括WSR-1/A、WSR-3和WSR-4在内的各种AN/APS-2F军用雷达改装的天气雷达覆盖南部和中部平原的龙卷探测和预警网络的最终目标[198]。1956年4月5日，上述某一部雷达被得克萨斯农工大学的气象学者在得克萨斯州布莱恩和学院站附近首次基于雷达的龙卷预警[8,198]。结合迈阿密飓风预报中心的努力，这些观测及其社会效益被用于开发美国第一个大型天气雷达网：C波段WSR-57雷达系统，该系统由遍布美国的66部AN/FPS-41雷达组成[149,198]。

在接下来的1～2年里，WSR-57网观测到了多个龙卷，但龙卷预警缺少一个关键部分：速度数据。研究人员发现，并不是每个钩状回波都会产生龙卷，也不是每个龙卷都与钩状回波有关，这使得研究人员相信，要想有效地发出龙卷警报，还需要其他信息。天气用途的多种脉冲

① 1英里≈1.6 km，下同。

多普勒雷达在20世纪60年代和70年代被开发，但直到1973年5月24日，NSSL的S波段多普勒雷达才发现雷达和龙卷之间的另一个联系[39]。在那次风暴中，NSSL雷达探测到一个紧密耦合的相邻距离库的风切变（龙卷涡旋特征TVS），与俄克拉何马州联合城附近的龙卷处于同一位置。这次观测的结论是，至少在某些情况下，多普勒雷达观测到的低层强旋转切变的存在可能表明龙卷的存在。当然，雷达波束一般位于地面以上，所以旋转存在的高度可能也在地面以上，这种旋转可能对应一个强中气旋，而不是龙卷[33]。尽管如此，这次观测发现了文献[33]中所描述的龙卷涡旋特征（TVS），并进一步推动了在美国建立地基多普勒雷达网的计划。

在20世纪70年代和80年代初，天气雷达界发生了很多变化。美国在20世纪70年代中期首次升级其国家天气雷达网，采用无多普勒功能的C-和S-波段WSR-74雷达组合。WSR-74是在现有的WSR-57的基础上增加的，以创建一个遍布美国的由128部雷达组成的网。通过整合71部WSR-74雷达和1981—1985年间增加的12部雷达，美国国家气象局（NWS）的"本地预警"雷达总数达到83部[198]。此外，许多WSR-57雷达在20世纪90年代中期一直在工作。大约在WSR-74网发展的同时，由于其对公共安全的巨大益处，天气雷达网开始在世界其他国家出现[47,82]。

然而，部分由于1973年在联合城附近的龙卷观测，多普勒雷达逐渐出现，许多研究中心都在开发新的雷达技术。那个时候，专门为气象用途设计的多普勒雷达在国家强风暴实验室（NSSL）、国家大气研究中心（NCAR）、空军剑桥研究实验室和康奈尔航空实验室等开发[11]。在20世纪70年代中期，NSSL在俄克拉何马城附近增加了第二部多普勒雷达，从而具备双多普勒性能。由于单台雷达只能提供沿雷达波束的多普勒速度，因此需要两台多普勒雷达来反演二维风场。在俄克拉何马州中部收集了一系列双多普勒个例[29,32,173]，其中一些个例包括时间序列数据，这些数据被用于检验龙卷中的多普勒谱[11,220-222]。

3.3.2 首次便携式多普勒雷达观测

虽然在20世纪80年代末，用于气象观测的地基雷达已经发展得很好了，但很明显，大多数龙卷不会发生在固定站点雷达附近。尽管固定站点雷达对于可持续的预警网络很重要，但它们不能提供大量的高质量数据用于了解更多关于龙卷动力学、成因和生命周期的信息，在地面上获得高空间分辨率数据的最好方法是用一个移动平台去观测龙卷。由Howie Bluestein领导的俄克拉何马大学研究小组与洛斯阿拉莫斯国家实验室（LANL）合作，将连续波X波段多普勒雷达转换为便携式天气雷达系统，用于观测龙卷[11]。LANL系统的便携特性使研究人员能够非常接近龙卷，使低海拔处的多普勒谱（反射率因子、平均径向速度和速度谱宽）和超过120 m·s^{-1}的风速能够被采样[16]。LANL雷达还允许频率调制，从而能够确定跨越多个距离库的谱[13]，这是通过便携式多普勒雷达测风进一步获得龙卷动力学知识的第一步。

第二代LANL雷达是一个安装在卡车车厢上的W波段扫描多普勒雷达，由马萨诸塞大学阿默斯特分校（UMASS）建造，并再次被OU的Bluestein团队使用。采用W波段可能的好处是一个相对较小的蝶形天线可以安装在卡车上，同时仍然提供比0.6°更好的波束宽度（在1999年升级后比0.2°更好）[11,12]。然而，W波段的衰减严重，导致需要非常接近强龙卷。正如将在接下来的小节中看到的，移动雷达系统频率的选择对观测任务是非常重要的，而且不同选择的差异很大。在1999年的龙卷多发季节，高空间分辨率的W波段UMASS雷达能够识别龙卷中的弱回波洞(WEHs)或“眼”、螺旋带、多重涡旋和摆线运动[12,17,19,179]。从 20 世纪 90 年代中期到 2010 年,该雷达被用于 VORTEX2 实验,为雷达和灾害性风暴研究带来了许多突破性的观测和发现。

在 20 世纪 90 年代中期的同一时间段,OU 的 Josh Wurman 和他的团队完成了第一部 X 波段车载多普勒雷达(DOW)的工作[209,210]。在第一次龙卷起源验证实验(VORTEX)[147]中,DOW 首次收集了强龙卷中反射率因子并估计了三维风场[206]。DOW 第一次收集的数据使人们更好地了解了龙卷中的离心碎片[58],以及风场的结构及其与龙卷带来的灾情之间的关系[1,205]。在第一个 DOW 成功实现后不久,类似于最初 NSSL 的双多普勒网,第二个 DOW 被建造用于双多普勒研究。DOW2 具备在多个地点放置移动雷达的能力,从而提供了更好地靠近风暴的数据收集机会[41]以及在前所未有的空间和时间分辨率下进行双多普勒研究的能力[3,123,211,212]。DOWs 在 2000 年代早期到中期经过多次迭代更新,由于 X 波段雷达是空间分辨率和便携式平台上的天线尺寸之间的相对可接受的折中方案,X 波段雷达成为移动多普勒天气雷达的主要设备。

3.3.3　当前的移动多普勒雷达系统

为了观测对流风暴尺度的过程和缓解严重的衰减问题,其他频率在 21 世纪初得到使用。共享大气研究和教学移动雷达(SMART-R 项目)是 OU、得克萨斯州农工大学、得克萨斯州(简称得州)理工大学和 NSSL 的合作成果[7]。SMART-Rs 是由 OU 的 Mike Biggerstaff 和他的团队从 2000 年代末开始运行的,它由一对 C 波段雷达组成,采集强对流风暴[188]、飓风[88]和其他现象[140]中的双多普勒风场。由于它们的波长更长,所以需要相对较大的天线,这使得搭载 SMART-R 的卡车相当大。虽然 SMART-Rs 不像其他移动平台那样灵活,无法接近龙卷,但在 VORTEX2 期间,SMART-Rs 在超级单体和龙卷附近的对流风暴尺度数据收集中发挥了重要作用[170,171,214]。在 2010 年代早期,其中一部车载 SMART-R 被改装为双偏振雷达,使其成为美国目前使用的唯一一部双偏振功能的 C 波段移动多普勒天气雷达。

另一个实现偏振观测的系统是 UMASS 偏振 X 波段雷达[20](图 3.4)。OU 的 Bluestein 团队使用了该雷达,该雷达推动了 X 波段偏振天气雷达的激增。许多关键偏振研究结果得到了 UMASS 的 X 波段雷达支持,包括超级单体的偏振特征(在 X 波段;Ryzhkov、Kumjian 等人以前在 S 波段报道过这样的观测)[176]。此外,该雷达还观测到了第一个 EF5 型龙卷(基于增强的藤田尺度),形成了一个长持续时间数据集,为认识周期性龙卷到长轨迹龙卷的过渡提供了线索[180]。

在频谱的另一方面,得州理工的 Ka 波段雷达(TTUKa)被设计用于收集超级单体和龙卷中极高空间分辨率的数据[78,197]。Ka 波段雷达的 3-dB 波束宽度为 0.33°,它提供了比大多数 X 波段移动雷达高三倍的方位分辨率和非常高的晴空探测灵敏度。由 Chris Weiss 和他在德州理工的团队运作的这两部 Ka 波段雷达,类似于 SMART-Rs 和 DOWs,在任何可能的情况下协同工作,收集双多普勒观测数据。Ka 波段雷达的一个显著缺点是衰减明显,在强降雨时可导致信号完全消失。

到 2010 年代初,大量的移动偏振 X 波段雷达成为观测主力。NSSL NOXP 是一个 1°波束宽度的 X 波段偏振测量系统,用于 VORTEX2、夜间平原高架对流(PECAN)[70]和许多较小的外场项目。主要由 Don Burgess 和 Ted Mansell 牵头，NOXP 主要在 VORTEX2 期间[42,164]观测了龙卷动力结构和龙卷生成，并且是在几个龙卷和非龙卷超级单体钩状回波中确定雨滴大小的数据源[65]。DOWs 的多次更新迎来了 DOWs3-4 [203]，并最终产生 DOW6 和 DOW7 [213]，偏振 X 波段系统也被用于 VORTEX2、PECAN 以及雷达观测龙卷和雷暴试验(ROTATE)。在灾害天气研究中心(CSWR)Josh Wurman 的领导下，VORTEX2 中超级单体和龙卷观测个例主要由 DOWs 6 和 DOWs 7 观测得到[2,92,125,190,191,215]。CSWR 的 DOWs 为强对流风暴和龙卷研究，以及天气雷达学术圈做出了非常多的贡献，为很多文献提供了观测个例[193,207,216]。

图 3.4　UMASS X 偏振移动雷达正在扫描龙卷。照片由 Jeff Snyder 提供

在同一时期，又开发了不同形式和不同用途的 X 波段偏振测量系统。OU 的先进雷达研究中心(ARRC)开发了 PX-1000，一种可移动的 X 波段偏振天气雷达，波束宽度为 1.8° [44]。在 OU 的 Boon Leng Cheong 的领导下，PX-1000 参与了 PECAN 和其他各种较小的外场项目，同时在灾害性风暴季节作为“固定”平台驻扎在俄克拉何马州的诺曼。由于 PX-1000 是一个基于拖车的平台，它不适合追逐龙卷；然而，它是 2013 年 5 月 20 日俄克拉何马州穆尔市 EF5 龙卷的主要观测雷达，提供了关于龙卷碎片抛射和“未能到达的”锢囚的新发现[102]。PX-1000 还用作天气雷达脉冲压缩技术的试验，该技术现在被用于多种以龙卷尺度强烈天气探测为目标的移动和固定的天气雷达和其他世界各地的天气雷达[101]。阿拉巴马大学亨茨维尔分校(UAH)移动式阿拉巴马 X 波段(MAX)雷达是另一种用于飓风和龙卷研究的双偏振 X 波

段雷达[142]。MAX 主要用于闪电、夜间高架对流风暴(PECAN 期间)和龙卷研究(VORTEX-SE 期间)。

对龙卷研究人员至关重要的一点是极其迅速的体扫更新速度。例如,目前的 WSR-88D 新一代天气雷达网在体积扫描模式(VCP)12 [37] 中以 4.2 min 的速度完成它的最快体扫,而龙卷的平流时间尺度在秒级。在很多情况下,龙卷甚至不会持续 4.2 min,因此其整个演变过程就被错过了。由于这个原因,已经发展了以扫描速度为关键设计标准的一小部分移动雷达系统,具有秒级的扫描能力,目前有四个雷达属于这一类,都在 X 波段工作。首先开发的是 CSWR Rapid DOW[208],最初称为 DOW5(图 3.5)。该雷达利用 9.3～9.75 GHz 的频率范围内同时独立发射的 6 条笔形波束(0.8°～0.9°宽度)实现了 7 s 体扫。它可以通过在第二次扫描中增加第二组 6 条笔形波束来提高垂直分辨率,使 12 条波束的体扫更新时间达到 14 s;结合其 11 m 的距离分辨率,CSWR Rapid DOW 的时空分辨率可以对龙卷进行三维多重精细尺度研究[91,193,215]。用于龙卷研究的第二种快速扫描移动雷达是 MWR-05XP,这是一部由海军研究生院改造的雷达,由 Howie Bluestein 在 2000 年代末到 2010 年代中期在该领域使用[21]。MWR-05XP 在每个维度获得 2.0°有效波束宽度的 7 s 90°×20°体扫,和 24 s 360°×20°的体扫具有相同的空间分辨率。虽然它的空间分辨率(150 m)比其他快速扫描雷达略低,但它具有快速扫描雷达组的最高灵敏度,在 10 km 范围内具有－15 dBZ 的灵敏度。MWR-05XP 已被用于研究超级单体中龙卷涡旋特征 TVSs 的垂直发展[63,64]。

图 3.5 灾害性天气研究中心的快速 DOW 观测 2009 年 6 月 5 日怀俄明州歌珊县的龙卷
(图片由 Paul C. Robinson 提供)

第三种被开发的快速扫描移动雷达是快速 X 波段偏振雷达,简称 RaXPol [141]。作为快速扫描系统组中唯一的偏振雷达,Howie Bluestein 和他的团队自 2011 年以来一直使用 RaXPol 在非常近的距离观察龙卷的偏振特征极其迅速的变化。RaXPol 是一个基于抛物面天线的平台,可以 180°/s 的速度旋转,使 360°×20°扫描时间为 40 s(仰角的间隔和数量可以改变,以创建更快的扫描)。为了处理由于天线快速旋转而导致的有效波束宽度的扩大,RaXPol 利用跳频来更快地产生独立样本。到目前为止，RaXPol 已经对 2011 年[79]和 2013 年[192,193]的埃尔雷诺龙卷进行了分析，包括观测到类似于 PX-1000 的碎片尾和抛射物。

最后一种快速扫描移动雷达是大气成像雷达（AIR；图 3.6），由 Brad Isom 和几位 ARRC 合作者构建，并由 Jim Kurdzo 和 David Bodine 及其在 OU 的团队自 2012 年以来使用[81]。AIR 采用了一种被称为数字波束形成或雷达成像的技术[129]，虽然成像技术在大气风廓线研究领域已经使用过[43,129,137]，但 AIR 是首次尝试对强对流风暴进行成像。AIR 在垂直维度上传输一个宽波束（20°高，1°宽），并使用一个垂直的 36 单元相控阵，逐个接收宽度为 1°的波束。在这种方式下，每个脉冲形成单独的距离-高度显示（RHIs），天线只需要在方位维度上旋转，不需要用笔形波束进行垂直扫描。AIR 能够在 6 s 内形成 90°×20°的体扫，还可以在每个方位收集 RHI 形式的数据。AIR 已被用于分析强对流风暴和龙卷[103]以及阵风锋和密度流[115]。

图 3.6　俄克拉何马大学大气成像雷达（AIR）正在扫描龙卷

3.3.4　NEXRAD 和其他固定站点雷达

WSR-88D，也通常被称为 NEXRAD（“新一代雷达”），在 20 世纪 80 年代末和 90 年代初，将分布广泛的、地基、定点的多普勒雷达覆盖到美国大陆的绝大多数地区。特别是在美国的平原、中西部和东南部地区，在人口较多区域相对密集的多普勒天气雷达网，因其对强对流风暴和龙卷的先进预警和预报能力而挽救了不计其数的生命[143]。随着 WSR-88D 网的安装，龙卷预警次数和龙卷预警平均提前时间几乎增加了一倍，而预期伤亡人数下降了 40%～45%[169]。WSR-88D 是 S 波段雷达，短脉冲峰值功率 750 kW，0.9°波束宽度，体扫更新时间介于 4.2 min 和 10 min。这种功率、多普勒测速性能和时空分辨率的结合致使多年来众多算法得到了增强，包括风暴单体识别和追踪[83]、中气旋探测[105,178]和龙卷探测[131]算法。

尽管 WSR-88D 网的寿命已接近 30 年[219]，但对扫描策略、算法和性能进行了一系列重大更新，包括一个主要的使用寿命延长项目[49]。WSR-88D 采用了一系列针对不同天气情况的体积扫描模式 VCPs。从灾害性风暴的角度来看，以适当的速度扫描获取可接受的数据质量，加上理想的更新时间和仰角切割，对灾害性对流风暴算法的性能至关重要[34,35]。VCP 12[37]的实现对 WSR-88D 网尤其重要，因为它扫描速度更快，在较低仰角的垂直采样密度更大。WSR-88D 的另一个重要更新是 2000 年代中后期使用的超分辨率过采样技术。该技术被用于更好地探测中气旋和龙卷[36,202]，并增强了水文学的各个领域应用[165]。在快速更新速度方面，诸如自动体扫评估和终止（AVSET）和补充自适应体积内低层扫描（SAILS）等技术的改进使得在灾害性对流风暴扫描中更频繁地顾及较低仰角[49,50]。与之前描述的移动系统类似，WSR-88D 最关键的更新是在 2010 年代早期的双偏振观测能力[55]。发射倾斜 45°的脉冲并在水平和垂直通道上接收信号的能力为降雨量估计[157]、水凝物类型确定[139,159]和龙卷探测[158]开辟了新的可能性。

除了 WSR-88D，各种 S、C 和 X 波段定点雷达在 20 世纪 90 年代和 21 世纪初被引入美国各地的研究机构和大学，并作为国家科学基金会和其他机构资助的大规模研究计划的一部分。阿拉巴马大学亨茨维尔分校采购了一部被称为 ARMOR 的偏振 C 波段雷达，该雷达曾观测过许多大型冰雹和龙卷个例[89]，并参与了 VORTEX-SE 外场观测计划。俄克拉何马大学在 2009 年首次使用了双偏振 C 波段 OU-PRIME 雷达，一年之后，它以 0.5°波束宽度近距离采集了 1 次 EF4 龙卷过程的观测数据[138]。OU-PRIME 参与了超级单体钩状回波的偏振特性[93]以及有关龙卷碎片特征[23]的研究。此外，NSSL 和雷达运行中心（ROC）在过去几十年里运行了多个定点雷达，包括 KOUN（NSSL）和 KCRI（ROC），它们都是 WSR-88D 的偏振版本，位于俄克拉何马州诺曼的俄克拉何马大学韦斯特海默机场。这两部雷达在它们的历史上都观测到了许多强烈或猛烈的龙卷，并被用于各种相关的研究[22,98,158,181,195,217,218]。最后，机场多普勒天气雷达（TDWR）是一个由 45 部 C 波段天气雷达组成的网络，位于美国各地的主要机场，于 20 世纪 90 年代部署，用于应对微下击暴流和风切变引起的民航交通灾害[59,130]。Vasiloff[187]讨论了将 TDWRs 作为 WSR-88D 的补充，用于龙卷探测和预警。

在 WSR-88D 时代，许多概念已经显示出它们作为美国国家天气雷达网未来加强或替代的潜在优势。在这些优势中，龙卷预报员和研究人员关注的两个主要领域是：与龙卷的距离（即低层扫描能力）和时间分辨率[31]。这些概念中的第一个已经作为大气协同自适应遥测（CASA）项目的一部分[128]。CASA 计划在美国各地使用大量（可能是数千台）X 波段双偏振雷达，以满足美国对天气雷达的需求。拥有这么多雷达的好处是可以对龙卷附近进行低层扫描[100]，并有可能对预报预警过程进行多种多普勒雷达数据同化[177]。第一个 CASA 测试网部署在俄克拉何马州西南部，第二个网目前部署在达拉斯地区。一系列的龙卷个例促进了关于灾害性天气业务和数据同化的研究[114,145,160]。

美国下一代天气雷达网的第二个概念是多功能相控阵雷达（MPAR）[196,224]。MPAR 的概念认为，相控阵天线可以快速扫描风暴，因为不需要在方位角和仰角上操纵大型天线。作为 MPAR 研究的一部分，NSSL 采购了一套淘汰的 SPY-1A 美国海军相控阵雷达；它被命名为国家天气雷达试验台（NWRT；图 3.7）。NWRT 使用电子操控的笔形波束在 30～60 s 内完成一个体扫[76,77]，并观测到多个灾害性的风暴和龙卷个例[104,134,181]。在相控阵雷达创新遥感试验（PARISE）期间，预报员被邀请到灾害性天气试验平台，对 NWRT PAR 数据与 NEXRAD 数据进行比较，并评估更高时间分辨率的雷达数据对其预报的影响。预报员指出，快速扫描数据帮助他们更早地识别灾害性天气的前兆，并提高了龙卷预警的准确性[26,27]。

图 3.7　位于俄克拉何马州诺曼的国家天气雷达试验台(NWRT)
(该试验台将于 2018 年被 MPAR 先进技术演示器取代)

3.3.5　使用雷达的主要龙卷外场项目

在 20 世纪 90 年代中期之前，大多数涉及龙卷观测的外场工作都集中在目测[71,132]以及风暴追逐者和外场研究人员进行的现场观测[4]上，他们试图更多地了解强对流风暴的结构。在 NSSL 的龙卷拦截计划中进行的一些目测与前几节提到的同一次联合城风暴的雷达数据[133]相结合。Lemon 和 Doswell[109]报告了目测和雷达数据的进一步结合，Bluestein[10]对此进行了总结。在龙卷拦截计划期间，联合多普勒业务项目(JDOP)也致力于将目测与 NSSL 雷达特征[10]相结合。现场观测主要由可携带的龙卷观测站(TOTO)的数据组成，TOTO 是一个放置在龙卷路径上的设备，用于收集风速和风向、气压、温度和电晕放电的观测数据[9,10]。此外，还通过将无线电探空仪释放进入超级单体和龙卷中进行了额外补充的实地测量[14,15,153-155,189,200]。

如前所述，Howie Bluestein 和他的学生在 20 世纪 80 年代末和 90 年代初通过 LANL 系统首次对龙卷进行了地基便携式雷达观测[13]。到 20 世纪 90 年代中期，Bluestein 的研究团队也在使用 UMASS 的 W 波段雷达，但除了国家科学基金会(NSF)为 OU 的研究团队提供持续的资助外，还从未有组织地使用多部雷达来协同观测龙卷。在 1994 年和 1995 年进行的第一次龙卷起源验证试验(VORTEX)是在外场首次尝试多雷达(和其他仪器)实时协同研究龙卷[147]。在 VORTEX 的第一年，也就是 1994 年，只有 LANL 和 UMASS 的 W 波段可供研究人员使用[10]。其他的观测平台包括无线电探空仪、机载雷达、实地龙卷观测探头和自动站网。VORTEX 是第一次联合使用所有这些平台来认识强对流风暴和龙卷在运动学和热力学方面的联系。在 VORTEX 的第二年(1995 年)，第一部 DOW 加入了观测仪器的行列。DOW 在 1995 年 VORTEX 期间收集了第一批相对高分辨率的基于雷达的龙卷数据集，包括 5 月 16

日[209]、6 月 2 日和 6 月 8 日的个例。作为一个大型外场项目的一部分，这些首次收集的雷达观测资料在文献中进行了总结[10]。

在 VORTEX 项目之后，出现了一系列研究龙卷的小型外场项目，许多是由 OU 领导的，最终是强烈天气研究中心（CSWR）领导的实验。在 20 世纪 90 年代末，第一部 DOW 和第二部 DOW 的协同观测实现了对龙卷的第一次双多普勒研究。两个相关但规模较小的后续项目，SUBVORTEX（1998）和 VORTEX-99（1999）使用了类似的仪器，包括 1999 年 5 月 3 日在俄克拉何马州穆尔市龙卷期间收集的数据。Howie Bluestein 的小组在 20 世纪 90 年代到 2010 年代初用 UMASS 的 X 波段偏振雷达和 W 波段雷达进行了试验，DOWs 也在同一时间段参与了每年的灾害性天气外场试验。尽管大规模、有组织的外场试验利用现有的仪器提供了关于龙卷形成和龙卷动力学的更全面综合的认识，但这些小尺度的移动雷达研究同样对龙卷及其母风暴形成了有价值的理解。特别是，这些研究显著增加了移动雷达观测到的龙卷和超级单体个例的样本数量，因为与龙卷相关的重大外场项目每 10～20 年才进行一次。这些试验还纳入了几种新兴的移动雷达技术，包括偏振和快速扫描雷达，为超级单体的微物理特征提供了新的见解，并通过大幅提升的时间分辨率在理解龙卷形成和龙卷动力学方面取得了重大进展。

在 VORTEX 及其后续项目之后，该项目的第二阶段计划 VORTEX2 在 2009 年和 2010 年完成[214]。这一次，11 部移动雷达阵列（SMART-R1、SMART-R2、DOW 6、DOW 7、RS-DOW、NOXP、UMASS X-Pol、TTU Ka 1 和 2、UMASS-W 和 MWR-05XP；图 3.8）和众多的定点雷达（WSR-88D 雷达、NWRT、KOUN 和 CASA）参与了该项目。许多其他的仪器也参与其中，从便携式龙卷地面观测系统到三脚架观测系统、无人机系统、移动无线电探空仪、自动站、雨滴谱仪和摄影测绘团队。VORTEX2 被称为史上最大的龙卷现场项目，它收集了多个开创性的数据集。其中最引人注目的是 2009 年 6 月 5 日发生在怀俄明州东南部的戈申（Goshen）/拉格兰奇龙卷[2,92,121,125,190,191]，几乎所有的仪器小组都收集到了这个大型长路径龙卷的数据。多个具备多普勒测速功能的、具有不同分辨率、位于不同观测距离、性能各异的多普勒天气雷达，捕捉了到龙卷的发生、龙卷的维持以及龙卷消散阶段，进行了对流风暴尺度和龙卷尺度的详尽观测。戈申个例被广泛认为是历史上研究最多、观测最充分的超级单体和龙卷。凭借前所未有的仪器和雷达设备，VORTEX2 实验确定了许多用于在龙卷附近设备科学部署的现代策略，特别是仪器部署的位置。

图 3.8 参与 VORTEX2 的一些移动雷达

VORTEX2 之后，美国主要基于雷达有组织的龙卷研究项目是由 CSWR 的 Josh Wurman 和 Karen Kosiba 领导的“龙卷和雷暴的雷达观测试验”(ROTATE)。ROTATE 主要聚焦龙卷的最低层，需要特别近距离地观测。ROTATE 还关注了被龙卷卷入空中的碎片以及强龙卷和弱龙卷之间的区别。使用了 DOWs 6、7 和 8，以及一些移动自动站和便携式龙卷地面观测系统进行现场探测。ROTATE 项目做出了多个发现，包括现场和遥感探测结合的数据集[215]，新的龙卷精细尺度观测[207]，以及多涡旋龙卷的观测[216]。在 VORTEX2 之前，ROTATE 也收集了大量的数据集。除了 ROTATE，OU 和 TTU 的团队在 2010 年代每年都开展小规模的外场项目。Chris Weiss 继续领导 TTU Ka 波段雷达的年度工作，而 Howie Bluestein 则在每一季主要负责 RaXPol 雷达。OU 的 ARRC 管理着大气成像雷达，2010 年代早期由 Jim Kurdzo 负责，自 2016 年以来由 David Bodine 和 Casey Griffin 领导。Mike Biggerstaff 偶尔使用偏振 SMART-R 进行外场项目，而 Don Burgess 在每年对流的风暴预报试验工作中负责 NOXP 雷达。普渡大学的 Robin Tanamachi 在对流风暴季节仍然使用 UMASS 的 X 波段雷达。NWRT、KOUN、KCRI、PX-1000 和其他雷达在春季大平原的灾害性对流天气中经常以快速扫描模式运行。最后，从 2016 年春天开始，CSWR 从 ROTATE 转移到 TWIRL(Tornadic Winds: Insitu and Radar obseruations at Low levels；龙卷：低空实地和雷达观测)；TWIRL 旨在更好地了解龙卷是如何形成的，以及最低层的风是如何造成破坏的。目前三辆现有 DOW 卡车以及一组移动自动站和便携式龙卷地面观测系统也参与其中。

撰写本章时，正在进行的有组织的大规模龙卷外场项目是 VORTEX-SE(VORTEX-SE)，它是聚焦 2016 年至 2017 年美国东南部龙卷，是 VORTEX 的一个分支。VORTEX-SE 所在的地域独特，与大平原地区龙卷多发生在春季的午后形成对照，该区域龙卷在夜间盛行，且该地区具有非常多变的地形。由于在这种环境下部署移动雷达很困难，在风暴形成之前，VORTEX-SE 的大部分移动雷达设备都位于固定位置(图 3.9)，导致风暴形成和维持过程中雷达覆盖“网”很大。VORTEX-SE 还试图理解和关注龙卷警报相关的社会问题，特别是在夜间和能见度较低的时候。

图 3.9　在 VORTEX-SE 期间，RaXPol 被部署在阿拉巴马州东北部的沙山高原，并收集了经过高原的超级单体雷暴的数据

3.4 地基雷达的科学进展

本节介绍了自 20 世纪 70 年代多普勒雷达问世以来，地基雷达观测所取得的成果。本节讨论的主题包括龙卷探测、雷达分析技术、龙卷成因和龙卷动力学，以及龙卷的双偏振和快速扫描雷达观测。

3.4.1 龙卷探测

第一次龙卷探测研究是在 20 世纪 70 年代用 NSSL 多普勒雷达观测进行的。NSSL 多普勒雷达观测到的第一个龙卷涡旋特征（TVS）是在 1973 年 5 月 24 日俄克拉何马州联合城附近[38,39]。他们注意到，TVS 在高度和时间上是连续的，并与地表破坏轨迹密切对应。最重要的是，他们指出，TVS 在龙卷形成前 23 min 就出现了，因此获得了改善龙卷预警时间的机会。TVS 首先发生在中间层，并在龙卷发生前下降到最低仰角。在 1973 年至 1976 年期间，NSSL 多普勒雷达观测到 10 个 TVSs，其中 8 个与龙卷相关，2 个无关[33]。TVSs 经常出现在龙卷形成前 10 min 以上，因此 TVS 的探测为龙卷预警提前时间增加提供了更有力的线索。这些早期研究还指出了一些挑战。即便是对于目前的雷达系统，这些挑战也使龙卷探测更加复杂。首先，龙卷的直径必须比雷达的方位和距离分辨率大。因此，对于直径较小或距离较远的龙卷，雷达分辨率可能不足以观察到 TVS。在更远的距离，雷达波束高度太高，无法观测龙卷。

在这些初步研究揭示了 TVSs 经常与龙卷搭配后，龙卷探测研究集中于开发探测中气旋和龙卷的自动化技术。NSSL 的研究人员在 20 世纪 80 年代开发了一种中气旋自动检测算法（MDA），该算法可以计算中气旋的大小和方位角切变[223]。该算法为在 NEXRAD 雷达网[48]上实现的第一个 MDA 奠定了基础。20 世纪 90 年代末，对 MDA 进行了升级，消除了原始算法中包含某些阈值，并在不同尺度的中气旋生命周期中提供了更稳健的识别[178]。升级后的 MDA 对中气旋的检出率（POD）为 99％以上，而原始 MDA 的检出率为 50％～75％。

龙卷探测的早期工作集中在 TVS 和评估相邻距离库的风切变。在 NEXRAD 网上实现的第一个 TVS 算法采用了严格的条件，导致 POD 较低[131]。为了解决这个问题，NSSL 龙卷检测算法（TDA）在 20 世纪 90 年代被开发出来[131]。TDA 聚焦于研究 TVS 的三维特征，并开发了跟踪龙卷涡旋和表征其强度的新方法。与原始算法相比，NSSL TDA 的虚警率 FAR 更低并且检出率 POD 更高。

利用相邻距离库风切变来探测和表征龙卷存在一定的局限性。最大相邻距离库风切变的大小和位置取决于雷达波束相对于涡旋中心的位置[201]。如果龙卷的采样很差，那么相对于龙卷的雷达波束位置变化会引起 TVS 强度或位置的波动。通常，雷达探测到的旋转速度随着到龙卷或中气旋的距离增加而减小，这是峰值速度的空间采样较粗和离地面高度增加的结果。

3.4.2 雷达分析技术

这一节主要讨论了对龙卷和其母雷暴至关重要的雷达分析技术。这些技术包括双多普勒雷达三维风场反演和地基速度追踪显示分析。

3.4.2.1 双多普勒

双多普勒或多多普勒分析采用不同的技术，结合两部或多部雷达的多普勒速度数据来获得三维风场。雷达观测到的多普勒速度或径向速度是雷达方位角 θ、仰角 ϕ、距离 r、笛卡尔风分量(u、v、w)和散射体下落末速度 w_t 的函数。

$$v_r = u\cos\phi\sin\theta + v\cos\phi\cos\theta + (w - w_t)\sin\phi \tag{3.1}$$

从式(3.1)可以看出，雷达观测的是风沿雷达波束的径向分量。当两个或多个雷达观测到足够不同的分量时(例如，通常当雷达波束之间的角度在 30°～150°时)，可以利用连续方程作为附加约束来反演三维风。

双多普勒分析需要仔细的数据后处理技术以解决几个误差源，同时保持运动的可分辨尺度。由于脉冲多普勒雷达系统只能在 Nyquist 区间内或最大不模糊速度 v_a(式 3.2)内分析多普勒速度，因此在灾害性雷暴中多普勒速度常常是折叠的。

$$v_a = \frac{\lambda}{4T_s} \tag{3.2}$$

式中雷达波长为 λ，脉冲重复时间为 T_s。要获得 Nyquist 区间以外的多普勒速度，需要对多普勒速度数据进行手动或自动退模糊或去折叠处理。退模糊之后，多普勒速度数据被客观分析(即空间滤波)到一个共同的网格。选择客观分析参数来获得最小的可分辨尺度，同时减小随机误差[116,183]。两次巴恩斯客观分析技术是最常用的方法，且可在保持分辨波长的同时有效地抑制噪声[116]。最后，由于风暴的移动和演变是在体扫期间进行的，所以两次体扫之间风暴(平流)的移动必须得到纠正。通常假设风暴运动是恒定和线性演化的[69]，但超级单体和龙卷可能包含大量的非线性平流和演化，导致平流修正后误差较大。为了减轻非线性运动误差，开发了一种空间变化的平流校正技术[167,168]，以减少非线性平流误差[144]。使用快速扫描雷达的双多普勒分析也有助于减少非线性平流误差。

3.4.2.2 GBVTD

双多普勒分析需要位于特定几何位置的两个或多个多普勒雷达来获取三维风场，这对于近距离的龙卷探测来说很难获得。此外，双多普勒分析的空间分辨率受到消除噪声数据所需的客观分析的限制，因此，科学家们探索了单多普勒技术来反演三维风场。20 世纪 90 年代末，开发了地基速度跟踪显示(GBVTD)[107]，用于反演热带气旋中不同波数的平均径向和切向速度及其分量。GBVTD 在轴对称假设下对多普勒速度数据进行傅里叶分解。双多普勒分析不需要这种假设，因此气流的不对称会给 GBVTD 分析带来误差。GBVTD 已扩展到龙卷的地基雷达研究[18,90,106,179]，包括通过对连续性方程积分来反演垂直速度。GBVTD 的简化版本也被开发了出来，以获得零阶流[58]或计算角动量平衡[146]。GBVTD 的一个优点是它可以应用于原始(退模糊的)多普勒速度数据，因此能够比双多普勒分析提供更高的分辨率。

从 GBVTD 和双多普勒分析中获得精确的三维风场反演存在一些挑战。多普勒雷达风场分析假设雷达测量的速度是风速；然而，雷达探测的是散射体速度的反射率因子加权平均值，而不是空气速度。在龙卷中，散射体由碎片和水凝物组成，由于碎片的离心作用它们与空气相比以不同的速度移动[58]。随着散射体的大小或密度的增加，空气速度和碎片速度之间的差异增大，可能导致几十米/秒的误差[25,58]。由于碎片相对于径向大气运动的向外离心，带有碎片的多普勒雷达测量结果低层辐散过强，因此会导致异常强烈的下沉气流[25,135]。然而，对于较短波长的雷达（例如 W 或 Ka 波段），碎片离心误差得到了缓解，因为即使在存在大碎片时，雷达后向散射信号由较小的散射体主导[25]。

3.4.3　龙卷形成和龙卷动力学

尽管科学家们已经论述了令人信服的龙卷形成理论，但龙卷的形成过程仍然没有被完全理解。特别是，科学家们一直无法确定为什么有些雷暴会产生龙卷，而有些雷暴不会产生龙卷。例如，在预报参数倾向于发生龙卷的地区，在龙卷爆发期间会形成大量超级单体，但有些超级单体会产生龙卷，有些则不会。由于龙卷的发生频率相对较低，因此很难观测到龙卷的形成，为了综合了解母风暴中发生的动力学、热力学和微物理过程，必须收集大量的观测资料[214]。多普勒雷达能够精细观测龙卷形成过程中发生的动态过程，并且最近，双偏振雷达提供了有关母风暴中发生的微物理过程的信息[97,138]。

单多普勒和双多普勒分析为 20 世纪 70 年代和 80 年代的龙卷成因提供了基本的概念理解。早期研究发现了一个下落的 TVS，形成于中间层并向地表下移[29,33]。早期的双多普勒分析帮助建立了超级单体的概念模型[109]，确定了超级单体龙卷在上升气流和后侧下沉气流的交界附近的上升气流区内形成，并确定了后侧阵风锋在锢囚形成中的重要性[29,148]。在后侧下沉气流发展并增强垂直速度的水平梯度之后[30]，水平涡旋的倾斜和随后的拉伸被认为是低层涡旋增强的机制。这些双多普勒测量结果提供了有价值的分析，以便与第一次超级单体三维数值模拟进行比较[85-87,152]。结合雷达分析，这些观测和模拟研究阐明了超级单体风暴的基本动力学特征。

早期的雷达研究也注意到一个龙卷形成的循环过程，通常出现在强龙卷爆发期间的一些超级单体中。20 世纪 80 年代，Don Burgess 利用多普勒雷达观测首次分析了循环型中气旋形成。在中气旋形成过程中，涡度极大值沿着后侧阵风锋发展，靠近上升气流的涡度极大值被拉伸，导致低层旋转更加强烈。在消散过程中，中气旋向后平流，并沿后侧阵风锋形成一个新的涡度极大值，这一过程重复[56,57]。在锢囚期，中气旋的上升气流与驱动上升气流的带有正浮力暖湿空气隔绝，龙卷消散。当龙卷长时间处于母风暴的上升气流之下时，就会产生最强烈的龙卷（2 型龙卷形成模式），而当龙卷更快地向后移动时，就会产生较弱的龙卷（1 型龙卷形成模式）[56,57,62,180]。对于循环型龙卷超级单体（CTSs），2 型龙卷通常发生在 CTS 生命周期的后期，导致较弱、短暂的龙卷之后出现长路径龙卷。在锢囚期，龙卷直径通常减小，然而龙卷可以像成熟期一样强烈[62,102]。

尽管许多早期研究确定了超级单体结构在龙卷生命周期的不同阶段的演变方式，但低层旋转加强和龙卷发展的机制尚未完全了解。水平涡度沿风暴引起的温度梯度斜压被认为是涡度的一个重要来源[85,120]，尽管在许多龙卷个例中已经观测到后侧下沉气流和前侧下沉气流中地面附近温度梯度[119,166]是比较弱的。数值模拟发现，在不产生斜压涡度的情况下，通过扰乱低空入流并从高空输送更高角动量的空气[112]，或通过下落的雨幕[52]从高空输送高角动量到地面，可以启动龙卷的形成。最近，对 2003 年 5 月 8 日龙卷超级单体的高分辨率数值模拟发现，与没有地表摩擦的模拟相比，内部动量激增期间的摩擦产生的水平涡度，以及随后的倾斜，促进了低层旋转的增强，并产生了更强烈的龙卷[161]。因此，针对 2003 年 5 月 8 日个例的模拟结果表明，斜压产生的涡度弱于摩擦产生的涡度。相反，其他研究发现在理想的和真实的超级单体数值模拟中，摩擦涡度的产生弱于斜压涡度的产生[117,127]。

由于缺乏三维热力学观测来补充雷达获得的三维风场数据，仅从观测数据来探索龙卷形成过程具有挑战性。尽管如此，近年来的龙卷形成研究已经取得了实质性的进展，主要集中在后侧阵风锋(RFGF)或内部动量涌流对龙卷形成和消散的作用[92,108,170]，以及了解龙卷生命周期中的涡度和角动量演变[124,213]。高分辨率双多普勒三维风场反演数据已被用于记录水平涡度的倾斜和垂直涡度的拉伸形成龙卷[211]，证实了数值模拟预期的特征[199]。利用在 VORTEX2 期间怀俄明州戈申市龙卷个例的雷达数据，双多普勒分析被用来确定风暴产生的涡度(可能是斜压的)主要来自于前侧的下沉气流，而不是后侧的下沉气流[122]。怀俄明州戈申市龙卷的形成也与次级后侧阵风锋的形成相吻合，其调节了水平涡度的倾斜和拉伸[92]。

在 2010 年 5 月 10 日 2223 UTC 对俄克拉何马州穆尔市龙卷的双多普勒分析观测到了一个典型的超级单体龙卷特征，该分析使用了 OU-PRIME 和 KOUN 雷达在龙卷发生时间附近的数据。双多普勒分析如图 3.10 所示，包括叠加在 OU-PRIME 雷达反射率因子上的水平风场(图 3.10a)，以及叠加在垂直速度和垂直涡度等值线上的水平风场(图 3.10b)。强烈的后侧阵风锋激涌明显出现在龙卷以南，伴有强烈的东风和下沉气流(>10 m/s)。后侧阵风锋东南边缘的垂直速度超过 20 m/s，且横跨龙卷存在强烈的上升-下沉气流环流(垂直涡度最大值标注为“T”)。因此，垂直速度的较大水平梯度使得水平涡度倾斜成垂直涡度。在反射率上，强降水出现在龙卷东北侧，形成细长的钩状回波。

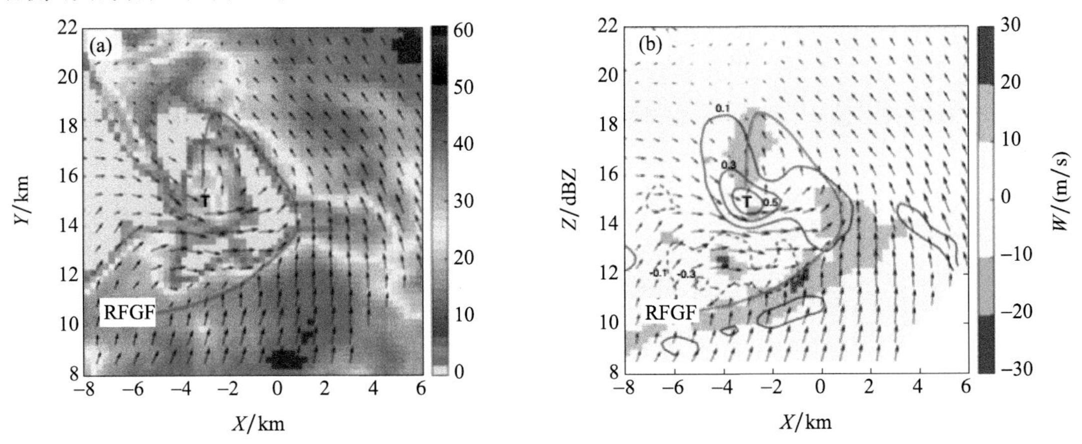

图 3.10 2010 年 5 月 10 日 2223 UTC，750 m AGL 处的双多普勒分析，着色部分为(a)雷达反射率因子和(b)垂直速度。双多普勒分析使用 OU-PRIME 和 KOUN 雷达数据。符号“T”表示龙卷的位置，灰色实线表示后侧阵风锋(RFGF)。红色实线表示垂直涡度在 0.1/s、0.3/s 和 0.5/s 处的等值线，而红色虚线表示垂直涡度在 −0.1/s 和 −0.3/s 处的等值线

20 世纪 90 年代的移动多普勒雷达系统使通过多普勒雷达详细了解龙卷动力学和结构成为可能。“车轮上的多普勒”(DOW)[209]和 UMASS W 波段雷达[12]观测到了龙卷的几个特征。龙卷移速超过 70 m/s，并观测到局部强烈的方位切变区域[12,209]。雷达反射率因子在地表附近最大，随高度的增加而减小。在高空，高反射率伴随着一个在龙卷中心出现反射率的极小值区，名为弱回波区(WEH)，被归因于碎片的离心作用[25,58]。在某些情况下弱回波洞的半径随高度扩大[206,209]，而在其他情况下 WEH 在近地表处半径最大[19]。这些观测还指出，地表辐合的缺乏[206]可能是由于碎片离心作用(导致碎片辐散)引起的观测到的多普勒速度的偏差[25,58]。

利用单多普勒和双多普勒风场反演方法对龙卷的三维风场结构进行了诊断。多普勒雷达研究中观测到的一系列龙卷旋流比与使用 GBVTD 分析[91,106,206]的实验室研究结果一致,并间接通过方位切变的局部强烈区域识别龙卷子涡旋[204,207]。GBVTD 分析也经常显示龙卷的中心下沉气流[106,191]以及低层入流[90,106]。然而,与数值模拟相比,单多普勒技术反演的入流速度要低得多。多部多普勒雷达分析也被用于分析龙卷尺度的下沉气流和子涡旋[74]。使用 GBVTDs 研究一个值得注意的特点是,它们反演的龙卷直径偏大。因为大多数移动雷达没有足够的方位分辨率充分表征较小直径的龙卷(例如,直径小于 200 m),除了近距离在高频运行的雷达(W 或 Ka 波段)。

3.4.4 龙卷的双偏振雷达特征

龙卷和龙卷风暴中会出现清晰的双偏振雷达特征,一些研究已经确定了超级单体雷暴的双偏振雷达特征。关于双偏振雷达理论和应用的概述,包括超级单体雷暴的偏振雷达特征,读者可以参考最新的偏振雷达理论和观测综述[94-96]。由于本章的重点是龙卷的地基观测,因此本节特别强调龙卷的偏振雷达特征,而不是雷暴母体的偏振特征。

利用 KOUN 雷达对龙卷进行了第一次双偏振雷达观测,第一次揭示了龙卷碎片特征(TDS)。空中的龙卷碎片产生了一种独特的偏振雷达特征,即中等到高的雷达反射率因子(Z_H)、低的同偏振相关系数(ρ_{HV})和接近零的差分反射率(Z_{DR})[156,158]。TDS 还应伴随龙卷涡旋特征(TVS),但如果空间分辨率不足以分辨 TVS,TDS 也可能单独存在[172]。Z_H 是碎片大小和浓度的函数,因此碎片浓度较高或直径较大往往产生较高的 Z_H。最近的研究观测到了范围更大的 Z_H,在 20～70 dBZ[22,163]。ρ_{HV}测量水平和垂直后向散射雷达信号之间的相关性,低 ρ_{HV}出现的情况有:大粒子导致的米散射、取样体积内不同的粒子类型和不规则的粒子形状。在龙卷中,由于碎片的大小、形状和组成的不同,被抛起的碎片会导致相关性的减小或低 ρ_{HV} 值。Z_{DR}是水平和垂直反射率因子的比值,对于瑞利散射,当粒子的长轴水平(垂直)排列时,Z_{DR}趋于正(负)值。由于碎片取向杂乱,Z_{DR}趋于 0 dB 附近。尽管如此,人们观测到的正和负 Z_{DR}的一致区域表明龙卷中碎片的取向存在一定程度的相似[23,45]。图 3.11 显示了 2013 年 5 月 20 日俄克拉何马州穆尔市 EF-5 龙卷期间 OU PX-1000 雷达观测到的 TDS 个例,它显示了一个明显的低 ρ_{HV}圆形区域,并伴随一个 TVS 和两处向南的碎片抛射。

龙卷碎片特征使龙卷可以实现业务化遥感探测 [20,97,158],特别是,由于 Z_{DR}在降水区可能为正,ρ_{HV}已被确定为检测 TDSs 最有用的偏振变量[20,22]。利用 TDSs 探测龙卷的能力在夜间或被雨水围绕的龙卷中尤其重要,因为此时龙卷很难被肉眼观察到。最近的研究开发了模糊逻辑龙卷自动检测算法,纳入了 TDSs 的偏振特征,改进了龙卷检测的统计结果[175,194]。当 TVS 没有被检测到或没有被充分识别时,TDS 可以添加额外的检测能力,例如,2014 年 8 月 31 日准线性对流系统(QLCS)中的一些龙卷产生了 TDSs,但没有产生 TVSs,因为方位分辨率不足以识别 TVS[172]。

虽然 TDSs 可以为龙卷探测提供一种有价值的工具,但有几点需要注意。一项利用 744 个已报道的龙卷个例进行的 TDS 研究发现,在对流风暴报道中只有 16%的龙卷产生了 TDS[186]。然而,该研究指出,100%的 EF4 级或以上猛烈龙卷和 58%的 EF2 级或以上强龙卷都产生了

TDS，与弱龙卷相比，这些特征在较长时间内都很明显。这是因为许多弱龙卷无法将足够多的碎片抛到雷达波束所在高度以供雷达探测，这在遥远的观测距离变得更具挑战性[97,186]。最后，在龙卷消散后，龙卷碎片可能会继续上升，因此 TDS 不一定是龙卷消散期间和消散后代表龙卷正在发生的可靠指标[22,163]。综上所述，TDS 可以证实有破坏性的龙卷，但是，没有 TDS 并不排除正在发生龙卷的可能性。

对龙卷碎片特征的研究验证了 TDSs 和龙卷灾情之间的关系，表明 TDS 参数可以为估计灾情严重程度提供信息[22,158,162]。TDS 参数包括 TDS 高度或体积，第 90 百分位反射率或第 10 百分位的 ρ_{HV}。TDS 的体积可达几十立方千米，并延伸到风暴顶部[22]。通过对 TDS 个例的大样本研究发现，TDS 高度越高的龙卷，其增强藤田指数 EF 越高[185,186]。这些发现与远距离碎片传输研究一致，即在 EF 等级较高的龙卷中，碎片会被运输得较远[73]。

TDS 特征与 EF 等级的相关性表明，TDS 与龙卷动力学之间可能存在某种关系。来自移动和固定平台/测站的高分辨率雷达数据已被用于研究 TDSs 和龙卷动力学之间的关系。龙卷尺度的 TDS 特征包括龙卷次级涡旋的 ρ_{HV} 较低，次级涡旋边缘的 Z_{DR} 为负[74]。在 TDS 的中心，同时观测到低 ρ_{HV} 和较高 ρ_{HV}。低 ρ_{HV} 可能是由于散射体数量较少且信噪比较低[180,192]，而较高 ρ_{HV} 可能是由于雨滴或其他瑞利散射体没有受到龙卷中心的离心力作用或雨滴没有被中央下沉气流回收[74]。TDS 似乎也受限于 35～40 m/s 的风速 [74,80]。

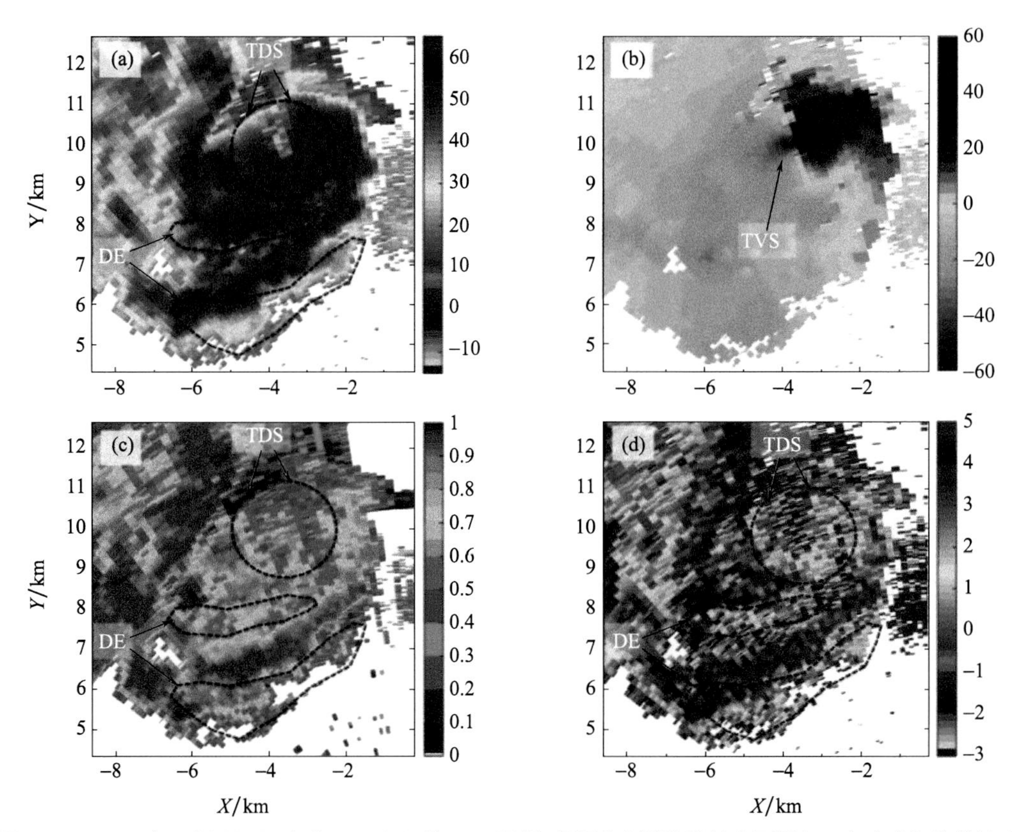

图 3.11 2013 年 5 月 20 日，来自 PX-1000 的 TDS 示例，显示(a)雷达反射率因子(Z_{HH})；(b)多普勒速度(v_r)；(c)同偏振相关系数 ρ_{HV}；(d)差分反射率(Z_{DR})。主要 TDS 以 TVS 为中心，伴随龙卷中心的最小 Z_{HH}。两处碎片抛射(标记为"DE")位于主要 TDS 的南部

Howie Bluestein 的研究小组在 2011 年和 2013 年使用 RaXpol 获得了近距离、快速扫描移动雷达观测的偏振雷达特征。在 2011 年 5 月 24 日俄克拉何马厄尔里诺 EF-5 龙卷形成之前，观测到一个 TDS，这是由发展的龙卷环流抛起与来自之前的龙卷和/或地表的碎片形成的[80]。在成熟阶段，龙卷和母风暴抛射的碎片高度通常超过 5 km（观测到的最大高度）。在短暂的减弱期间，TDS 高度下降，并注意到有大范围碎片沉降，这与没有速度测量进行龙卷尺度的早先预测一致[22]。从等值面分析中，他们确定了 TDS 中向上传播超过 30 $m\ s^{-1}$ 的波状结构，并假设这些结构可能与离心波有关。从 PPI 扫描中，他们识别出了与观测中发现的灰尘流入带相似的低 ρ_{HV}"锯齿状"凸出。RaXpol 雷达数据也用于研究 2013 年俄克拉何马州厄尔里诺 EF-3 龙卷[192]。当龙卷经过一所房屋造成 EF-3 程度的破坏时，他们注意到与局部地区被抛起的碎片相关的高雷达反射率因子。他们还注意到"碎片悬垂"的存在，即低 ρ_{HV} 伴随高仰角的 WEH，并认为这与碎片被抛入风暴尺度的上升气流相对应。

风暴尺度气流通过强上升气流和下沉气流影响 TDS 的三维结构。碎片被再循环到风暴尺度的上升气流中[22,192]，可能被抛到强烈超级单体的中上层。强下沉气流也可能有助于龙卷碎片的沉降[138]。针对 2013 年俄克拉何马州穆尔市 EF-5 龙卷[102]，研究了在龙卷环流和主上升气流基础上，后侧下沉气流（RFD）对"抛射"碎片的作用。在速度数据中观察到一系列的后侧阵风锋激涌（RFGFSs），伴随着同偏振相关系数显著降低的相同区域（表明存在碎片）。图 3.11c 展现了两个碎片抛射的例子，明显能看到两条低 ρ_{HV} 带。该研究将这些碎片抛射归因于 RFD 内碎片沉降结合 RFGF（后侧阵风锋）重新抛起碎片。此外，RFD 被认为在作者所称的"没有达成"锢囚中发挥了主要作用，导致龙卷轨迹形成一个环路，而没有发生完全锢囚。在 2011 年俄克拉何马州厄尔里诺龙卷中，X 波段雷达也观测到了类似的碎片抛射[79]。摄影测量分析也揭示了低 ρ_{HV} 区域与沿后侧阵风锋悬浮着尘埃和碎片的区域[193]，以及其他没有明显辐合的区域相关。

双偏振 TDSs 的一个新兴应用是表征龙卷内的散射体。需要这样的工作来纠正离心力导致的碎片辐散对空气微元多普勒速度正确测量的影响，以及提供与龙卷有关的灾害信息。一些研究假设较大的悬浮粒子会导致较高的雷达反射率因子和较低的 ρ_{HV}[22,158]，双波长雷达观测和数值模拟也表明，较大的碎片可能与较大的双波长比有关[23,25]。研究 TDS 如何变化取决于龙卷碎片类型，Boon Leng Cheong 以及一个科学家和工程师团队开发了一种双偏振雷达模拟器，为碎片和 LES 模式输入雷达截面数据，产生实际的双偏振 TDS 特征，包括低 ρ_{HV} 和负 Z_{DR} 特征[45]。该模拟还被用于探索偏振谱密度，它用来表示多普勒速度函数的双偏振雷达变量。由于水凝物和碎片以不同的速度移动，因此预计可以使用偏振谱密度在多普勒谱中识别水凝物和碎片，并随后过滤由碎片离心作用引起的空气微元碎片速度偏差[184]。

3.4.5 快速扫描雷达应用

在快速扫描雷达技术开发之前，Howie Bluestein 采用只扫描低仰角的策略实现快速扫描，每 10～15 s 扫描一次最低仰角，并观察到龙卷风场在几十秒或更短的时间内演化[18]。在他们的 GBVTD 分析中，他们观察到切向风速在 30～60 s 的时间尺度上振荡，最强的风出现在龙卷直径最小的时候。这些初步分析有助于证明对当前快速扫描雷达的需求。

Howie Bluestein 的研究小组已经使用 MWR-05XP 和 RaXpol 对龙卷的形成进行了几次体积快速扫描研究。这些研究主要集中在垂直方向上，探索自下而上或自上而下的龙卷形成方式。通过角动量从中气旋高度到地面的径向辐合产生的动力管道效应[110,182]可能导致自上而下的龙卷形成方式。利用 MWR-05XP 资料研究了 4 个龙卷形成的个例，这些个例的分析表明，中气旋尺度的涡旋特征是从地面向上发展[63]。采用时间分辨率与 NEXRAD 雷达相似的模拟 TVSs，很明显，TVS 演化的时间采样不足可能会呈现龙卷是自上而下的发生的假象，即使实际上正在发生的是龙卷自下而上地形成。利用一次 EF-5 型龙卷发生过程的 RaXpol 资料，发现在龙卷发生前 3 min，低层中气旋增强，直径减小，而在高空中气旋强度保持不变。在龙卷发生前，高空强度增强，约 30 s 内垂直柱面的 TVSs 快速增强[79]。在一次中低层逆温层内(1.5～2 km)，他们观察到逆温层以上区域相比其以下区域，龙卷的加强存在滞后。从这种快速扫描的小样本观测来看，缺乏自上而下的龙卷形成过程表明，动力管道效应对龙卷的发生可能并不重要。

从快速扫描雷达数据中发现了龙卷风场结构和雷达反射率特征的快速变化。快速扫描雷达揭示了龙卷在不同尺度上的结构变化[216]，龙卷轨迹和风速的突变[64,102]，以及次级涡旋的演化[174,193,216]。在俄克拉何马州肖尼市的 EF-4 龙卷期间，AIR 数据显示明显的在龙卷轨迹右侧局部碎片的上升，因为反射率最大值随时间上升。利用 AIR 数据，观察到 WEH 宽度为 1 km 的大直径龙卷(图 3.12)仅在 20 s 内快速转变为 WEH 无序的小直径龙卷[103]。快速扫描 DOW 捕捉到一个持续次级涡旋包括几个环路的复杂轨迹，且次级涡旋移动速度超过 80 m/s[216]。最近，利用快速扫描 DOW 和 RaXpol 数据进行了第一次快速扫描双多普勒分析[193]，揭示了单胞和大涡旋、多涡结构之间的过渡，该分析还提供了足够的空间分辨率来分辨次级涡旋中 95 m/s 的风。

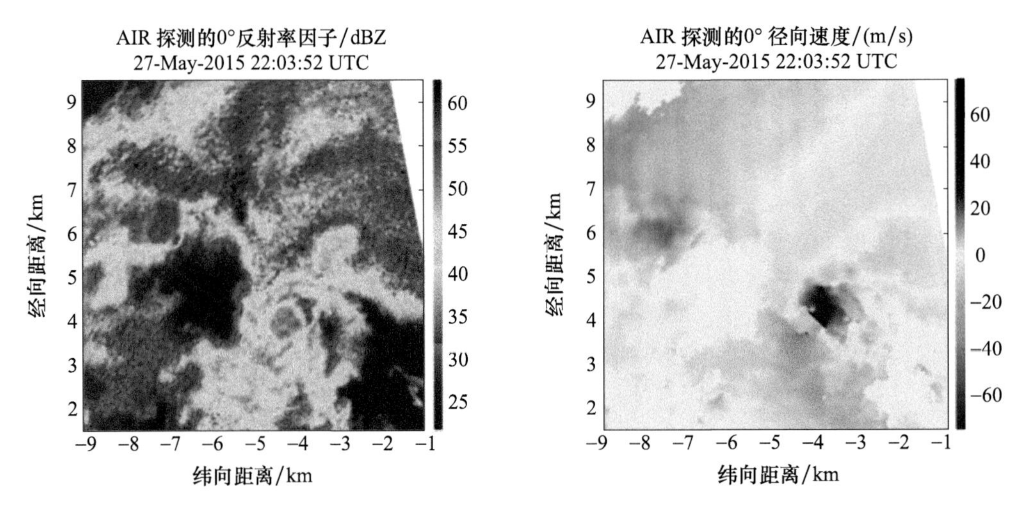

图 3.12　反射率因子(左)和径向速度(右)。来自 2015 年 5 月 27 日得克萨斯州加拿大龙卷期间大气成像雷达(AIR)的数据。在此个例中，AIR 收集了 6-s 一次的体扫数据

快速扫描雷达已观测到龙卷和中气旋之间的复杂相互作用，导致龙卷和中气旋强度的变化。MWR-05XP 数据的分析记录了两个 TVS 强度相似的龙卷的合并[63,66]。在合并前，原始龙卷减弱并远离先前的上升气流，而在后侧阵风锋与前侧阵风锋交会处附近，第二个龙卷在其

西北部强辐合区内形成。在合并的过程中，TVSs 以螺旋式彼此围绕旋转（所谓的 Fujiwhara 效应）。合并后，单一 TVS 比先前两个 TVSs 更为强烈。这种相互作用发生在不到 1 min 的时间内，如果没有亚分钟级时间分辨率的雷达数据，它的演变不可能被记录下来[66]。利用 AIR 资料详细描述了一次中气旋的快速发展及其与初始中气旋的相互作用[103]。当第二个中气旋从西侧靠近初始中气旋时，初始中气旋减弱。当两个中气旋靠近并相互旋转时，初始中气旋向北移动；当两个中气旋距离增加时，初始中气旋再次增强。

MWR-05XP 和 RaXpol 也被用来记录龙卷的消散，并在四个龙卷个例中观察到类似的模式。两次由 MWR-05XP 观测的龙卷消散过程均为"由内而外"，先在 1.5 km 处消散，然后在 1.5 km 以上，最后在 1.5 km 之下的高度内消散[64]。他们注意到 1.5 km 以下和以上 TVS 运动的差异，TVS 分别移动到风暴的右侧和左侧，而龙卷消散首先出现在不同 TVS 运动之间的界面上。他们推测 1.5 km 以下 TVS 移动到风暴右侧是由于 1 个次级 RFGFS（后侧阵风锋）波涌的增强造成的，且初始消散高度似乎在 RFGFS 的顶部。龙卷的上部随后减弱，因为入流被切断，随后在接近地表处出现了完全锢囚。这种消散过程大约持续了 3 min。利用来自 2011 年厄尔里诺龙卷的高分辨率 RaXpol 数据，记录了一次 EF-3 龙卷的消散[79]。他们观测到龙卷在 3 min 内以"由内而外"的方式减弱，最初减弱发生在 1～3 km 处。随后，整个龙卷的消散发生在 30 s 左右，在此期间，所有高度的龙卷强度都下降。

3.5 尚未解决的龙卷研究问题与未来雷达技术

本节讨论几个尚未解决的龙卷研究领域和需要改进的雷达分析技术。该部分最后介绍了未来的雷达系统，它们将引领新的科学和工程进展，以处理这些研究领域和改善龙卷预报。

3.5.1 未解决的龙卷研究问题

迄今为止，虽然龙卷生成的研究已经取得了很大的进展，并从观测和数值模拟研究中发展出了一些引人注目的龙卷生成理论，但对龙卷如何形成的科学认识仍然不完善。因此，在相似的环境中，评估为什么有些超级单体会产生龙卷，而有些却不会，这是一个挑战。导致低层旋转增强的风暴尺度过程仍存在争议，可能在不同的时空尺度上涉及多个过程。特别是，斜压产生的水平涡度在多大程度上促进了龙卷的形成仍存在争议，包括超级单体的哪些区域产生大量由斜压产生的水平涡度。其他研究指出，即使没有斜压产生水平涡度，龙卷也可以继续发生。因此，为了验证这些假设，关键是要将从雷达数据获得的三维运动场与来自其他补充观测（例如，可移动式自动站、飞机或无人机、探空设备、雨滴谱仪）的热力或微物理信息相结合，以了解这些动力、热力、微物理过程在龙卷形成中是如何演变的。

龙卷的典型特征在不同个例中有很大的不同，包括它们的直径、速度和持续时间。因此，可以推测不同个例中的龙卷形成过程不同，随后的增强和消散机制也因个例而异。虽然一个单一的、统一的龙卷过程理论可能是人们所希望的，但龙卷的形成、加强、维持和消散可能涉及一系列具体的和不同的物理过程。因此，值得探讨的是，具体某一次龙卷可能涉及的具体物理过程如何根据这些特征而变化（例如，弱龙卷相对强龙卷）。到目前为止，快速扫描雷达观测的样本量很小，因此需要快速扫描雷达提供更多的龙卷个例，以确定龙卷形成、增强和消散过程中的全谱特征。快速扫描的偏振雷达可以帮助研究龙卷形成过程中微物理过程的演变，以及偏振雷达特征和动力特征之间的关系。这样的研究可以阐明龙卷和非龙卷风暴之间的差异，这可能有助于为预报员提供更准确和及时的龙卷预警。

数值模拟的进步使雷达数据和超级单体内其他测量的高分辨率数据、风暴尺度数据同化成为可能。由于偏振雷达的探测值与水凝物的粒子谱分布密切相关，偏振数据同化可以改善超级单体的微物理分析以及评估超级单体和龙卷生命周期不同阶段的微物理演化。对于这类研究，必须仔细考虑数值天气预报(NWP)模型中使用的微物理方案[99]，并且可能需要双参数或三参数微物理方案，甚至计算耗时的水凝物分档计算的微物理参数化方案，因为过分简化的水凝物分布表达(例如，单参数)不能充分代表超级单体中独特的水凝物谱分布。其他观测资料的同化，如移动自动站、无人机、探空等，也将有助于提供更真实的超级单体模拟，以研究龙卷相关过程。目前，先进的 NWP 模型能够进行嵌套网格模拟，足以通过 25～100 m 的分辨率粗略解析龙卷尺度过程[161]。超级计算能力的持续发展将带来更多的超级单体模拟研究，并具有足够的分辨率来解析龙卷尺度过程，最终这些非常高分辨率的模型可能用于业务预报模式。

大涡模拟模式的龙卷模拟分辨率最高，解析尺度小至 1～3 m。虽然一方面通过非常高分辨率的龙卷模拟，另一方面通过移动雷达数据，对龙卷形成和演变的过程已经了解了很多，但将两者结合起来更能利用它们各自的优势。例如，多普勒雷达数据不能轻易解析龙卷与地表或角流区域的相互作用，但可以表征真实情况和三维不对称。到目前为止，龙卷 LES 模式试验仅限于理想化的轴对称初始化边界条件，因此这些试验无法解释真实的三维龙卷雷暴母体。因此，需要将高分辨率模式雷达观测数据纳入 LES 模式，以实现“真实的”LES 试验。这种试验可以从轴对称分析(如 GBVTD)开始，初始化 LES 模式的侧边和上边界条件(图 3.3)。然后，可以使用移动雷达数据进行数据同化试验。

为了提高用于研究超级单体和龙卷的雷达分析技术的准确性，还需要进行一些改进。超级单体和龙卷需要非常精确的三维风场来了解龙卷的形成和动力结构，并进行轨迹分析来评估涡度的产生。到目前为止，用于获得超级单体雷暴(偶尔还有龙卷)三维风场的双多普勒雷达风场反演方法尚未得到广泛验证，因此，人们对它们的准确性知之甚少。低空质量通量和非线性平流的探测不充分测量和估计是两个具有挑战性的问题，可能会造成较大的误差，尽管快速扫描雷达的双多普勒探测可能有助于解决后者。垂直指向雷达可能有助于探测超级单体的垂直运动。特别是，利用 AIR 的宽大发射波束进行的垂直指向探测可能特别有助于获得垂直运动的体积测量。垂直指向调频连续波(FMCW)雷达网也可能提供一种相对廉价的方法来获取超级单体中的多个垂直速度剖面，同时还可以改进微物理参数的反演。

3.5.2 未解决的龙卷研究问题和未来雷达技术

尽管目前美国有大量的探测龙卷尺度强烈天气的快速扫描雷达，但为了开发下一代天气雷达，壁垒正在不断被打破。在撰写本节时，目前有三种主要的快速扫描雷达正在开发中：PX-10000(或 PX-10k；升级版的 PX-1000)、先进技术演示器(ATD)和偏振大气成像雷达(PAIR)。其中预计第一个完成的是 PX-10k，计划在 2017 年年中完成。PX-10k 将是一个安装在小型拖车上的 X 波段平台，带有用于远程操作的集成发电机。PX-10k 具有 100 MHz 的可用线性调频带宽用于脉冲压缩，以及800 W的固态发射机用于发射偏振波，将能够为天气雷达提供引人瞩目的距离分辨率，同时在50 km距离保持10 dBZ的较高灵敏度。与PX-1000相比，PX-10k的抛物面碟形天线稍大一些，产生的3-dB波束宽度为1.4°(运行/工作nominal中心频率为9.25 GHz)。此外，该天线将能够达到180°/s或更快的旋转速率，使得单仰角更新在2 s以内。PX-10k还能将中心频率调整到8.9～9.6 GHz。当考虑到线性调频带宽时，该性能可以使PX-10k根据其相对于其他X波段雷达的位置及频率可以进行动态地适应性调整，PX-10k移动雷达的特别灵活性非常适合外场观测项目。

预期的第二个系统ATD(图3.13)，是NSSL获得的一个定点雷达，用于替代俄克拉何马州诺曼市的NWRT。ATD由麻省理工学院林肯实验室、NSSL和通用动力公司联合设计，并将于2018年年中由林肯实验室交付NSSL安装完成。ATD的特点为在S波段工作的有源相控阵，是未来双偏振相控阵天气雷达的试验场。ATD将被用于在气象和航空联合雷达中研究偏振测量和多功能性，作为国家频谱高效监测雷达(SENSR)计划的一个潜在解决方案(MPAR)。ATD较宽一面的波束宽度为1.7°，使用76块64单元面板形成一个4 m×4 m的阵列，使其成为美国未来全尺度天气雷达一个相对合适的工具。总共由4864个单元构成，每个单元峰值功率为6 W，将产生大约30 kW的总峰值功率(不含阵列加权)，因此可能需要脉冲压缩来获得ATD所需的灵敏度和距离分辨率。由于其相控阵设计，具有自适应能力的快速扫描和交叉任务将成为可能。

图3.13 多功能相控阵雷达(MPAR)和先进技术演示器(ATD)概念图。图由MIT林肯实验室提供

最后，由NSF和OU资助，OU ARRC牵头完成的PAIR将是一种C波段移动偏振成像雷达，它吸取了当前AIR的经验教训，并将偏振能力和不同的频率应用到新的设计中。该PAIR预计将于2019年完成，并将配备一个峰值功率6.4 kW的固态发射机，具有同时或交替双偏振探测的能力。该天线阵列将足够大，以容纳C波段大致2°的波束宽度，但天线转动有限，只有非常少的波束调整。将有足够的线性调频带宽来实现优于10 m的距离分辨率，并有足够的灵敏度在10 km距离处探测到−17 dBZ，90°×20°的更新速度大约为2.5 s。PAIR的C波段设计将明显改善现有X波段快速扫描雷达的衰减特性，还将为龙卷的双多普勒偏振观

测提供第二个频率，为通过不同波长的雷达探测理解龙卷碎片开辟新的科学前沿。C波段还将帮助补充SMART-R雷达，它们中只有一部是双偏振雷达，并可开展包括飓风观测等其他任务类型。

快速扫描X波段雷达经常用于研究龙卷，这些系统的1°波束宽度限制了它们观测直径较小的龙卷和近地表风的能力。例如，在5 km距离上，X波段的雷达波束宽度约为87 m。在数值模拟中，拐角流风场最低仅几十米，而X波段雷达很少获取该区域的样本；还有更精细波束宽度的雷达，比如UMASS的W波段（波长约3 mm）雷达和德州理工大学的Ka波段（波长约8 mm）雷达，它们的波束宽度在5 km距离上分别为16 m和28 m，因此，这些较精细波束宽度的雷达很可能对龙卷的拐角流区域进行采样。然而，它们目前缺乏快速扫描和偏振能力，并且非常容易衰减，甚至信号由于衰减而完全消失。为了解决这一需求，快速扫描Ka-或W波段雷达的开发将通过满足此类研究所需的时间和空间分辨率，为龙卷动力学提供更完整的理解。较短的波长对龙卷碎片的敏感度较低，对水凝物的敏感度较高，因此碎片离心误差较小，风速测量更准确[25]。最后，新的技术进步可以利用双频或多频雷达探测来研究龙卷，提供关于散射体特征[23]更详细的信息，并绘制碎片离心偏差的三维结构[25]。双频或多频雷达探测也将提供额外的信息，以获得超级单体内水凝物的特征和分布。

尽管这些技术进步目前仅限于研究使用的雷达系统，但最终目标是向一个全国范围内的快速扫描和双偏振的业务雷达网络迈进，以利用这些系统所带来的科学和工程上的进步。科学研究将提高对龙卷相关过程（如龙卷生成）的理解，并有望减少龙卷误报率和增加龙卷预警提前时间。源自这些雷达的科学认识让我们得以一窥未来快速扫描偏振雷达网的能力。

致谢　David Bodine由美国国家科学基金会AGS-1303685资助。James Kurdzo目前是MIT林肯实验室的员工；该实验室资金或资源没有被用于产生本出版作品中的结果/发现。作者感谢Casey Griffin提供了2010年5月10日的双多普勒数据，并审阅了本章的早期草稿。

参考文献*

[1] Alexander, C. R., and J. Wurman. 2005. The 30 May 1998 Spencer, South Dakota, storm. Part I: The structural evolution and environment of the tornadoes. *Monthly Weather Review* 133(1): 72-97.

[2] Atkins, N. T., A. McGee, R. Ducharme, R. M. Wakimoto, and J. Wurman. 2012. The LaGrange tornado during VORTEX2. Part II: Photogrammetric analysis of the tornado combined with dual-Doppler radar data. *Monthly Weather Review* 140(9): 2939-2958.

[3] Beck, J. R., J. L. Schroeder, and J. M. Wurman. 2006. High-resolution dual-Doppler analyses of the 29 May 2001 Kress, Texas, cyclic supercell. *Monthly Weather Review* 134(11): 3125-3148.

[4] Bedard, A. J., and C. Ramzy. 1983. Surface meteorological observations in severe thunderstorms. Part I: Design details of TOTO. *Journal of Climate and Applied Meteorology* 22(5): 911-918.

* 参考文献沿用原版书中内容，未改动

[5] Benjamin, T. B. 1962. Theory of the vortex breakdown phenomenon. *Journal of Fluid Mechanics*14: 593-629.

[6] Bieringer, P. , and P. S. Ray. 1996. A comparison of tornado warning lead times with and without NEXRAD Doppler radar. *Weather and Forecasting* 11:47-52.

[7] Biggerstaff, M. I. , L. J. Wicker, J. Guynes, C. Ziegler, J. M. Straka, E. N. Rasmussen, A. Doggett, L. D. Carey, J. L. Schroeder, and C. Weiss. 2005. The shared mobile atmospheric research and teaching radar: a collaboration to enhance research and teaching. *Bulletin of the American Meteorological Society*86(9):1263-1274.

[8] Biglerm, S. G. 1956. A note on the successful identification and tracking of a tornado by radar. *Weatherwise*9(6):198-201.

[9] Bluestein, H. B. 1983. Surface meteorological observations in severe thunderstorms. Part II: Field experiments with TOTO. *Journal of Climate and Applied Meteorology* 22(5):919-930.

[10] Bluestein, H. B. 1999. A history of severe-storm-intercept field programs. *Weather and Forecasting*14(4): 558-577.

[11] Bluestein, H. B. 2013. *Severe Convective Storms and Tornadoes :Observations and Dynamics*. Berlin: Springer.

[12] Bluestein, H. B. , and A. L. Pazmany. 2000. Observations of tornadoes and other convective phenomena with a mobile, 3-mm wavelength, Doppler radar: the spring 1999 field experiment. *Bulletin of the American Meteorological Society*81(12):2939-2951.

[13] Bluestein, H. B. , and W. P. Unruh. 1989. Observations of the wind field in tornadoes, funnel clouds, and wall clouds with a portable Doppler radar. *Bulletin of the American Meteorological Society*70(12):1514-1525.

[14] Bluestein, H. B. , and G. R. Woodall. 1990. Doppler-radar analysis of a low-precipitation severe storm. *Monthly Weather Review* 118(8):1640-1665.

[15] Bluestein, H. B. , E. W. McCaul, G. P. Byrd, and G. R. Woodall. 1988. Mobile sounding observations of a tornadic storm near the dryline: the Canadian, Texas storm of 7 May 1986. *Monthly Weather Review*116 (9):1790-1804.

[16] Bluestein, H. B. , J. G. Ladue, H. Stein, D. Speheger, and W. F. Unruh. 1993. Doppler radar wind spectra of supercell tornadoes. *Monthly Weather Review* 121(8):2200-2222.

[17] Bluestein, H. B. , C. C. Weiss, and A. L. Pazmany. 2003. Mobile Doppler radar observations of a tornado in a supercell near Bassett, Nebraska, on 5 June 1999. Part I: Tornadogenesis. *Monthly Weather Review*131 (12):2954-2967.

[18] Bluestein, H. B. , W. C. Lee, M. Bell, C. C. Weiss, and A. L. Pazmany. 2003. Mobile Doppler radar observations of a tornado in a supercell near Bassett, Nebraska, on 5 June 1999. Part II: Tornado-vortex structure. *Monthly Weather Review* 131:2968-2984.

[19] Bluestein, H. B. , C. C. Weiss, and A. L. Pazmany. 2004. The vertical structure of a tornado near Happy, Texas, on 5 May 2002: high-resolution, mobile, W-band, Doppler radar observations. *Monthly Weather Review*132(10):2325-2337.

[20] Bluestein, H. B. , M. M. French, R. L. Tanamachi, S. Frasier, K. Hardwick, F. Junyent, and A. L. Pazmany. 2007. Close-range observations of tornadoes in supercells made with a dualpolarization, X-band, mobile Doppler radar. *Monthly Weather Review* 135(4):1522-1543.

[21] Bluestein, H. B. , M. M. French, I. PopStefanija, R. T. Bluth, and J. B. Knorr. 2010. A mobile, phased-array

Doppler radar for the study of severe convective storms. *Bulletin of the American Meteorological Society*91(5):579-600.

[22] Bodine, D. J., M. R. Kumjian, R. D. Palmer, P. L. Heinselman, and A. V. Ryzhkov. 2013. Tornado damage estimation using polarimetric radar. *Weather and Forecasting* 28(1):139-158.

[23] Bodine, D. J., R. D. Palmer, and G. Zhang. 2014. Dual-wavelength polarimetric radar analyses of tornadic debris signatures. *Journal of Applied Meteorology and Climatology* 53:242-261.

[24] Bodine, D. J., T. Maruyama, R. D. Palmer, C. J. Fulton, H. B. Bluestein, and D. C. Lewellen. 2016. Sensitivity of tornado dynamics to debris loading. *Journal of the Atmospheric Sciences* 73:2783-2801.

[25] Bodine, D. J., R. D. Palmer, T. Maruyama, C. J. Fulton, Y. Zhu, and B. L. Cheong. 2016. Simulated frequency dependence of radar observations of tornadoes. *Journal of Atmospheric and Oceanic Technology* 33(9):1825-1842.

[26] Bowden, K. A., and P. L. Heinselman. 2016. A qualitative analysis of NWS forecasters' use of phased-array radar data during severe hail and wind events. *Weather and Forecasting* 31:43-55.

[27] Bowden, K. A., P. L. Heinselman, D. M. Kingfield, and R. P. 2015. Thomas. Impacts of phasedarray radar data on forecaster performance during severe hail and wind events. *Weather and Forecasting*30:389-404.

[28] Bradford, M. 2001. *Scanning the Skies: A History of Tornado Forecasting*. Norman: University of Oklahoma Press.

[29] Brandes, E. A. 1977. Flow in severe thunderstorms observed by dual-Doppler radar. *Monthly Weather Review*105(1):113-120.

[30] Brandes, E. A. 1984. Vertical vorticity generation and mesocyclone sustenance in tornadic thunderstorms: the observational evidence. *Monthly Weather Review* 112:2253-2269.

[31] Brotzge, J., and Donner, W. 2013. The tornado warning process: a review of current research, challenges, and opportunities. *Bulletin of the American Meteorological Society* 94(11):1715-1733.

[32] Brown, R. A., D. W. Burgess, J. K. Carter, L. R. Lemon, and D. Sirmans. 1975. NSSL dual-Doppler radar measurements in tornadic storms: a preview. *Bulletin of the American Meteorological Society*56(5):524-526.

[33] Brown, R. A., L. R. Lemon, and D. W. Burgess. 1978. Tornado detection by pulsed Doppler radar. *Monthly Weather Review* 106:29-38.

[34] Brown, R. A., J. M. Janish, and V. T. Wood. 2000. Impact of WSR-88D scanning strategies on severe storm algorithms. *Weather and Forecasting* 15(1):90-102.

[35] Brown, R. A., V. T. Wood, and D. Sirmans. 2000. Improved WSR-88D scanning strategies for convective storms. *Weather and Forecasting* 15(2):208-220.

[36] Brown, R. A., V. T. Wood, and D. Sirmans. 2002. Improved tornado detection using simulated and actual WSR-88D data with enhanced resolution. *Journal of Atmospheric and Oceanic Technology*19(11):1759-1771.

[37] Brown, R. A., V. T. Wood, R. M. Steadham, R. R. Lee, B. A. Flickinger, and D. Sirmans. 2005. New WSR-88D volume coverage pattern 12: results of field tests. *Weather and Forecasting* 20:385-393.

[38] Burgess, D. W. 1976. Single Doppler radar vortex recognition: Part I. Mesocyclone signatures. In17*th Radar Meteorology Conference*, Seattle, WA, 97-103. Boston: American Meteorological Society(Preprints).

[39] Burgess, D. W., L. R. Lemon, and R. A. Brown. 1975. Tornado characteristics revealed by Doppler radar. *Geophysical Research Letters* 2(5):183-184.

[40] Burgess, D. W. , V. T. Wood, and R. A. Brown. 1982. Mesocyclone evolution statistics. In *12th Conference on Severe Local Storms*, San Antonio, TX, 422-424. Boston: American Meteorological Society(Preprints).

[41] Burgess, D. W. , M. A. Magsig, J. Wurman, D. C. Dowell, and Y. Richardson. 2002. Radar observations of the 3 May 1999 Oklahoma City tornado. *Weather and Forecasting* 17(3):456-471.

[42] Burgess, D. W. , E. R. Mansell, C. M. Schwarz, and B. J. Allen. 2010. Tornado and tornadogenesis events seen by the NOXP, X-band, dual-polarization radar during VORTEX2 2010. In *25th Conference on Severe Local Storms*. Boston: American Meteorological Society.

[43] Cheong, B. L. , R. D. Palmer, C. D. Curtis, T. Y. Yu, and D. S. Zrni'c, and D. Forsyth. 2008. Refractivity retrieval using the phased-array radar: first results and potential for multimission operation. *IEEE Transactions on Geoscience and Remote Sensing* 46(9):2527-2537.

[44] Cheong, B. L. , R. Kelley, R. D. Palmer, Y. Zhang, M. Yeary, and T. Y. Yu. 2013. PX-1000: a solid-state polarimetric X-band weather radar and time-frequency multiplexed waveform for blind range mitigation. *IEEE Transactions on Instrumentation and Measurement* 62(11):3064-3072.

[45] Cheong, B. L. , D. J. Bodine, C. J. Fulton, S. M. Torres , T. Maruyama, and R. D. Palmer. 2017. SimRadar: a polarimetric radar time-series simulator for tornadic debris studies. *IEEE Transactions on Geoscience and Remote Sensing* 55:2858-2870.

[46] Church, C. R. , J. T. Snow, G. L. Baker, and E. M. Agee. 1979. Characteristics of tornado-like vortices as a function of swirl ratio: a laboratory investigation. *Journal of the Atmospheric Sciences* 36:1755-1776.

[47] Collier, C. G. , C. A. Fair, and D. H. Newsome. 1988. International weather-radar networking in western Europe. *Bulletin of the American Meteorological Society* 69:16-21.

[48] Crum, T. , and Alberty, R. 1993. The WSR-88D and the WSR-88D operational support facility. *Bulletin of the American Meteorological Society* 74(9):1669-1687.

[49] Crum, T. , S. Smith, J. Chrisman, R. Vogt, M. Istok, R. Hall, B. Saffle. 2013. WSR-88D radar projects 2013 update. In *29th Conference on EIPT*, Austin, TX

[50] Daniel, A. E. , J. N. Chrisman, S. D. Smith, and M. W. Miller. 2014. New WSR-88D operational techniques: responding to recent weather events. In *30th Conference on Environmental Information Processing Technologies*, Atlanta, GA. Boston: American Meteorological Society.

[51] Davies-Jones, R. P. 1973. The dependence of core radius on swirl ratio in a tornado simulator. *Journal of the Atmospheric Sciences* 30:1427-1430.

[52] Davies-Jones, R. 2008. Can a descending rain curtain in a supercell instigatetornadogenesis barotropically? *Journal of the Atmospheric Sciences* 65:2469-2497.

[53] Davies-Jones, R. , R. J. Trapp, H. B. Bluestein. 2001. Tornadoes and tornadic storms. In*Severe Convective Storms*, ed. C. A. Doswell, 167-222. Boston: American Meteorological Society

[54] Doviak, R. J. , and D. S. Zrni'c. 1993. *Doppler Radar and Weather Observations*. 2nd ed. New York: Dover Publications.

[55] Doviak, R. J. , V. Bringi, A. Ryzhkov, A. Zahrai, and D. Zrni' c. 2000. Considerations for polarimetric upgrades to operational WSR-88D radars. *Journal of Atmospheric and Oceanic Technology* 17:257-278.

[56] Dowell, D. C. , and H. B. Bluestein. 2002. The 8 June 1995 Mclean, Texas, storm. Part I: Observations of cyclic tornadogenesis. *Monthly Weather Review* 130(11):2626-2648.

[57] Dowell, D. C. , and H. B. Bluestein. 2002. The 8 June 1995 Mclean, Texas, storm. Part II: Cyclic tornado formation, maintenance, and dissipation. *Monthly Weather Review* 130(11):2649-2670.

[58] Dowell, D. C., C. R. Alexander, J. M. Wurman, and L. J. Wicker. 2005. Centrifuging of hydrometeors and debris in tornadoes: radar-reflectivity patterns and wind-measurement errors. *Monthly Weather Review* 133(6):1501-1524.

[59] Evans, J., and D. Turnbull. 1989. Development of an automated windshear detection system using Doppler weather radar. *Proceedings of the IEEE* 77:1661-1673.

[60] Fiedler, B. H. 1998. Wind-speed limits in numerically simulated tornadoes with suction vortices. *Quarterly Journal of the Royal Meteorological Society* 124:2377-2392

[61] Fiedler, B. H., and R. Rotunno. 1986. A theory for the maximum windspeeds in tornado-like vortices. *Journal of the Atmospheric Sciences* 43:2328-2340.

[62] French, M. M., H. B. Bluestein, D. C. Dowell, L. J. Wicker, M. R. Kramar, and A. L. Pazmany. 2008. High-resolution, mobile Doppler radar observations of cyclic mesocyclogenesis in a supercell. *Monthly Weather Review* 136(12):4997-5016.

[63] French, M. M., H. B. Bluestein, I. PopStefanija, C. A. Baldi, and R. T. Bluth. 2013. Reexamining the vertical development of tornadic vortex signatures in supercells. *Monthly Weather Review* 141(12), 4576-4601.

[64] French, M. M., H. B. Bluestein, I. PopStefanija, C. A. Baldi, and R. T. Bluth. 2014. Mobile, phased-array, Doppler radar observations of tornadoes at X band. *Monthly Weather Review* 142(3):1010-1036.

[65] French, M. M., D. W. Burgess, E. R. Mansell, and L. J. Wicker. 2015. Bulk hook echo raindrop sizes retrieved using mobile, polarimetric Doppler radar observations. *Journal of Applied Meteorology and Climatology* 54(2):423-450.

[66] French, M. M., P. S. Skinner, L. J. Wicker, and H. B. Bluestein. 2015. Documenting a rare tornado merger observed in the 24 May 2011 El Reno-Piedmont, Oklahoma, supercell. *Monthly Weather Review* 143(8): 3025-3043.

[67] Fujita, T. T. 1974. Jumbo tornado outbreak of 3 April 1974. *Weatherwise* 27(3), 116-126.

[68] Fujita, T. T., D. L. Bradbury, and C. F. V. Thullenar. 1970. Palm Sunday tornadoes of April 11, 1965. *Monthly Weather Review* 98:29-69.

[69] Gal-Chen, T. 1982. Errors in fixed and moving frame of reference: application for conventional and Doppler radar analysis. *Journal of the Atmospheric Sciences* 39:2279-2300.

[70] Geerts, B., D. Parsons, C. L. Ziegler, T. M. Weckwerth, M. I. Biggerstaff, R. D. Clark, M. C. Coniglio, B. B. Demoz, R. A. Ferrare, W. A. Gallus, K. Haghi, J. M. Hanesiak, P. M. Klein, K. R. Knupp, K. Kosiba, G. M. McFarquhar, J. A. Moore, A. R. Nehrir, M. D. Parker, J. O. Pinto, R. M. Rauber, R. S. Schumacher, D. D. Turner, Q. Wang, X. Wang, Z. Wang, J. Wurman. 2017. The 2015 plains elevated convection at night field project. *Bulletin of the American Meteorological Society* 98(4):767-786.

[71] Golden, J. H., and B. J. Morgan. 1972. NSSL Notre Dame tornado intercept program, Spring 1972. *Bulletin of the American Meteorological Society* 53(12):1178-1180.

[72] Goliger, A. M., and R. V. Milford. 1998. A review of worldwide occurrences of tornadoes. *Journal of Wind Engineering and Industrial Aerodynamics* 74-76:111-121.

[73] Grazilus, T. 1993. Significant Tornadoes 1680-1991. The Tornado Project of Environmental Films.

[74] Griffin, C. B., D. J. Bodine, and R. D. Palmer. 2017. Kinematic and polarimetric radar observations of the 10 May 2010, Moore-Choctaw, Oklahoma tornadic debris signature. *Monthly Weather Review* 145: 2723-2741.

[75] Groenemeijer, P., and T. Kühne. 2014. A climatology of tornadoes in Europe: results from the European

severe weather database. *Monthly Weather Review* 142(12):4775-4790.

[76] Heinselman, P. L. , and S. M. Torres. 2011. High-temporal-resolution capabilities of the national weather radar testbed phased-array radar. *Journal of Applied Meteorology and Climatology* 50(3):579-593.

[77] Heinselman, P. L. , D. L. Priegnitz, K. L. Manross, T. M. Smith, and R. W. Adams. 2008. Rapid sampling of severe storms by the national weather radar testbed phased array radar. *Weather and Forecasting* 23 (5):808-824.

[78] Hirth, B. D. , J. L. Schroeder, S. W. Gunter, and J. G. Guynes. 2012. Measuring a utility scale turbine wake using the TTUKa mobile research radars. *Journal of Atmospheric and Oceanic Technology* 29 (6): 765-771.

[79] Houser, J. B. , H. B. Bluestein, and J. C. Snyder. 2015. Rapid-scan, polarimetric, Doppler-radar observations of tornadogenesis and tornado dissipation in a tornadic supercell: the "El Reno, Oklahoma" storm of 24 May 2011. *Monthly Weather Review* 143(7):2685-2710.

[80] Houser, J. L. , H. B. Bluestein, and J. C. Snyder. 2017. A fine-scale radar examination of the tornadic debris signature and weak-echo reflectivity band associated with a large, violent tornado. *Monthly Weather Review* 144:4101-4130.

[81] Isom, B. , R. Palmer, R. Kelley, J. Meier, D. Bodine, M. Yeary, B. L. Cheong, Y. Zhang, T. Y. Yu, and M. I. Biggerstaff. 2013. The atmospheric imaging radar: simultaneous volumetric observations using a phased array weather radar. *Journal of Atmospheric and Oceanic Technology* 30(4):655-675.

[82] Joe, P. , and S. Lapczak. 2002. Evolution of the Canadian operational radar network. In *Proceedings of ERAD*, Delft, Netherlands, 370-382.

[83] Johnson, J. T. , P. L. MacKeen, A. Witt, E. D. W. Mitchell, G. J. Stumpf, M. D. Eilts, and K. W. Thomas. 1998. The storm cell identification and tracking algorithm: an enhanced WSR-88D algorithm. *Weather and Forecasting* 13(2):263-276.

[84] Karstens, C. D. , T. M. Samaras, B. D. Lee Jr. , W. A. Gallus, C. A. Finley. 2010. Near-ground pressure and wind measurements in tornadoes. *Monthly Weather Review* 138(7):2570-2588.

[85] Klemp, J. B. , and R. Rotunno. 1983. A study of the tornadic region within a supercell thunderstorm. *Journal of the Atmospheric Sciences* 40:359-377.

[86] Klemp, J. B. , and R. B. Wilhelmson. 1978. The simulation of three-dimensional convective storms dynamics. *Journal of the Atmospheric Sciences* 35:1070-1096.

[87] Klemp, J. B. , R. B. Wilhelmson, and P. S. Ray. 1981. Observed and numerically simulated structure of a mature supercell thunderstorm. *Journal of the Atmospheric Sciences* 38:1558-1580.

[88] Knupp, K. R. , J. Walters, and M. Biggerstaff. 2006. Doppler profiler and radar observations of boundary layer variability during the landfall of Tropical Storm Gabrielle. *Journal of the Atmospheric Sciences* 63 (1):234-251.

[89] Knupp, K. R. , T. A. Murphy, T. A. Coleman, R. A. Wade, S. A. Mullins, C. J. Schultz, E. V. Schultz, L. Carey, A. Sherrer, E. W. McCaul, B. Carcione, S. Latimer, A. Kula, K. Laws, P. T. Marsh, K. Klockow. 2014. Meteorological overview of the devastating 27 April 2011 tornado outbreak. *Bulletin of the American Meteorological Society* 95(7):1041-1062.

[90] Kosiba, K. , and J. Wurman. 2010. The three-dimensional axisymmetric wind field structure of the Spencer, South Dakota, 1998 tornado. *Journal of the Atmospheric Sciences* 67(9):3074-3083.

[91] Kosiba, K. A. , and J. Wurman. 2013. The three-dimensional structure and evolution of a tornado boundary

layer. *Weather and Forecasting* 28(6):1552-1561.

[92] Kosiba, K., J. Wurman, Y. Richardson, P. Markowski, P. Robinson, and J. Marquis. 2013. Genesis of the Goshen County, Wyoming, tornado on 5 June 2009 during VORTEX2. *Monthly Weather Review* 141(4): 1157-1181.

[93] Kumjian, M. R. 2011. Precipitation properties of supercell hook echoes. *Electronic Journal of Severe Storms Meteorology* 6(5):1-21.

[94] Kumjian, M. R. 2013. Principles and applications of dual-polarization weather radar. Part I: Description of the polarimetric radar variables. *Journal of Operational Meteorology* 1:226-242.

[95] Kumjian, M. R. 2013. Principles and applications of dual-polarization weather radar. Part II: Warm-and cold-season applications. *Journal of Operational Meteorology* 1:243-264.

[96] Kumjian, M. R. 2013. Principles and application of dual-polarization weather radar. Part III: Artifacts. *Journal of Operational Meteorology* 1:265-274.

[97] Kumjian, M. R., and A. V. Ryzhkov. 2008. Polarimetric signatures in supercell thunderstorms. *Journal of Applied Meteorology and Climatology* 48:1940-1961.

[98] Kumjian, M. R., A. V. Ryzhkov, V. M. Melnikov, and T. J. Schuur. 2010. Rapid-scan superresolution observations of a cyclic supercell with a dual-polarization WSR-88D. *Monthly Weather Review* 138(10): 3762-3786.

[99] Kumjian, M. R., Z. J. Lebo, and H. C. Morrison. 2015. On the mechanisms of rain formation in an idealized supercell storm. *Monthly Weather Review* 143:2754-2773

[100] Kurdzo, J. M., and R. D. Palmer. 2012. Objective optimization of weather radar networks for low-level coverage using a genetic algorithm. *Journal of Atmospheric and Oceanic Technology* 29(6):807-821.

[101] Kurdzo, J. M., B. L. Cheong, R. D. Palmer, G. Zhang, and J. B. Meier. 2014. A pulse compression waveform for improved-sensitivity weather radar observations. *Journal of Atmospheric and Oceanic Technology* 31(12):2713-2731.

[102] Kurdzo, J. M., D. J. Bodine, B. L. Cheong, and R. D. Palmer. 2015. High-temporal resolution polarimetric X-band Doppler radar observations of the 20 May 2013 Moore, Oklahoma, tornado. *Monthly Weather Review* 143(7):2711-2735.

[103] Kurdzo, J. M., F. Nai, D. J. Bodine, T. A. Bonin, R. D. Palmer, B. L. Cheong, J. Lujan, A. Mahre, A. D. Byrd. 2017. Observations of severe local storms and tornadoes with the atmospheric imaging radar. *Bulletin of the American Meteorological Society* 98(5):915-935.

[104] Kuster, C. M., P. L. Heinselman, and T. J. Schuur. 2016. Rapid-update radar observations of downbursts occurring within an intense multicell thunderstorm on 14 June 2011. *Weather and Forecasting* 31(3): 827-851.

[105] Lee, R. R., and A. White. 1998. Improvement of the WSR-88D mesocyclone algorithm. *Weather and Forecasting* 13(2):341-351.

[106] Lee, W. C., and J. Wurman. 2005. Diagnosed three-dimensional axisymmetric structure of the Mulhall tornado on 3 May 1999. *Journal of the Atmospheric Sciences* 62(7):2373-2393.

[107] Lee, W. C., B. J. D. Jou, P. L. Chang, and S. M. Deng. 1999. Tropical cyclone kinematic structure retrieved from single-Doppler radar observations. Part I: Doppler velocity patterns and the GBVTD technique. *Monthly Weather Review* 127:2419-2439.

[108] Lee, B. D., C. A. Finley, and C. D. Karstens. 2012. The Bowdle, South Dakota, cyclic tornadic supercell of

22 May 2010: surface analysis of rear-flank downdraft evolution and multiple internal surges. *Monthly Weather Review* 140(11):3419-3441.

[109] Lemon, L. R., and C. A. Doswell. 1979. Severe thunderstorm evolution and mesocyclone structure as related to tornadogenesis. *Monthly Weather Review* 107(9):1184-1197.

[110] Leslie, L. M. 1971. The development of concentrated vortices: a numerical study. *Journal of Fluid Mechanics*48(1):1-21.

[111] Lewellen, D. C., and W. S. Lewellen. 2007. Near-surface intensification of tornado vortices. *Journal of the Atmospheric Sciences*64:2176-2194.

[112] Lewellen, D. C., and W. S. Lewellen. 2007. Near-surface vortex intensification through corner flow collapse. *Journal of the Atmospheric Sciences* 64:2195-2209.

[113] Lewellen, D. C., W. S. Lewellen, and J. Xia. 2000. The influence of a local swirl ratio on tornado intensification near the surface. *Journal of the Atmospheric Sciences* 57:527-544.

[114] Mahale, V. N., J. A. Brotzge, and H. B. Bluestein. 2014. The advantages ofa mixed-band radar network for severe weather operations: a case study of 13 May 2009. *Weather and Forecasting* 29(1):78-98.

[115] Mahre, A., T. Y. Yu, R. Palmer, and J. Kurdzo. 2017. Observations of a cold front at high spatiotemporal resolution using an X-band phased array imaging radar. *Atmosphere* 8(2):30.

[116] Majcen, M., P. Markowski, Y. Richardson, D. Dowell, and J. Wurman. 2008. Multipass objective analyses of Doppler radar data. *Journal of Atmospheric and Oceanic Technology* 25(10):1845-1858.

[117] Markowski, P. M. 2016. An idealized numerical simulation investigation of the effects of surface drag on the development of near-surface vertical vorticity in supercell thunderstorms. *Journal of the Atmospheric Sciences*73:4349-4385.

[118] Markowski, P. M., and Y. P. Richardson. 2014. What we know and don't know about tornado formation. *Physics Today* 67(9):26-31.

[119] Markowski, P. M., J. M. Straka, and E. N. Rasmussen. 2002. Direct surface thermodynamic observations within the rear-flank downdrafts of nontornadic and tornadic supercells. *Monthly Weather Review*130(7):1692-1721.

[120] Markowski, P., E. Rasmussen, J. Straka, R. Davies-Jones, Y. Richardson, and R. J. Trapp. 2008. Vortex lines within low-level mesocyclones obtained from Psuedo-Dual-Doppler radar observations. *Monthly Weather Review* 136:3513-3535.

[121] Markowski, P., Y. Richardson, J. Marquis, J. Wurman, K. Kosiba, P. Robinson, D. Dowell, E. Rasmussen, and R. Davies-Jones. 2012. The pretornadic phase of the Goshen County, Wyoming, supercell of 5 June 2009 intercepted by VORTEX2. Part I: Evolution of kinematic and surface thermodynamic fields. *Monthly Weather Review* 140(9):2887-2915.

[122] Markowski, P. M., Y. Richardson, J. Marquis, R. Davies-Jones, J. Wurman, K. Kosiba, P. Robinson, E. Rasmussen, and D. Dowell. 2012. The pretornadic phase of the Goshen County, Wyoming, supercell of 5 June 2009 intercepted by VORTEX2. Part II: Intensification of Low-Level Rotation. *Monthly Weather Review* 140:2916-2938.

[123] Marquis, J., Y. Richardson, J. Wurman, and P. Markowski. 2008. Single-and dual-Doppler analysis of a tornadic vortex and surrounding storm-scale flow in the Crowell, Texas, supercell of 30 April 2000. *Monthly Weather Review* 136(12):5017-5043.

[124] Marquis, J., Y. Richardson, P. Markowski, D. Dowell, and J. Wurman. 2012. Tornado maintenance inves-

tigated with high-resolution dual-Doppler and EnKF analysis. *Monthly Weather Review*140(1):3-27

[125] Marquis, J., Y. Richardson, P. Markowski, D. Dowell, J. Wurman, K. Kosiba, P. Robinson, G. Romine. 2014. An investigation of the Goshen County, Wyoming, tornadic supercell of 5 June 2009 using EnKF assimilation of mobile mesonet and radar observations collected during VORTEX2. Part I: Experiment design and verification of the EnKF analyses. *Monthly Weather Review*142(2):530-554.

[126] Maruyama, T. 2011. Simulation of flying debris using a numerically generated tornado-like vortex.. *Journal of Wind Engineering and Industrial Aerodynamics* 99:249-256.

[127] Mashiko, W., H. Niino, and T. Kato. 2009. Numerical simulation of tornadogenesis in an outer-rainband minisupercell of Typhoon Shanshan on 17 September 2006. *Monthly Weather Review*137:4238-4260.

[128] McLaughlin, D., D. Pepyne, B. Philips, J. Kurose, M. Zink, D. Westbrook, E. Lyons, E. Knapp, A. Hopf, A. Defonzo, R. Contreras, T. Djaferis, E. Insanic, S. Frasier, V. Chandrasekar, F. Junyent, N. Bharadwaj, Y. Wang, Y. Liu, B. Dolan, K. Droegemeier, J. Brotzge, M. Xue, K. Kloesel, K. Brewster, F. Carr, S. Cruz-Pol, K. Hondl, and P. Kollias. 2009. Short-wavelength technology and the potential for distributed networks of small radar systems. *Bulletin of the American Meteorological Society* 90(12):1797-1817.

[129] Mead, J. B., G. Hopcraft, S. J. Frasier, B. D. Pollard, C. D. Cherry, D. H. Schaubert, and R. E. McIntosh. 1998. A volume-imaging radar wind profiler for atmospheric boundary layer turbulence studies. *Journal of Atmospheric and Oceanic Technology* 15(4):849-859.

[130] Michelson, M., W. Shrader, and J. Wieler. 1990. Terminal Doppler Weather Radar. *Microwave Journal*33:139-148.

[131] Mitchell, E. D. W., S. V. Vasiloff, G. J. Stumpf, A. Witt, M. D. Eilts, J. T. Johnson, and K. W. Thomas. 1998. The national severe storms laboratory tornado detection algorithm. *Weather and Forecasting* 13(2):352-366.

[132] Moller, A. R.: The improved NWS storm spotters' training program at Ft. Worth, Tex. 1978. *Bulletin of the American Meteorological Society*59(12):1574-1582.

[133] Moller, A., C. Doswell, J. McGinley, S. Tegtmeier, and R. Zipser. 1974. Field observations of the Union City tornado in Oklahoma. *Weatherwise* 27(2):68-79.

[134] Newman, J. F., and P. L. Heinselman. 2012. Evolution of a quasi-linear convective system sampled by phased array radar. *Monthly Weather Review* 140(11):3467-3486.

[135] Nolan, D. S. 2013. On the use of Doppler radar-derived wind fields to diagnose the secondary circulations of tornadoes. *Journal of the Atmospheric Sciences* 70:1160-1171.

[136] Nolan, D. S., N. A. Dahl, G. H. Bryan, and R. Rotunno. 2017. Tornado vortex structure, intensity, and surface wind gusts in large-Eddy simulations with fully developed turbulence. *Journal of the Atmospheric Sciences*74(5):1573-1597.

[137] Palmer, R. D., B. L. Cheong, M. W. Hoffman, S. J. Frasier, and F. J. López-Dekker. 2005. Observations of the small-scale variability of precipitation usingan imaging radar. *Journal of Atmospheric and Oceanic Technology*22(8):1122-1137.

[138] Palmer, R. D., D. Bodine, M. Kumjian, B. Cheong, G. Zhang, Q. Cao, H. B. Bluestein, A. Ryzhkov, T. Y. Yu, and Y. Wang. 2011. Observations of the 10 May 2010 tornado outbreak using OU-PRIME: potential for new science with high-resolution polarimetric radar. *Bulletin of the American Meteorological Society* 92(7):871-891.

[139] Park, H. S., A. V. Ryzhkov, D. S. Zrni'c, and K. E. Kim. 2009. The hydrometeor classification algorithm

for the polarimetric WSR-88D:description and application to an MCS. *Weather and Forecasting* 24(3): 730-748.

[140] Payne, C. D. , T. J. Schuur, D. R. MacGorman, M. I. Biggerstaff, K. M. Kuhlman, and W. D. Rust. 2009. Polarimetric and electrical characteristics of a lightning ring in a supercell storm. *Monthly Weather Review*138(6):2405-2425.

[141] Pazmany, A. L. , J. B. Mead, H. B. Bluestein, J. C. Snyder, and J. B. Houser. 2013. A mobile rapid-scanning X-band polarimetric(RaXPol)Doppler radar system. *Journal of Atmospheric and Oceanic Technology* 30(7):1398-1413.

[142] Petersen, W. A. , K. R. Knupp, D. J. Cecil, J. R. Mecikalski, C. Darden, and J. Burks. 2007. The university of Alabama Huntsvillethor center instrumentation: research and operational collaboration. In *33rd International Conference on Radar Meteorology*.

[143] Polger, P. D. , B. S. Goldsmith, R. C. Przywarty, and J. R. Bocchieri. 1994. National Weather Service warning performance based on the WSR-88D. *Bulletin of the American Meteorological Society*75(2): 203-214.

[144] Potvin, C. K. , L. J. Wicker, and A. Shapiro. 2012. Assessing errors in variational dual-Doppler wind syntheses of supercell thunderstorms observed by storm-scale mobile radars. *Journal of Atmospheric and Oceanic Technology*29(8):1009-1025.

[145] Putnam, B. J. , M. Xue, Y. Jung, N. Snook, and G. Zhang. 2014. The analysis and prediction of microphysical states and polarimetric radar variables in a mesoscale convective system using double-moment microphysics, multinetwork radar data, and the ensemble Kalman filter. *Monthly Weather Review*142(1):141-162.

[146] Rasmussen, E. N. , and J. M. Straka. 2007. Evolution of low-level angular momentum in the 2 June 1995 Dimmitt, Texas tornado cyclone. *Journal of the Atmospheric Sciences* 64:1365-1378.

[147] Rasmussen, E. N. , J. M. Straka, R. Davies-Jones, C. A. Doswell, F. H. Carr, M. D. Eilts, and D. R. MacGorman. 1994. Verification of the origins of rotation in tornadoes experiment: VORTEX. *Bulletin of the American Meteorological Society*75(6):995-1006.

[148] Ray, P. S. , R. J. Doviak, G. B. Walker, D. Sirmans, J. Carter, and B. Bumgarner. 1975. Dual-Doppler observations of a tornadic storm. *Journal of Applied Meteorology* 14:1521-1530.

[149] Rockney, V. 1958. The WSR-57 radar. In *Proceedings of Seventh Conference on Weather Radar*, Miami Beach, FL, F14-F20. Boston: American Meteorological Society.

[150] Rotunno, R. 1979. A study in tornado-like vortex dynamics. *Journal of the Atmospheric Sciences*36:140-155.

[151] Rotunno, R. 2013. The fluid dynamics of tornadoes. *Annual Review of Fluid Mechanics* 45:59-84.

[152] Rotunno, R. , and J. B. Klemp. 1982. The influence of the shear-induced pressure gradient on thunderstorm motion. *Monthly Weather Review* 110:136-151.

[153] Rust, W. D. 1989. Utilization of a mobile laboratory for storm electricity measurements. *Journal of Geophysical Research*94(D11):13305-13311.

[154] Rust, W. D. , and T. C. Marshall. 1989. Mobile, high-wind, balloon-launching apparatus. *Journal of Atmospheric and Oceanic Technology*6(1):215-217.

[155] Rust, W. D. , D. W. Burgess, R. A. Maddox, L. C. Showell, T. C. Marshall, and D. K. Lauritsen. 1990. Testing a mobile version of a cross-chain LORAN atmospheric(M-CLASS)sounding system. *Bulletin of*

the American Meteorological Society 71(2):173-180.

[156] Ryzhkov, A. , D. Burgess, D. Zrni'c, T. Smith, and S. Giangrande. 2002. Polarimetric analysis of a 3 May 1999 tornado. In 21*th Conference on Severe Local Storms*, San Antonio, TX. Boston: American Meteorological Society(Preprints).

[157] Ryzhkov, A. V. , S. E. Giangrande, and T. J. Schuur. 2005. Rainfall estimation with a polarimetric prototype of WSR-88D. *Journal of Applied Meteorology and Climatology* 44:502-515.

[158] Ryzhkov, A. V. , T. J. Schurr, D. W. Burgess, and D. Zrni'c. 2005. Polarimetric tornado detection. *Journal of Applied Meteorology and Climatology*44:557-570.

[159] Ryzhkov, A. V. , T. J. Schuur, D. W. Burgess, P. L. Heinselman, S. E. Giangrande, and D. S. Zrni'c. 2005. The joint polarization experiment: Polarimetric rainfall measurements and hydrometeor classification. *Bulletin of the American Meteorological Society* 86(6):809-824.

[160] Schenkman, A. D. , M. Xue, A. Shapiro, K. Brewster, and J. Gao. 2011. Impact of CASA radar and Oklahoma mesonet data assimilation on the analysis and prediction of tornadic mesovortices in an MCS. *Monthly Weather Review* 139(11), 3422-3445.

[161] Schenkman, A. D. , M. Xue, and M. Hu. 2014. Tornadogenesis in a high-resolution simulation of the 8 May 2003 Oklahoma City supercell. *Journal of the Atmospheric Sciences* 71:130-154.

[162] Schultz, C. J. , L. D. Carey, E. V. Schultz, B. C. Carcione, C. B. Darden, C. C. Crowe, P. N. Gatlin, D. J. Nadler, W. A. Petersen, and K. R. Knupp. 2012. Dual-polarization tornadic debris signatures Part I: Examples and utility in an operational setting. *Electronic Journal of Operational Meteorology*13:120-137.

[163] Schultz, C. J. , S. E. Nelson, L. D. Carey, L. Belanger, B. C. Carcione, C. B. Darden, T. Johnstone, A. L. Molthan, G. J. Jedlovec, E. V. Schultz, C. C. Crowe, and K. R. Knupp. 2012. Dualpolarization tornadic debris signatures Part II: Comparisons and caveats. *Electronic Journal of Operational Meteorology* 13:138-158.

[164] Schwarz, C. M. , and D. W. Burgess. 2010. Verification of the origins of rotation in tornadoes experiment, part 2(VORTEX2): data from the NOAA(NSSL) X-band dual-polarized radar. In25*th Conference on Severe Local Storms*. Boston: American Meteorological Society.

[165] Seo, B. C. , and W. F. Krajewski. 2010. Scale dependence of radar rainfall uncertainty: initial evaluation of NEXRAD's new super-resolution data for hydrologic applications. *Journal of Hydrometeorology*11(5):1191-1198.

[166] Shabbott, C. J. , and P. M. Markowski. 2006. Surface in situ observations within the outflow of forward-flank downdrafts of supercell thunderstorms. *Monthly Weather Review* 134:1422-1441.

[167] Shapiro, A. , K. M. Willingham, and C. K. Potvin. 2010. Spatially variable advection correction of radar data. Part I: Theoretical considerations. *Journal of the Atmospheric Sciences* 67(11):3445-3456

[168] Shapiro, A. , K. M. Willingham, and C. K. Potvin. 2010. Spatially variable advection correction of radar data. Part II: Test results. *Journal of the Atmospheric Sciences* 67(11):3457-3470.

[169] Simmons, K. M. , and D. Sutter. 2005. WSR-88D radar, tornado warnings, and tornado casualties. *Weather and Forecasting* 20(3):301-310.

[170] Skinner, P. S. , C. C. Weiss, M. M. French, H. B. Bluestein, P. M. Markowski, and Y. P. Richardson. 2014. VORTEX2 observations of a low-level mesocyclone with multiple internal rear-flank downdraft momentum surges in the 18 May 2010, Dumas, Texas supercell. *Monthly Weather Review*142(8):2935-2960.

[171] Skinner,P. S. ,C. C. Weiss,L. J. Wicker,C. K. Potvin,and D. C. Dowell. 2015. Forcing mechanisms for an internal rear-flank downdraft momentum surge in the 18 May 2010 Dumas,Texas supercell. *Monthly Weather Review* 143(11):4305-4330.

[172] Skow,K. D. ,and C. Cogil. 2017. A high-resolution aerial survey and radar analysis of quasilinear convective system surface vortex damage paths from 31 August 2014. *Weather and Forecasting* 32:441-467.

[173] Smull,B. F. ,and R. A. Houze. 1987. Dual-Doppler radar analysis of a midlatitude squall line with a trailing region of stratiform rain. *Journal of the Atmospheric Sciences* 44(15):2128-2149.

[174] Snyder,J. C. ,and H. B. Bluestein. 2014. Some considerations for the use of high-resolution mobile radar data in tornado intensity determination. *Weather and Forecasting* 29(4):799-827.

[175] Snyder,J. C. ,and A. V. Ryzhkov. 2015. Automated detection of polarimetric tornadic debris signatures using a hydrometeor classification algorithm. *Journal of Applied Meteorology and Climatology* 54:1861-1870.

[176] Snyder,J. C. ,H. B. Bluestein,V. Venkatesh,and S. J. Frasier. 2013. Observations of polarimetric signatures in supercells by an X-band mobile Doppler radar. *Monthly Weather Review* 141(1):3-29.

[177] Stensrud, D. J. , L. J. Wicker, K. E. Kelleher, M. Xue, M. P. Foster, J. T. Schaefer, R. S. Schneider, S. G. Benjamin,S. S. Weygandt,J. T. Ferree,and J. P. Tuell. 2009. Convective-scale warn-onforecast system. *Bulletin of the American Meteorological Society* 90:1487-1499.

[178] Stumpf, G. J. , A. Witt, E. D. Mitchell, P. L. Spencer, J. T. Johnson, M. D. Eilts, K. W. Thomas, and D. W. Burgess. 1998. The National Severe Storms Laboratory mesocyclone detection algorithm for the WSR-88D. *Weather and Forecasting* 13(2):304-326.

[179] Tanamachi, R. L. , H. B. Bluestein, W. C. Lee, M. Bell, A. Pazmany. 2007. Ground-based velocity track display(GBVTD)analysis of W-band Doppler radar data in a tornado near Stockton,Kansas,on 15 May 1999. *Monthly Weather Review* 135(3):783-800.

[180] Tanamachi, R. L. , H. B. Bluestein, J. B. Houser, S. J. Frasier, and K. M. Hardwick. 2012. Mobile, X-band,polarimetric Doppler radar observations of the 4 May 2007 Greensburg,Kansas,tornadic supercell. *Monthly Weather Review* 140(7):2103-2125

[181] Tanamachi,R. L. ,P. L. Heinselman,and L. J. Wicker. 2015. Impacts of a storm merger on the 24 May 2011 El Reno,Oklahoma,tornadic supercell. *Weather and Forecasting* 30(3):501-524.

[182] Trapp,R. J. ,and R. Davies-Jones. 1997. Tornadogenesis with and without a dynamic pipe effect. *Journal of the Atmospheric Sciences* 54(1):113-133.

[183] Trapp,R. ,and C. A. Doswell. 2000. Radar data objective analysis. *Journal of Atmospheric and Oceanic Technology* 17:105-120.

[184] Umeyama,A. Y. ,S. M. Torres,and B. L. Cheong. 2017. Bootstrap dual-polarimetric spectral density estimator. *IEEE Transactions on Geoscience and Remote Sensing* 55:2299-2312.

[185] Van Den Broeke,M. S. 2015. Polarimetric tornadic debris signature variability and debris fallout signatures. *Journal of Applied Meteorology and Climatology* 54:2389-2405.

[186] Van Den Broeke,M. S. ,and S. T. Jauernic. 2014. Spatial and temporal characteristics of polarimetric tornadic debris signatures. *Journal of Applied Meteorology and Climatology* 53(10):2217-2231.

[187] Vasiloff,S. V. 2001. Improving tornado warnings with the Federal Aviation Administration's terminal Doppler weather radar. *Bulletin of the American Meteorological Society* 82(5):861-874.

[188] Vasiloff,S. V. ,and K. W. Howard. 2009. Investigation of a severe downburst storm near Phoenix,Arizo-

na, as seen by a mobile Doppler radar and the KIWA WSR-88D. *Weather and Forecasting* 24(3): 856-867.

[189] Wakimoto, R. M., and J. W. Wilson. 1989. Non-supercell tornadoes. *Monthly Weather Review* 117(6): 1113-1140.

[190] Wakimoto, R. M., N. T. Atkins, and J. Wurman. 2011. The LaGrange tornado during VORTEX2. Part I: Photogrammetric analysis of the tornado combined with single-Doppler radar data. *Monthly Weather Review* 139(7):2233-2258.

[191] Wakimoto, R. M., P. Stauffer, W. C. Lee, N. T. Atkins, and J. Wurman. 2012. Finescale structure of the LaGrange, Wyoming, tornado during VORTEX2: GBVTD and photogrammetric analyses. *Monthly Weather Review* 140(11):3397-3418.

[192] Wakimoto, R. M., N. T. Atkins, K. M. Butler, H. B. Bluestein, K. Thiem, J. Snyder, and J. Houser. 2015. Photogrammetric analysis of the 2013 El Reno tornado combined with mobile X-band polarimetric radar data. *Monthly Weather Review* 143(7):2657-2683.

[193] Wakimoto, R. M., N. T. Atkins, K. M. Butler, H. B. Bluestein, K. Thiem, J. C. Snyder, J. Houser, K. Kosiba, and J. Wurman. 2016. Aerial damage survey of the 2013 El Reno tornado combined with mobile radar data. *Monthly Weather Review* 144(5):1749-1776.

[194] Wang, Y., and T. Y. Yu. 2015. Novel tornado detection using an adaptive neuro-fuzzy system with S-band polarimetric weather radar. *Journal of Atmospheric and Oceanic Technology* 32:195-208.

[195] Wang, Y., T. Y. Yu, M. Yeary, A. Shapiro, S. Nemati, M. Foster, D. L. Andra, and M. Jain. 2008. Tornado detection using a neuro-fuzzy system to integrate shear and spectral signatures. *Journal of Atmospheric and Oceanic Technology*25(7):1136-1148.

[196] Weber, M. E., J. Y. N. Cho, J. S. Herd, Flavin, J. M., Benner, W. E., Torok, G. S. 2007. The next-generation multimission U. S. surveillance radar network. *Bulletin of the American Meteorological Society*88: 1739-1751.

[197] Weiss, C. C., J. L. Schroeder, J. Guynes, P. S. Skinner, and J. Beck. 2009. The TTUKa mobile Doppler radar: Coordinated radar and in situ measurements of supercell thunderstorms during Project VORTEX2. In34*th Conference on Radar Meteorology*, Williamsburg, VA. 11B. 2. Boston: American Meteorological Society(Preprints).

[198] Whiton, R. C., P. L. Smith, S. G. Bigler, K. E. Wilk, and A. C. Harbuck. 1998. History of operational use of weather radar by U. S. weather services. Part I: The pre-NEXRAD era. *Weather and Forecasting*13: 219-243.

[199] Wicker, L. J., and R. B. Wilhelmson. 1995. Simulation and analysis of tornado development and decay within a three-dimensional thunderstorm. *Journal of the Atmospheric Sciences* 52:2675-2703.

[200] Wilson, J. W., G. B. Foote, N. A. C. rook, J. C. Fankhauser, C. G. Wade, J. D. Tuttle, C. K. Mueller, and S. K. Krueger. 1992. The role of boundary-layer convergence zones and horizontal rolls in the initiation of thunderstorms: a case study. *Monthly Weather Review* 120(9):1785-1815.

[201] Wood, V. T., and R. A. Brown. 1997. Effects of radar sampling on single-Doppler velocity signatures of mesocyclones and tornadoes. *Weather and Forecasting* 12:928-938.

[202] Wood, V. T., R. A. Brown, and D. Sirmans. 2001. Technique for improving detection of WSR-88D mesocyclone signatures by increasing angular sampling mesocyclone signatures by increasing angular sampling. *Weather and Forecasting* 16(1):177-184.

[203] Wurman,J. 2001. The DOW mobile multiple-Doppler network. In30*th International Conference on Radar Meteorology*.

[204] Wurman,J. 2002. The multiple-vortex structure of a tornado. *Weather and Forecasting* 17(3):473-505.

[205] Wurman,J. ,and C. R. Alexander. 2005. The 30 May 1998 Spencer,South Dakota,storm. Part II:Comparison of observed damage and radar-derived winds in the tornadoes. *Monthly Weather Review*133(1):97-119.

[206] Wurman,J. ,and S. Gill. 2000. Finescale radar observations of the Dimmitt,Texas(2 June 1995),tornado. *Monthly Weather Review* 128(7):2135-2164.

[207] Wurman,J. ,and K. Kosiba. 2013. Finescale radar observations of tornado and mesocyclone structures. *Weather and Forecasting* 28(5):1157-1174.

[208] Wurman,J. ,and M. Randall. 2001. An inexpensive,mobile,rapid-scan radar. In30*th International Conference on Radar Meteorology*,Munich,Germany. 3. 4. Boston:American Meteorological Society(Preprints).

[209] Wurman,J. ,J. M. Straka,and E. N. Rasmussen. 1996. Fine-scale Doppler radar observations of tornadoes. *Science* 272:1774-1777.

[210] Wurman,J. ,J. Straka,E. Rasmussen,M. Randall,and A. Zahrai. 1997. Design and deployment of a portable,pencil-beam,pulsed,3-cm Doppler radar. *Journal of Atmospheric and Oceanic Technology*14(6):1502-1512.

[211] Wurman,J. ,Y. Richardson,C. Alexander,S. Weygandt,and P. F. Zhang. 2007. Dual-Doppler analysis of winds and vorticity budget terms near a tornado. *Monthly Weather Review* 135 (6):2392-2405.

[212] Wurman,J. ,Y. Richardson,C. Alexander,S. Weygandt,and P. F. Zhang,2007. Dual-Doppler and single-Doppler analysis of a tornadic storm undergoing mergers and repeated tornadogenesis. *Monthly Weather Review* 135(3):736-758.

[213] Wurman,J. ,K. Kosiba,P. Markowski,Y. Richardson,D. Dowell,and P. Robinson. 2010. Finescale single-and dual-Doppler analysis of tornado intensification,maintenance,and dissipation in the Orleans,Nebraska,supercell. *Monthly Weather Review* 138(12):4439-4455.

[214] Wurman,J. ,D. Dowell,Y. Richardson,P. Markowski,E. Rasmussen,D. Burgess,L. Wicker,and H. B. Bluestein. 2012. The second verification of the origins of rotation in tornadoes experiment:VORTEX2. *Bulletin of the American Meteorological Society* 93(8):1147-1170.

[215] Wurman,J. ,K. Kosiba,and P. Robinson. 2013. In situ,Doppler radar,and video observations of the interior structure of a tornado and the wind-damage relationship. *Bulletin of the American Meteorological Society*94(6):835-846.

[216] Wurman,J. ,K. Kosiba,P. Robinson,and T. Marshall. 2014. The role of multiple-vortex tornado structure in causing storm researcher fatalities. *Bulletin of the American Meteorological Society*95(1):31-45.

[217] Xue,M. ,S. Liu,and T. Y. Yu. 2007. Variational analysis of oversampled dual-Doppler radial velocity data and application to the analysis of tornado circulations. *Journal of Atmospheric and Oceanic Technology*24(3):403-414.

[218] Yu,T. Y. ,Y. Wang,A. Shapiro,M. B. Yeary,D. S. Zrni′c,and R. J. Doviak. 2007. Characterization of tornado spectral signatures using higher-order spectra. *Journal of Atmospheric and Oceanic Technology*24(12):1997-2013.

[219] Yussouf, N. , and D. J. Stensrud. 2008. Impact of high temporal frequency radar data assimilation on storm-scale NWP model simulations. In24*th Conference on Severe Local Storms*, 9B. 1. Boston: American Meteorological Society.

[220] Zrni'c, D. , and M. Istok. 1979. Wind speeds in two tornadic storms and a tornado, deduced from Doppler spectra. *Journal of Applied Meteorology* 19(12):1405-1415.

[221] Zrni'c, D. S. , and R. J. Doviak. 1975. Velocity spectra of vortices scanned with a pulse-Doppler radar. *Journal of Applied Meteorology* 14(8):1531-1539.

[222] Zrni'c, D. S. , R. J. Doviak, D. W. Burgess, and R. A. Brown. 1977. Probing tornadoes with a pulse Doppler radar. *Quarterly Journal of the Royal Meteorological Society* 103(438):707-720.

[223] Zrni'c, D. S. , D. W. Burgess, and L. D. Hennington. 1985. Automatic detection of mesocyclonic shear with Doppler radar. *Journal of Atmospheric and Oceanic Technology* 2:425-438.

[224] Zrni'c, D. S. , J. F. Kimpel, D. E. Forsyth, A. Shapiro, G. Crain, R. Ferek, J. Heimmer, W. Benner, T. J. McNellis, and R. J. Vogt. 2007. Agile-beam phased array radar for weather observations. *Bulletin of the American Meteorological Society* 88:1753-1766.

第4章　利用地基天气雷达研究雷暴

LucaBaldini，Nicoletta Roberto，Mario Montopoli，Elisa Adirosi

4.1 引言

雷暴是由积雨云产生的短时局地风暴，伴随着起电现象，但并非总是伴随着强风、大雨和冰雹，在早期阶段，一个显著的特征是强烈的上升气流。相反，在降水柱中的强下沉气流是消散阶段的标志。

雷暴的独特之处在于闪电活动。因此，对雷暴的研究包括研究闪电现象和与云特性随时间的演化有关的电荷机制。

雷暴的高度通常达到15 km或更高，高于民航飞机在巡航时达到的高度。因此，它们与空中交通的一些特定风险有关，必须及时发现它们在飞行路线上的存在，以确保飞机和乘客的安全。雷暴经常与由高降水率引起的几种水文地质灾害有关，这种高降水率可产生滑坡和山洪暴发，特别是在水文响应非常快的小流域。

由于雷暴周期短范围小，遥感仪器特别是天气雷达对于观测、研究和预报这类现象是必要的。雷达(无线电探测和测距)是一种使用无线电波探测和定位"目标物"的设备，对天气雷达来说，"目标物"是云和降水粒子，这些粒子的物理性质是根据它们向雷达散射返回波的一些特征来估计的，比如振幅、相位和偏振。天气雷达波束的空间位置通过相对于发射脉冲的时间延迟和天线指向角来确定。为了获得足够的空间分辨率，考虑到物理和技术的限制，天线波束宽度在大多数情况下被限制在1°。许多国家气象局管理由S波段或C波段雷达组成的网络，以提供无缝隙测量。小型的、有时是移动的X波段系统也越来越多地以联网的方式用于集中监测从小流域到人口密集的大都市等相对较小的区域，这些区域特别容易受到恶劣天气的影响[43]。一般来说，在平均发射功率和天线波束宽度相同的情况下，频率越高，系统的灵敏度越高，天线尺寸越小。这意味着较高频率的系统更便宜且更易于管理。另一方面，高于S波段(在该波段衰减效应可以忽略)的频率意味着通过降水路径传播造成的气象回波衰减增强。然而，由于双偏振技术也提供了一种减轻路径衰减效应的方法，C波段和X波段双偏振系统也可以用来研究具有显著衰减效应的高降水率相关的雷暴。本章重点讨论S和C波段系统收集的观测数据，这些系统的特征与业务服务使用的系统相似。

早期天气雷达提供了降水云的反射率因子，这仍然是每个天气雷达的基本观测量[1]。探测风暴径向多普勒速度的系统始于20世纪70年代。1988年，美国在主要机场部署了称为

WSR-88D的基于多普勒S波段系统的全国雷达网，以及C波段终端多普勒天气雷达（TD-WR）。后来，具有多普勒功能的雷达成为业务气象服务的标准。自20世纪60年代以来，对天气雷达回波偏振信息的利用进行了研究，在所有可能的方案中，20世纪70年代提出的在水平（H）和垂直（V）态（雨介质的本征偏振态）运行的线性正交方案被认为是一种便捷的方案，可用于业务实现[57]。除了一些例外情况外，在业务系统中采用这种方案（现在称为“双偏振”）主要是在2000年以后。2017年，全球业务天气雷达的一个重要组成部分是双偏振多普勒系统（例如，自2013年以来，所有WSR-88D系统都是双偏振的），几乎所有地方都在进行双偏振升级计划。双偏振技术使得新的雷达探测量被定义，不仅提高了定量估计降水的准确性，而且有助于揭示降水云的其他特性。关于天气雷达测量原理和技术的更多细节，包括多普勒和双偏振系统，读者可以参考的主要教科书如Doviak和Zrnic[16]以及Bringi和Chandrasekar[10]。

当用雷达观测资料分析雷暴特性时，通常第一步是对流区域的识别。根据反射率的相对峰值，在雷达反射率数据的水平剖面上识别对流单体[59]。它们通常是短暂的（20～30 min），而雷达探测的快速更新速度（通常小于5～10 min）可以推断它们在空间和时间上的演化[67]，并研究它们的结构与天气条件的关系[2]。这是天气雷达网相对于气象卫星等其他天气监测系统的一个关键优势。此外，天气雷达扫描大气使用不同的仰角，这样就得到了风暴的三维图像，它可以用来获得诸如延伸高度和更常用的降水云垂直结构等信息。多普勒天气雷达估计沿径向与降水相关的质量的平均速度，称为平均多普勒速度。为了获得速度的其他两个分量，必须将覆盖重叠地区不同雷达的同步多普勒测量结果结合起来，从而获得雷暴内的三维通量[33]。随着双偏振被确认为提供先进基于雷达的天气产品的标准技术，新的探测量已经可用来识别水凝物并确定它们在降水云中的空间分布。在对流风暴中，这些探测量提供了与雷暴生命特定阶段相关的更清晰的特征，从而更精确地识别局地特征，如上升气流和冰雹轴的区域。先进的分析方法旨在从雷达探测中识别导致闪电的条件。应该指出的是，天气雷达无法捕捉所有可以描述对流触发机制的风暴前活动。然而，最近基于双偏振探测的研究旨在利用非气象散射体（如对空气运动敏感的昆虫移动）返回的信号来检测一些风暴前特征[45]。

本章的结构如下：第2节重点介绍了基于单偏振和多普勒雷达数据的方法来研究对流的内部结构和了解其随时间的演变；然后，第3节介绍了双偏振雷达探测量，并通过具体的事例解释了它们与雷暴的不同阶段直接相关的特征；最后，在建立由雷达（特别是双偏振系统）估计的云微物理与云起电过程之间的关系方面的一些进展将在第4节中介绍；第5节通过重点阐述当前雷达研究风暴起电的成果来结束本章。

4.2 单偏振、多普勒和双多普勒方法

自20世纪50年代末以来，基于单偏振天气雷达的对流风暴特性提取方法问世，尤其值得一提的是用于检测冰雹的方法[15]。单偏振冰雹检测方法是基于反射率垂直剖面的分析：高反射率等值线（如45 dBZ）高于冻结层的高度或雷达回波顶高度与地面冰雹发生有一定的关系[62]。后来，设计了一种相对简单的算法提取冰雹发生概率和冰雹尺度，该方法基于0 ℃等温线高度、雷达回波顶高度和温度导出量，并在WSR-88D雷达业务使用[69]。基于双偏振雷达

探测的水凝物分类系统(HCS)中的较新的冰雹识别算法(请参阅本章的下一节)已经明显优于早期基于反射率的方法[29]。

观测对流系统的另一个重要优势是用速度信息来表示。事实上,天气雷达系统中多普勒速度的可用性能够揭示沿着雷达波束的平均运动(即沿天线指向的径向方向)。这是因为天气雷达只在每个脉冲的一个方向上发送和接收能量,因此,探测到的运动速度大小必然只是能量发射方向上的速度大小,术语"平均运动"指的是将雷达波束内所有目标物的各种速度分量取平均,从而形成了速度产品。需要注意的是,天气雷达需要水凝物甚至其他散射物的存在,如昆虫[46,51]来追踪风的轨迹。雷达可以探测水凝物的速度,这要归功于多普勒效应(或多普勒频移),它是相对于波源移动的观察者看到波的频率或波长(或其他周期性事件)的变化。多普勒效应以1842年描述它的奥地利物理学家Christian Doppler的名字命名。对于天气雷达,一个目标物(例如大气中的一个水凝物粒子群)相对于波的发射源(即雷达)的运动将产生目标-源距离的连续时间变化,使后向散射波前沿到达的时间逐渐变长或变短,这取决于目标的运动方向。天气雷达通过测量接收信号相位的时间变化(即角频率)来测量目标-源距离的时间变化(即速度)。通常,天气雷达的目标物是液态和/或固态粒子,其运动由几个因素决定,包括重力、浮力和局地风。在雷暴中,上升/下沉气流和湍流运动控制着这些粒子的输送。因此,在速度域中,接收到的雷达信号可以被合理地视为具有不同速度的各种贡献的分布(通常称为频谱),并以整个目标运动的平均速度值为中心。速度谱的范围(即谱宽)是雷达波束内目标物湍流效应的一个度量。所有谱速度贡献的相干累积和就是天气雷达接收到的信号。

由于雷达通过测量接收信号相位的时间变化来估计速度,所以当每个方向的相位变化超过180°时就会产生相位折叠,导致速度模糊(或速度折叠)。也就是说,当被观测目标的运动速度足够快,产生的径向速度大于雷达可测最大值($V_{\max}$)时,就会出现速度模糊。在这种情况下,雷达仍然会在区间($-V_{\max}$,$V_{\max}$)内记录速度,该速度是模糊的(即错误的)。该最大速度由$V_{\max}=\mathrm{PRF}\times\lambda/4$给出,其中$\lambda$为雷达波长,PRF为发射脉冲的重复频率。例如,对$\lambda$的依赖表现为:X波段雷达比使用相同PRF的S波段系统更容易出现速度模糊。增加PRF会增大$V_{\max}$,但代价是降低雷达观测的距离覆盖,即降低最大不模糊距离$r_{\max}=\mathrm{c}/(2\times\mathrm{PRF})$,其中c为光速。距离和速度模糊之间的权衡是显而易见的,因为$r_{\max}V_{\max}=\mathrm{c}\cdot\lambda/8$。在S波段也需要速度去折叠技术来恢复不模糊速度场,同时保持测距能力不变。该技术有两种类型:基于对脉冲发射次序的修改(也称为发射编码技术)或基于后处理方法。第一类包括双PRF[14]和交错PRF[75]以及偏振[26]和相位编码[7]等方法。在双PRF模式下,用一个PRF发射一组脉冲,然后用另一个PRF发射第二组脉冲。两个PRF对应两个不同的速度,对它们进行估计和比较以获得不模糊速度。缺点是,在观测相同场景的时间相等时,信号的每个子部分(即与两个PRF相关的部分)收集的样本数量比使用单个PRF时要少。因此,双PRF速度估计具有更高的标准差。为了克服这一问题,可以采用交错PRF。将具有相同PRF的N个脉冲序列以不同的时移发射,产生一个连续脉冲的发射序列,其时间间隔为T_i($T_1<T_2<T_N$)。该方法的目标是找到T_1、T_2、T_N的最佳排列,给出最大不模糊速度和距离。较大的N提高了去折叠的成功率,但增加了从降水数据中去除杂波的最终滤波过程的复杂性。结合脉冲对分析得到的速度谱即使不模糊速度。

无论是哪种发射方式，后处理技术都是基于识别多普勒雷达探测的模糊速度场中不符合实际的梯度。模糊修正的完成基于：(1)利用环境风外部信息重建的速度场的连续性[16]；(2)依赖图像处理技术的自洽方法[74]；(3)速度方位显示(VAD)可以检查沿方位角的速度连续性[12]，或其众多变型，如扩展 VAD[58]、梯度 VAD[23]或 AR VAD[72]。

三维风矢量重建技术利用单个[36,61]或多个多普勒雷达系统[8,17]收集的速度径向分量测量数据。这种技术是基于对所研究的风场结构的简化假设，在最终重建的风产品中，往往必须接受局部气流的不完整描述。在机场附近，多普勒雷达被用来探测风的强度和方向随高度的强烈变化，通常称为风切变。就风切变而言，小尺度雷暴单体的下沉气流是最危险的，对空中安全预警系统的准确性提出了重大挑战[68]。

在灾害性个例分析的框架下，对雷暴中存在的旋转的量化是多普勒天气雷达的一个相关应用。多普勒雷达探测旋转的能力提高了气象学家观察雷暴内部并确定旋转成分存在的能力，旋转成分通常是龙卷发展的前兆[11,70]。2011 年 4 月 16 日，WSR-88D 在北卡罗来纳州罗利/达勒姆探测到一次超级单体雷暴，图 4.1 给出了一个值得关注的旋转示例，从图中可以看出一些有趣的特征。在雷达反射率方面(图 4.1a,c)，在图 4.1a 的左下中部有明显的钩状回波。钩状回波是一个强的后侧下沉气流加上一个强的入流区(见图 4.1c)。需要注意的是，入流区的反射率值较低，因为该区域的空气在上升，且没有明显的降雨。相反，在下沉气流占主导地位的区域，预计反射率值会较高。从多普勒特征图(图 4.1b、图 4.1d)可以看出，虚线右侧的中气旋内风向通常远离雷达(颜色较暖表示远离雷达的径向速度较大)，虚线左侧则相反，风向普遍朝向雷达(颜色较冷表示朝向雷达径向速度较大)。

实际上，图 4.1b 中虚线以东的红色模糊是没有修正多普勒速度模糊的个例。因此，存在一个紧凑的空间范围，分为风突然变化的两边。当这两个部分紧挨着时，意味着风在一小段距离内变化非常快，可以检测到一个中气旋特征(在例子中是逆时针方向)。值得注意的是，当风向垂直于雷达波束(虚线)时，我们无法判断发生了什么，除非我们有来自另一个位于不同位置的雷达的信息。

另一个以多普勒雷达数据为基础的重要气象应用是在天气预报模式中同化多普勒风场，以提高预报技术，尤其是针对恶劣天气[71]。最后，在不太主流的多普勒雷达应用中，鸟类追踪[24]和火山云表征[47]值得一提。

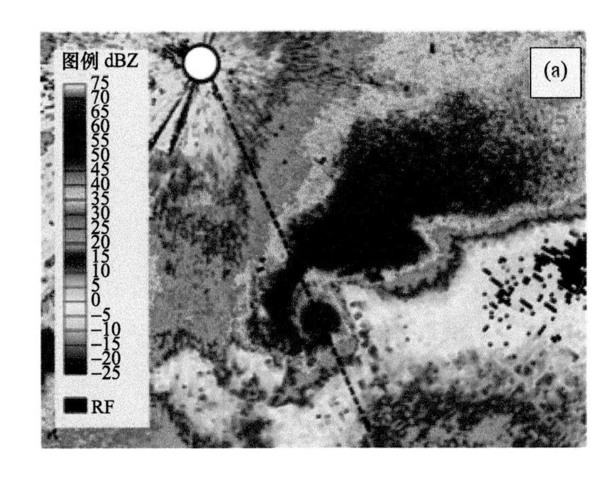

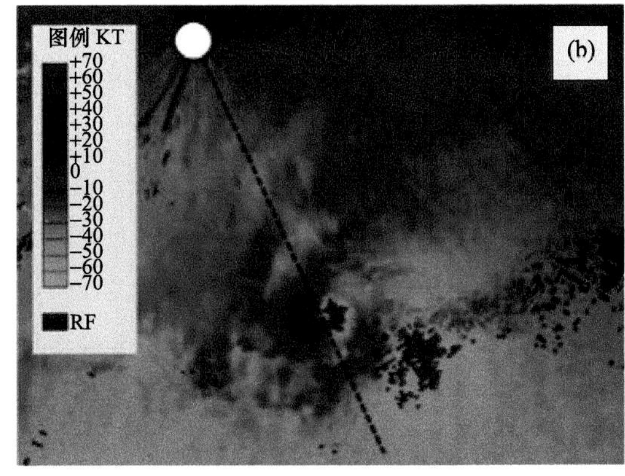

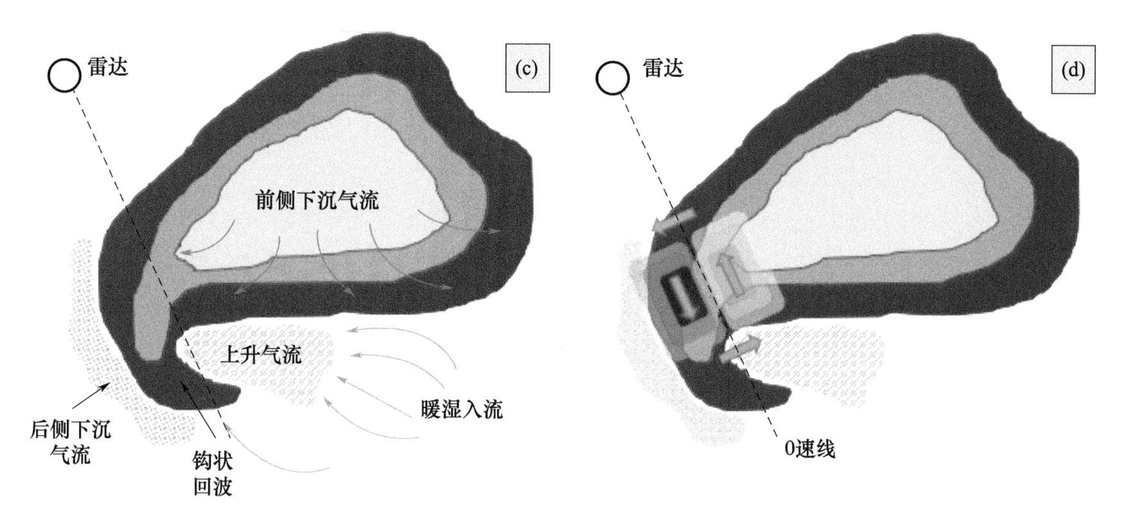

图 4.1 示例为 2011 年 4 月 16 日晚 20:40UTC,KRAX WSR-88D(北卡罗来纳州罗利市)在 0.5°仰角采集到的(a)雷达反射率图像和(b)相应的多普勒径向速度图像。图(a)和(b)中色标的单位分别为 dBZ 和(KT/h)。(c)和(d)分别展示了超级单体主要特征和预期雷达多普勒特征的示意图。关于这一个例的补充材料可以在文献[40]中找到

4.3 对流的双偏振雷达特征

双偏振技术是 20 世纪 70 年代末提出的一种研究成果,旨在证明云和降水粒子后向散射回波的偏振量中包含有用的信息。回波的偏振态主要是由粒子的性质决定的,例如形状(如雨滴在下落时呈扁圆形)、粒子谱分布、密度、形状、下落状态(如翻滚的冰雹)和温度。双偏振(线性正交)方案采用水平和垂直偏振态,对应于不随传播路径变化的雨介质的本征极化态。

直到 2000 年以后,这种方案才成为改进的业务天气雷达的参考标准。如今,大多数业务系统都采用所谓的 STAR(同时发射和接收)模式实现双偏振:它们同时发射水平偏振和垂直偏振脉冲,同时使用两个接收机接收 H 和 V 态的脉冲。实际上,在 STAR 方案中,作为 H 和 V 偏振态同时结合的结果,每个脉冲发射一个 45°线性偏振波。正如后面解释的那样,这对双偏振雷达提供的一些探测量的可用性造成了一些限制。

其他雷达系统设计,例如:双偏振交替脉冲发射和提供更多探测量的“全偏振”方案,现在主要用于研究型雷达。其他使用调制交替脉冲发射的雷达系统方案中,h 和 v 脉冲在发射过程中按照特定的脉冲次序进行切换。抑或是“全偏振”方案,其提供更大的测量数据,h 和 v 脉冲使用两个单独的发射通道发射,主要用于研究型雷达。

由双偏振雷达系统提供的基本探测量包括等效反射率因子 Z_H,通常是水平偏振(单位为 mm^6/m^3;更多的是取对数的 10 倍,单位为 dBZ),差分反射率 Z_{DR}(通常以分贝表示)定义为 h 和 v 偏振脉冲的共极返回信号之比(它实际上是在 H 和 V 接收通道上返回信号平均功率之比);同偏振相关系数(以下简称相关系数)(ρ_{HV}),定义为 H 和 V 偏振脉冲的相同偏振返回信号的复相关系数与传播差分相移的振幅,即复同偏振相关系数的相位。差分相移实际上是一

个传播变量，表示 h 和 v 偏振雷达波之间的传播相位差。差分传播相移距离廓线的距离导数的一半称为比差分相移(K_{DP},°/km)。进一步的探测量是线性退偏振比(L_{DR},dB)，定义为交叉偏振和同偏振平均功率电平的比值。但是，L_{DR}在 STAR 方案中不能直接使用，因为该方案中不存在交叉偏振探测。此外，如文献[63,30]所述，在交叉耦合效应(即后向散射信号功率从一个偏振态移动到产生交叉偏振分量的正交态)的情况下，STAR 方案测得的 Z_{DR}可能是不正确的。

每一次探测对所测粒子的某些特定物理性质都很敏感，因此可以使用双偏振测量来确定每个雷达取样体积中主要的水凝物类型。水凝物的识别已经取得了相当好的成功，在使用双偏振雷达的大多数业务网络中已应用了称为水凝物分类系统(HCS)。它们大多基于模糊逻辑(FL)方法，将现有的关于不同双偏振探测量识别水凝物能力的知识(经验的或基于理论的)集成到自动分类器中。

从根本上看，基于文献[37,76]所描述的方法，多年来提出了许多模糊逻辑 HCS。它们使用不同的输入测量集，将不同的水凝物类别作为输出，以及不同的隶属函数。后者是表示被测量变量(或变量向量)属于给定类别水凝物的程度的函数。最近，基于有监督、半监督和非监督方式的方法得到了改进，如聚类[4,28]、神经网络和支持向量机[53]。在对流风暴情况下，HCS 是支持气象学者快速有效地分析降水垂直结构或突出冰雹轴发生区域的基础。FL 方案的一个例外是文献[41]中描述的贝叶斯分类器。

双偏振技术的早期应用之一是为探测融化层以下的冰雹轴提供可能性[3]，即 0℃等温线以下的层结，厚几百米，其中冰粒子融化成雨滴，与周围的雨区形成对比。在融化层(ML)之上，双偏振测量的一个显著特征是正 Z_{DR}柱，其顶部可以延伸到 0℃层以上几千米。Z_{DR}柱揭示了强上升气流的区域，在那里雨滴被抬升到较冷的高度，然后冻结并失去稳定性，从而导致 Z_{DR}在最高高度的快速减小。直接探测表明，在正 Z_{DR}柱中，涉及水滴碰并的暖雨过程占主导地位[32]。由于某些水滴在上升气流中下落时的增长(即那些下落末速度超过上升气流速度的水滴，因此暴露在高液水含量的环境中)，粒子分选机制在很大程度上导致了柱内观察到的 Z_H 和 Z_{DR}的垂直结构。雨滴在这样的柱中可以长到很大，尽管它们的浓度很低。9 mm 直径可被认为是高空粒径的上限，但在文献[25]的测量中了地面上的最大雨滴直径为 9.7 mm。图 4.2 中的示例显示了 Z_{DR}柱的特征，展现了意大利 C 波段双偏振系统在距离雷达约 20 km 和 50 km 处观测到的两个对流单体。数据在 RHI(距离高度显示)扫描模式下收集，该模式下仰角变化，而方位角保持不变。虽然这种模式在业务雷达中不常见，但用这种方式收集的数据更能符合和代表降雨单体的瞬时发展，因此可以更好地在垂直剖面上揭示特征。注意衰减校正算法被应用于 Z_H 和 Z_{DR}，以合理地解释这些探测量。典型的 Z_H 对流特征如图 4.2(左上)所示，其高反射率核心延伸到 ML 上方(图 4.2 中虚线标记了从附近的无线电探空测得的 0℃等温线)。在最接近雷达的单体中，融化层以下，Z_{DR}大于 0 dB 的区域(图 4.2 右上)与高反射率区域(图 4.2 左上)几乎重合。该区域伴随下沉气流，Z_{DR}值高表明有较大的雨滴。由于上升气流对过冷水滴的作用，两个单体中均存在位于 ML 之上的正 Z_{DR}值。实际上，在较远的降雨单体中，ML 上方较窄的 Z_{DR}柱相对于 ML 下方较大的 Z_{DR}值左移了 2 km。同样的降雨单体中 Z_H 也有类似的变化。Z_{DR}柱以及 ML 之下较大的 Z_H 和 Z_{DR}分别表征了循环的上升和下降气流机制共存。相对于最近的降雨单体，远降雨单体在 ML 之下的 Z_{DR}值较低，可能是由于上升气流和下沉气流的强度不同(假定 C 波段衰减得到了适当补偿)。在对流单体的核心和 ML 上方(约 1.0 °/km)发现较高的 K_{DP}值(高达 2.5 °/km)，表明液态水的存在

(图 4.2 左下)。

通过应用水凝物分类算法,雷暴结构中包含的水凝物的位置变得容易探测,如图 4.2 右下图所示。可能的水凝物类别集为雨、冰雹、雹雨混合物、干雪、湿雪和霰-小冰雹,为每一距离库中的雷达三参数 Z_H、Z_{DR}和 K_{DP}进行分类的算法是基于文献[53]中描述的支持向量机方法。单体核心内的区域被划分为由霰组成的区域,霰是云起电机制中典型的水凝物。总体而言,Z_{DR}柱形成的早期探测可以揭示与对流相关的上升气流的强度,并且是极短期天气预报的重要工具[34]。混合相区从 0 ℃延伸到 10 ℃,由过冷雨滴、部分冻结雨滴和其他形式的湿冰水凝物组成,可以通过 LDR 的增大值检测到,它们通常在正 Z_{DR}柱上形成一个"帽"。在此个例中,它们还与较低的 ρ_{HV}值有关,当系统无法探测 LDR 时,可用作其替代。

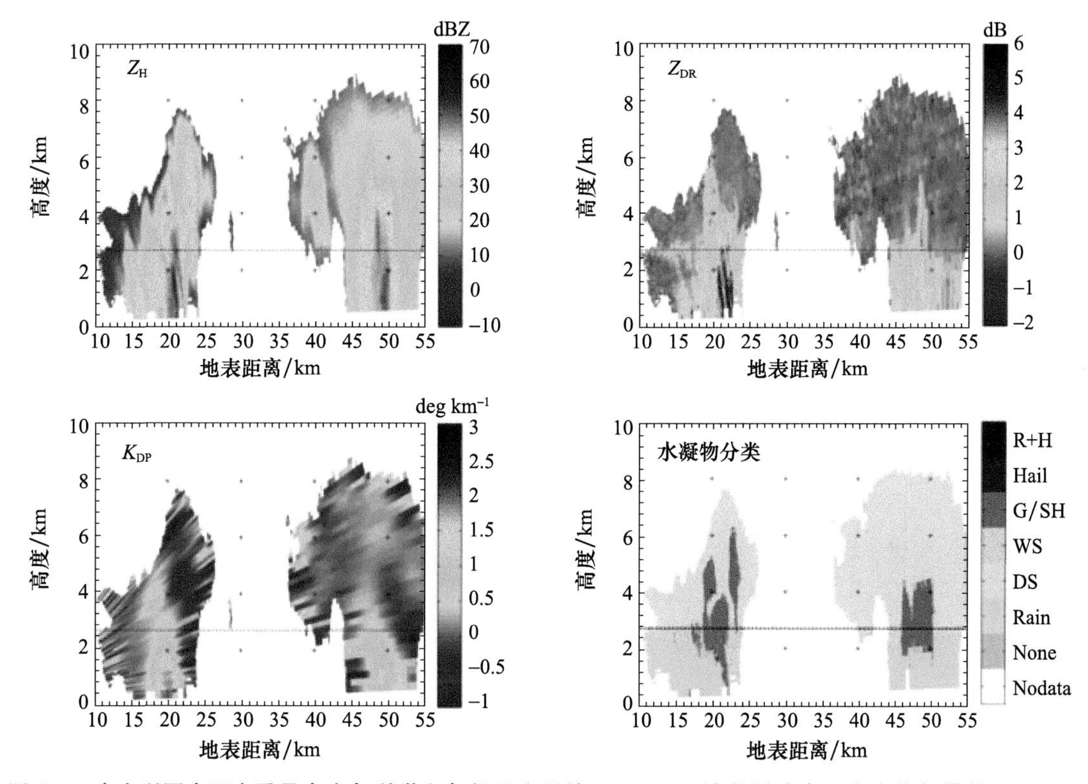

图 4.2 意大利国家研究委员会大气科学和气候研究所的 Polar 55 C 波段雷达在固定方位扫描的 RHI 显示。位于意大利罗马,41.84°N,12.65°E,海拔 130 m。数据采集于 2012 年 10 月 12 日 27°方位角。如图所示为 Z_H(左上)和 Z_{DR}(右上),K_{DP}(左下),以及水凝物分类结果(右下)。在水凝物分类的色标中,"DS""WS" "G/SH"和"R+H"分别表示干雪、湿雪、霰/小雹、冰雹混合物;"Nodata"和"none"表示没有测量值以及 HCS 无法提供可接受的结果。黑色虚线表示从 Pratica di Mare 机场(41.66°N,12.48°E)进行的最近无线电探空得出的 0℃层

为了进一步指出对流中一系列双偏振探测量的特性,用图 4.3 来进行讨论。它展示了不同的双偏振探测量,数据再次由 NASA NPOL S 波段雷达在爱荷华洪水研究地面验证(IFloodS)外场试验期间用 RHI 模式收集,该试验是 NASA/JAXA 全球降水测量任务地面验证计划的一部分[50]。反射率的探测结果(图 4.3 左上)突出显示了一个对流单体,它在雷达站上方延伸超过 14 km。在 0 ℃等温线以上(估计接近海拔 3900 m)发现了较高的反射率值,与近 0 dB 的 Z_{DR}值(右上)和低的共同偏振关系数(左下)相关。这些值表明混合相态翻滚粒子的存在。在液体粒子存在处(右下)探测到了较大的正 K_{DP}值(这种探测量对翻滚粒子不敏感)。一般来说,接近 0 ℃的正 K_{DP}区域表明液水被抬升到这些高度上。这些液水有助于下沉气流阶段的开启。对流风暴上层的冰晶可能被强电场力垂直排列,并与负 K_{DP}相关。1991 年在佛罗里达州对流和降水起电项目(CaPE)期间收集的探测数据在文献[31]中进行了分析,并呈现了这种特征。对于使用 STAR 模式的双偏振雷达,雷达波束穿过平均倾斜角度不为 0°的下落非球形水凝物,决定了返回的 H 和 V 偏振波之间的耦合。可以发现增强的正和负 Z_{DR}值的放射状条纹,这是由于冰晶的方向相对于 H 和 V 极化面旋转造成的耦合效应。

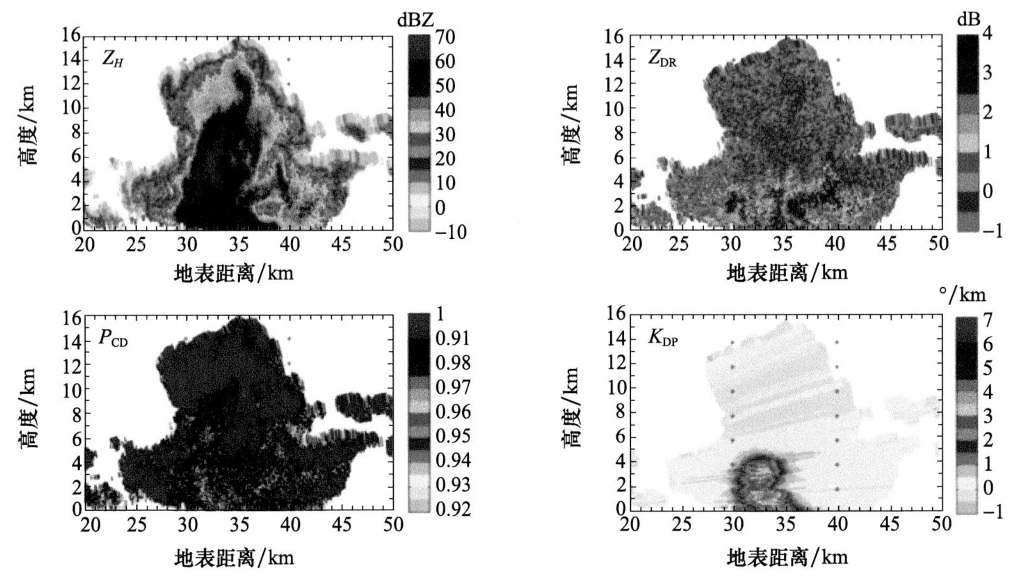

图 4.3 2013 年 5 月 20 日 02:16UTC,在 IfloodS 试验期间,NASA S 波段 NPOL 雷达沿 130.4°方位角收集的雷达探测数据(雷达位于 42.27°N,92.51°E。从左上角开始,顺时针方向依次为反射率因子、差分反射率、同偏振相关系数和比差分相移。高度是海拔高度)

在普通的(不产生冰雹的)风暴中,主要的冰相水凝物类型是由大冰晶在下落过程中吸附过冷云滴增长产生的霰粒子。过冷液滴与冰晶碰撞后在其上冻结(这种过程被称为“淞附”),导致霰的形成。这种粒子通常会呈锥形,如科罗拉多州立大学 CHILL 的雷达收集到的探测数据和装备了体积粒子采样探头的飞机收集的图像所示,如文献[18]中所述。霰粒子在下落过程中虽然表现出振荡和翻滚,但稳定下落模式表现为顶点垂直向上。这些粒子在雷暴中很重要,与下一节所示的闪电触发有关。

4.4 利用双偏振雷达观测估计云起电过程

闪电的形成取决于降水云的微物理性质。云中的电荷通常呈垂直分层的三极结构[66]:正电荷聚集在云的上部,负电荷聚集在云的中部,最后在云的底部发现一个低强度的正电荷中心(见图 4.4 右)。目前,用来解释云起电机理的最广为接受的理论是非感应起电(NIC)。NIC 理论解释了冰相水凝物的电荷是冰晶和霰在一个存在过冷水滴和冰晶的环境中碰撞的结果,然后促进了电荷分离。NIC过程获得的电荷分离可以产生一个能克服空气介电强度(即能在绝缘材料中形成导电路径的电场的最小值,也称为介电击穿水平)的强带电场(数百 kV/m^2),并产生与观测到的放电有关的电流。室内实验证实了NIC理论,认为影响电荷强度及其极性的主要因素是云液水含量和环境温度[60]。事实上,冰相水凝物的相对生长速度受云中过饱和度的影响[55]。综合而言,这一机制需要存在较大的冰相水凝物(即霰或冰雹丸),它们在悬浮着过冷水滴的环境中与冰晶碰撞。云上部气团的输送以及包含的电荷的垂直分布取决于上升气流的强度,当电荷分离的数量足够多时,触发大气的电击穿产生闪电。图4.4(左图)描绘了雷暴中有利于闪电形成的水凝物分布的典型配置。可以总结如下:

冻结阶段(大约－40 ℃),云体上部的冰晶带正电荷;

混合相态阶段Ⅰ(从约－15 ℃至－10 ℃),云体中部带负电荷的霰粒子;

混合相态阶段Ⅱ(－10 ℃之下),部分带正电荷的霰和冰雹在云底降落,因为在温度高于－10 ℃时,霰带正电荷[60]。

在雷达特征方面,下面的小节展示了双偏振雷达如何识别这些阶段。

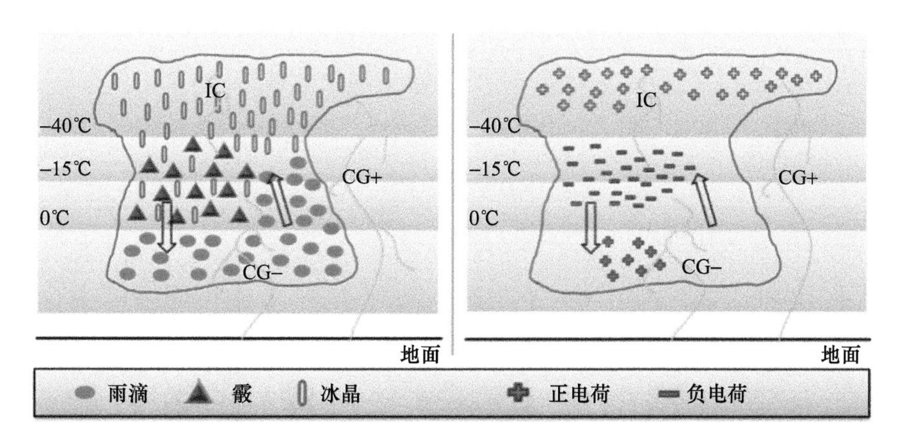

图4.4 一场雷暴中的水凝物分布(左图)和电荷分布概念图(右图)。云对地正和负(分别为CG＋和CG－)以及云内(IC)闪电用黄色曲线表示,红色箭头表示上升气流和下沉气流。注意冰晶是垂直排列的

4.1 闪电期间观测到的双偏振量

双偏振天气雷达能够以合适的空间和时间分辨率提供有关云起电机制的冰相水凝物信息。为了深入了解云中的闪电活动,自20世纪80年代以来,雷达探测与闪电探测数据被结合研究使用[27]。本地或全球覆盖的闪电网络(LN)可检测闪电的位置、强度和极性(正或负)。这些LN网络基于ELF(极低频)到VLF(甚低频)或VHF(甚高频)系统。那些工作在VHF以下频率的网络通常有很大的空间覆盖范围。事实上,它们可以探测到视线之外的闪电,利用电离层传播效应,让闪电产生的信号通过天波路径传播几千千米。另一方面,VHF系统对云内外的带电过程非常敏感,但空间覆盖范围有限。

通常情况下,频率位于LF/VLF的LNs探测云地闪(CG)的回击,但一些复杂的系统也能够检测云间(IC)闪和极性[5]。通过使用对发射能量偏振敏感的接收机来认识IC和CG之间的区别,通常IC和CG闪电分别是水平偏振和垂直偏振电场。闪电定位通常利用到达时间

(TOA)技术实现,该技术使用至少三个传感器检测到的信号时间间隔的三角形组合。另一种很少使用的定位技术是基于测向仪和基于干涉的方法

图 4.5 的上半部分显示了 LF/VHF 网络在意大利中部探测到的闪电,叠加在位于罗马的 C 波段天气雷达观测到的雷达反射率上的例子。可以注意到闪电和雷达反射率值之间的相关性:在雷达反射率值高的区域(大于 40 dBZ)上有大量的闪电事件。

一些结合地面双偏振雷达和闪电网络观测的工作表明,闪电与微物理特性之间存在良好的相关性[65]。例如,利用全球卫星观测资料[49]和 C 波段双偏振雷达收集的地基观测资料发现霰的含量与闪电密度之间存在线性关系[52]。这些相关性有不同的应用领域,如数值天气预报(NWP)模型和短临预报技术。通常情况下,NWP 使用的是由一个或多个在很短的时间间隔和很短空间距离内测得的回击构成的闪电数据。需要采用一些将回击数据转换为闪电的方法。国家闪电探测网络(NLDN)的阈值通常是最大时间间隔 0.5 s,连续闪电之间的最大横向距离 10 km。图 4.5 中的散点图(下图)显示了不同个例中与霰有关的柱状冰水含量(IWC)与闪电空间密度之间的线性关系[52]。文献[52]中报道的线性关系与其他文献的结果基本一致[22,49]。然而,IWC 与闪电密度的斜率在不同的个例中并不相同(见图 4.5 底部所示的表格)。这种斜率与双偏振雷达探测的其他雷暴特征(如上升气流强度)之间可能的关系是目前正在研究的课题。

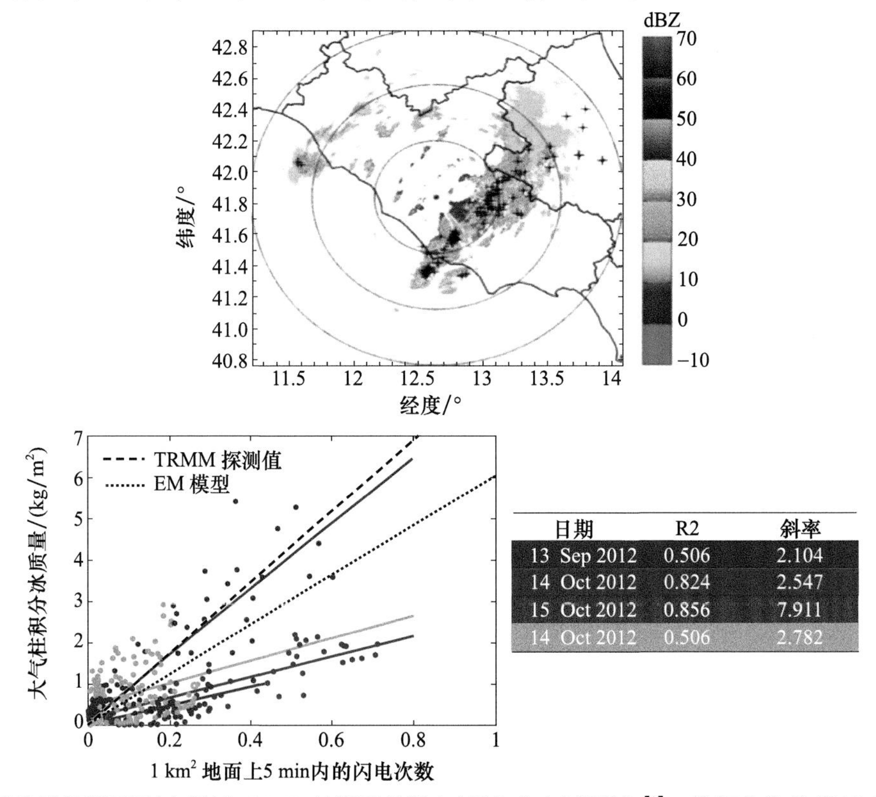

日期	R2	斜率
13 Sep 2012	0.506	2.104
14 Oct 2012	0.824	2.547
15 Oct 2012	0.856	7.911
14 Oct 2012	0.506	2.782

图 4.5 雷达反射率因子(上图)与 LINET 检测到的闪电(黑色十字)相叠加[6]。数据收集于 2015 年 10 月 14 日 15:05 UTC,雷达仰角 1.6°。下图显示 LINET 在 5 min 内记录的 1 km² 地面上的闪电次数和由 Polar 55C 收集的双偏振观测量估计的 IWC(kg/m²)之间的散点图[52]。作为参考,黑点线表示[22]大气柱积分霰冰水含量与最小阈值的线性关系,黑虚线表示[49]中大气柱积分冰质量与陆地闪电密度的线性关系。表中显示了不同个例中(每行的颜色对应散点图中的一个相同个例)的相关性(R^2)和线性关系的斜率

换句话说，从图4.5可以看出，相同数量的闪电密度对应着IWC的较大变化。因此，不仅IWC的准确计算对闪电密度的预测很重要，起电机制的效率也起着重要的作用。在这方面，对上升气流的深入研究可以解释图4.5所示的斜率变化。实际上，结合双偏振雷达观测和VHF LN测量结果，文献[65]发现当上升气流强度大于10 m/s且霰总量大于2000/km^{3}时，VHF源和CG数量增加，而文献[56]记录了在放电过程的最后一步闪电跳跃之前上升气流强度和霰总量的增加。另一个值得一提的重要方面是，IWC的估计可能存在很大的不确定性。在这方面，双偏振雷达探测有助于减少单偏振系统的不确定性，能够像本章之前所示的那样区分冰相区域和其他区域，然后更好地约束IWC的估计。

在过去的几十年里，双偏振雷达探测在确定与闪电活动相关的关键云微物理特征方面已经表现出了特有的能力。如上一节所述，Z_{DR}和K_{DP}探测的特征（特别是正Z_{DR}柱）能够识别对流的重要微物理过程。参考图4.4的方案，Z_{DR}特征在混合相层中占主导地位，而K_{DP}特征在冰相层中较重要[13,38,65]。Z_{DR}和K_{DP}值为正的层恰好位于融化层的上方，而高层中Z_{DR}和K_{DP}通常为负值。在起电过程中观测到的云的另一个特征是混合相层中存在圆锥形状的霰粒子，在下落过程中其顶点为垂直指向。这些水凝物的典型特征是负的Z_{DR}和K_{DR}值，伴随高的Z_H值[19,52]。在云的上部冰相区也发现了与闪电相关的特殊雷达偏振特征。K_{DP}和Z_{DR}均为明显负值的报道见[38]。

K_{DP}和Z_{DR}值的降低伴随着闪电密度的增加[21]，这些负的信号与垂直排列的冰粒子有关。一些研究表明，强电场可以在初始冰晶中诱导产生瞬时偶极子，可以排列冰晶[64]。而下落过程中主轴垂直指向的霰的存在（也存在于冻结层）可使K_{DP}为负值。

4.4.2 雷达和闪电联合探测的应用

一些重要的外场试验活动致力于探索与偏振雷达特征相关的带电风暴的主要特征，例如2000年5月至7月在美国开展的STEPs（强雷暴起电和降水研究）[35]，以及最近在CHUVA（巴西主要降水系统的云过程）项目框架下自2010年以来在巴西开展的试验[39]。特别是，在CHUVA项目中，开展专门的试验来研究云带电的特性：为X波段双偏振多普勒雷达设计了一个特别的扫描方式，以便以足够的分辨率探测偏振雷达探测量的垂直分布。雷达数据与安装在西南雷达区的甚高频传感器密集网络探测到的闪电测量数据相结合，通过对数百万条偏振雷达变量垂直廓线的观测，[42]中展现了在不同高度上有闪电活动和没有闪电活动的风暴的主要特征。他们确定了云垂直廓线中的偏振雷达特征，推断了云起电涉及的主要水凝物和过程。他们在不同的云层中发现了K_{DP}和Z_{DR}的变化。在温度范围为0至－15℃的层结中发现了增大的正K_{DP}值，这可能是由于抬升的过冷水滴造成的，而在温度范围为－15至－40℃（混合相I）的层结中发现K_{DP}和Z_{DR}值通常为负。他们将这些值归因于锥形霰的存在。最后发现，在闪电密度快速增加的冻结层（4 min内每千米14个VHF源），K_{DP}下降了10倍，变为负值。当闪电发生时，这样的K_{DP}低值与冰晶的排列现象相符。通过大量的统计数据，这项工作的结果支持了过去几十年在不同云层上发现的许多关于偏振雷达特征的结果，这些结果往往是通过分析少数个例研究得到的。

对流雷暴中的闪电对航空和电力基础设施管理等多种活动都有影响。因此，近几十年来，对闪电预报系统的开发进行了大量的投入。根据时间分辨率和空间尺度的不同，有不同的预报系统和预报模式。那些集中在短时间(几个小时)和有限区域的预报通常基于天气雷达数据[48]。雷达预报闪电的技术大多是基于对流初始前兆信号的识别,如冰水含量的估计、霰的识别和量化以及上升气流强度的估计等。基于卫星数据[44]或数值天气预报模式[20]的技术,通常可以获得高时空分辨率的闪电预报。后者基于闪电模拟模型,利用与精细起电方案相关的一维[22]或三维云结构再现云中的充电机制。通过深入了解云微物理过程及其与偏振雷达探测研究揭示的电活动的关系,闪电模型模拟可以获得有益的认识和重要的改进。此外,这些知识可被用于 NWP 模式,以改进微物理方案、云参数化和模式初始化。

4.5 总结

几十年来,天气雷达一直被用于探测和预测恶劣天气。在雷暴情况下,天气雷达有望揭示其重要特征,如风暴顶部高度、范围、风场特征、降雨率等,从而得到一些风暴激烈程度的指数。自 20 世纪 70 年代以来,多普勒雷达在识别对流系统的动力结构方面发挥了重要作用。双偏振雷达系统为获取雷暴结构及其时间演变的细节提供了基础性贡献。对双偏振探测量的具体特征进行分析,最终在自动分类算法的支持下,可以识别对流单体生命的不同阶段。云体上部的 Z_{dr} 柱和 K_{dp} 特征的识别在云起电机制的关系中起着相关的作用。当与其他仪器的测量协同使用时,雷达探测将变得更加有用。事实上,多传感器观测提供了一个从多个角度研究雷暴的机会。最近文献报道的结果强调了结合双偏振雷达和闪电网络观测来研究云结构和云充电机制相关的微物理特性的能力。尽管对云起电机制的理解取得了进展,但关键问题仍未解决。一是从风暴到雷暴云的过渡,即从首次雷达回波到第一次闪电时间的闪电形成过程的理解是对临近预报闪电的基础。雷达探测数据的快速更新(最终由使用的电子调控天线[73]实现),以及通过密集雷达网络结合临近预报工具[54]实现的高空间分辨率测量,有望在与云起电机制的触发有关的高密度冰粒子的识别、量化和预测方面带来进步。将天气雷达观测与可见光-红外星载仪器相结合,有望对探测闪电或对流前期信号做出重要贡献。例如从 2021 年开始,第三代气象卫星(MTG)将能够从云发展的早期阶段探测云体和起电,或通过全球导航卫星系统(GNSS)测量得出的水汽特征,可以用于预警对流的触发[9]。

致谢 作者感谢两位评审人(Roberto Cremonini 和一位匿名者),他们的建议对提高本章的质量和意义做出了贡献。作者感谢意大利民防部多年来对 CNR-ISAC 地中海地区对流系统研究的支持。

参考文献*

[1] Atlas, D., ed. 1990. *Radar in Meteorology*. Boston: American Meteorological Society. https://doi.org/

* 参考文献沿用原版书中内容,未改动

10. 1007/978-1-935704-15-7.

[2] Austin, P. M. , and R. A. Houze. 1972. Analysis of the structure of precipitation patterns in New England. *Journal of Applied Meteorology* 11: 926-935. https://doi. org/10. 1175/1520-0450 (1972) 011<0926:AOTSOP>2. 0. CO;2.

[3] Aydin, K. , T. Seliga, and V. Balaji. 1986. Remote sensing of hail with a dual linear polarization radar. *Journal of Climate and Applied Meteorology* 25: 1475-1484. https://doi. org/10. 1175/1520-0450 (1986)025%3C1475:RSOHWA%3E2. 0. CO;2.

[4] Bechini, R. , and V. Chandrasekar. 2015. Semisupervised robust hydrometeor classification method for dual-polarization radar applications. *Journal of Atmospheric and Oceanic Technology* 32: 22-47. https://doi. org/10. 1175/JTECH-D-14-00097. 1.

[5] Betz, H. -D. , K. Schmidt, P. Oettinger, and M. Wirz. 2004. Lightning detection with 3-D discrimination of intracloud and cloud-to-ground discharges. *Geophysical Research Letters* 31: L11108. https://doi. org/10. 1029/2004GL019821.

[6] Betz, H. -D. , K. Schmidt, and W. P. Oettinger. 2009. LINET-an international VLF/LF lightning detection network in Europe. In *Lightning: Principles, Instruments and Applications*, Chap. 5, ed. H. -D. Betz, U. Schumann, and P. Laroche. Dordrecht: Springer. https://doi. org/10. 1007/978-1-4020-9079-05.

[7] Bharadwaj, N. , and V. Chandrasekar. 2007. Phase coding for range ambiguity mitigation in dual-polarized Doppler weather radars. *Journal of Atmospheric and Oceanic Technology* 24: 1351-1363. https://doi. org/10. 1175/JTECH2061. 1.

[8] Bousquet, O. , and P. Tabary. 2014. Development of a nationwide real-time 3-D wind and reflectivity radar composite in France. *Quarterly Journal of the Royal Meteorological Society* 140: 611-625. https://doi. org/10. 1002/qj. 2163.

[9] Brenot, H. , J. Neméghaire, L. Delobbe, N. Clerbaux, P. De Meutter, A. Deckmyn, A. Delcloo, L. Frappez, and M. Van Roozendael. 2013. Preliminary signs of the initiation of deep convection by GNSS. *Atmospheric Chemistry and Physics* 13: 5425-5449. https://doi. org/10. 5194/acp-13-5425-2013.

[10] Bringi, V. N. , and V. Chandrasekar. 2001. *Polarimetric Doppler Weather Radar: Principles and Applications*. Cambridge: Cambridge University Press.

[11] Brown, R. A. , W. C. Bumgarner, K. C. Crawford, and D. Sirmans. 1971. Preliminary Doppler velocity measurements in a developing radar hook echo. *Bulletin of the American Meteorological Society* 52: 1186-1188. https://doi. org/10. 1175/1520-0477(1971)052<1186:PDVMIA>2. 0. CO;2.

[12] Browning, K. A. , and R. Wexler. 1968. The determination of kinematic properties of a wind field using Doppler radar. *Journal of Applied Meteorology* 7: 105-113. https://doi. org/10. 1175/1520-0450(1968) 007<0105:TDOKPO>2. 0. CO;2.

[13] Caylor, I. , and V. Chandrasekar. 1996. Time-varying ice crystal orientation in thunderstorms observed with multiparameter radar. *IEEE Transactions on Geoscience and Remote Sensing* 34: 847-858. https://doi. org/10. 1109/36. 508402.

[14] Dazhang, T. , S. G. Geotis, R. E. Passarelli Jr. , A. L. Hansen, and C. L. Frush. 1984. Evaluation of an alternating-PRF method for extending the range of unambiguous Doppler velocity. In *Proceedings of the 22nd Conference on Radar Meteorology*, Zurich, Switzerland, 10-13 Sept 1984, 523-527.

[15] Donaldson, R. J. 1961. Radar reflectivity profiles in thunderstorms. *Journal of Meteorology* 18: 292-305. https://doi. org/10. 1175/1520-0469(1961)018%3C0292:RRPIT%3E2. 0. CO;2.

[16] Doviak, R. J., and D. S. Zrni'c. 1994. *Doppler Radar and Weather Observations*. 2nd ed. Washington: Academic.

[17] Doviak, R. J., P. S. Ray, R. G. Strauch, and L. J. Miller. 1976. Error estimation in wind fields derived from dual-Doppler radar measurement. *Journal of Applied Meteorology* 15: 868-878. https://doi.org/10.1175/1520-0450(1976)015%3C0868:EEIWFD%3E2.0.CO;2.

[18] El-Magd, A., V. Chandrasekar, V. N. Bringi, and W. Strapp. 2000. Multiparameter radar and in situ aircraft observation of graupel and hail. *IEEE Transactions on Geoscience and Remote Sensing* 38: 570-578. https://doi.org/10.1109/36.823951.

[19] Evaristo, R., T. M. Bals-Elsholz, E. R. Williams, D. J. Smalley, M. F. Donovan, and A. Fenn. 2013. Relationship ofgraupel shape to differential reflectivity: theory and observations. Presented at 93rd American Meteorological Society Annual Meeting, Austin, TX, 5-10 Jan 2013.

[20] Federico, S., E. Avolio, M. Petracca, G. Panegrossi, P. San, D. Casella, and S. Dietrich. 2014. Simulating lightning into the RAMS model: implementation and preliminary results. *Natural Hazards and Earth System Sciences* 14: 2933-2950. https://doi.org/10.5194/nhess-14-2933-2014.

[21] Figueras i Ventura, F., F. Honor, and P. Tabary. 2013. X-band polarimetric weather radar observations of a hailstorm. *Journal of Atmospheric and Oceanic Technology* 30: 143-2151. https://doi.org/10.1175/JTECH-D-12-00243.1.

[22] Formenton, M., G. Panegrossi, D. Casella, S. Dietrich, A. Mugnai, P. San, F. Di Paola, H.-D. Betz, C. Price, and Y. Yair. 2013. Using a cloud electrification model to study relationships between lightning activity and cloud microphysical structure. *Natural Hazards and Earth System Sciences* 13: 1085-1104. https://doi.org/10.5194/nhess-13-1085-2013.

[23] Gao, J., K. K. Droegemeier, J. Gong, and Q. Xu. 2004. A method for retrieving mean horizontal wind profiles from single-Doppler radar observations contaminated by aliasing. *Monthly Weather Review* 132: 1399-1409. https://doi.org/10.1175/1520-0493-132.1.1399.

[24] Gasteren, H. V., I. Holleman, W. Bouten, E. V. Loon, and J. Shamoun-Baranes. 2008. Extracting bird migration information from C-band Doppler weather radars. *MonthlyWeather Review* 150: 674-686. https://doi.org/10.1111/j.1474-919X.2008.00832.x.

[25] Gatlin, P. N., M. Thurai, V. N. Bringi, W. Petersen, D. Wolff, A. Tokay, L. Carey, andM. Wingo. 2015. Searching for large raindrops: a global summary of two-dimensional videodisdrometer observations. *Journal of Applied Meteorology and Climatology* 54: 1069-1089. https://doi.org/10.1175/JAMC-D-14-0089.1.

[26] Golestani, Y., V. Chandrasekar, and R. J. Keeler. 1995. Dual polarized staggered PRT scheme for weather radars: analysis and application. *IEEE Transactions on Geoscience and Remote Sensing* 33 (2): 239-246. https://doi.org/10.1109/36.377923.

[27] Goodman, S. J., D. E. Buechler, P. D. Wright, and W. D. Rust. 1988. Lightning and precipitation history of a microburst-producing storm. *Geophysical Research Letters* 15: 1185-1188.

[28] Grazioli, J., D. Tuia, and A. Berne. 2015. Hydrometeor classification from polarimetric radar measurements: a clustering approach. *Atmospheric Measurement Techniques* 8: 149-170. https://doi.org/10.5194/amt-8-149-2015.

[29] Heinselman, P., and A. Ryzhkov. 2006. Validation of polarimetric hail detection. *Weather Forecast* 21: 839-850. https://doi.org/10.1175/WAF956.1.

[30] Hubbert, J. C., S. M. Ellis, M. M. Dixon, and G. G. Meymaris. 2010. Modeling, error analysis, and evaluation of dual-polarization variables obtained from simultaneous horizontal and vertical polarization transmit radar. Part II: experimental data. *Journal of Atmospheric and Oceanic Technology* 27: 1599-1607. https://doi.org/10.1175/2010JTECHA1337.1.

[31] Hubbert, J. C., S. M. Ellis, W.-Y. Chang, S. Rutledge, and M. Dixon. 2014. Modeling and interpretation of S-Band ice crystal depolarization signatures from data obtained by simultaneously transmitting horizontally and vertically polarized fields. *Journal of Climate and Applied Meteorology* 53: 1659-1677. https://doi.org/10.1175/JAMC-D-13-0158.1.

[32] Illingworth, A. J., J. W. F. Goddard, and S. M. Cherry. 1987. Polarization radar studies of precipitation development in convective storms. *Quarterly Journal of the Royal Meteorological Society* 113: 469-489. https://doi.org/10.1002/qj.49711347604.

[33] Kropfli, R. A., and L. J. Miller. 1975. Thunderstorm flow patterns in three dimensions. *Monthly Weather Review* 103: 70-71. https://doi.org/10.1175/1520-0493(1975)103<0070:TFPITD>2.0.CO;2.

[34] Kumjian, M., A. Khain, N. Benmoshe, E. Ilotoviz, A. Ryzhkov, and V. Phillips. 2014. The anatomy and physics of ZDR columns: investigating a polarimetric radar signature with a spectral bin microphysical model. *Journal of Applied Meteorology and Climatology* 53: 1820-1843. https://doi.org/10.1175/JAMC-D-13-0354.1.

[35] Lang, T. J., et al. 2004. The severe thunderstorm electrification and precipitation study. *Bulletin of the American Meteorological Society* 85: 1107-1125. https://doi.org/10.1175/BAMS-85-8-1107.

[36] Lhermitte, R. M., and D. Atlas. 1961. Precipitation motion by pulse Doppler radar. In *Proceedings of 9th Weather Radar Conference*, Kansas City, MO, 23-26 Oct 1961, 218-223.

[37] Liu, H., and V. Chandrasekar. 2000. Classification of hydrometeors based on polarimetric radar measurements: development of fuzzy logic and neuro-fuzzy systems, and in situ verification. *Journal of Atmospheric and Oceanic Technology* 17: 140-164. https://doi.org/10.1175/1520-0426(2000)017<0140:COHBOP>2.0.CO;2.

[38] López, R. E., and J.-P. Aubagnac. 1997. The lightning activity of a hailstorm as a function of changes in its microphysical characteristics inferred from polarimetric radar observations. *Journal of Geophysical Research* 102: 16799-16813. https://doi.org/10.1029/97JD00645.

[39] Machado, L. A., et al. 2014. TheChuva project: how does convection vary across Brazil? *Bulletin of the American Meteorological Society* 95: 1365-1380. https://doi.org/10.1175/BAMS-D-13-00084.1.

[40] Madaus, L. 2011. How to recognize rotation on Doppler radar. http://lukemweather.blogspot.it/2011/04/how-to-recognize-rotation-on-doppler.html. Accessed May 20, 2017.

[41] Marzano, F. S., D. Scaranari, M. Montopoli, and G. Vulpiani. 2008. Supervised classification and estimation of hydrometeors from C-band dual-polarized radars: a Bayesian approach. *IEEE Transactions onGeoscience and Remote Sensing* 46: 85-98. https://doi.org/10.1109/TGRS.2007.906476.

[42] Mattos, E. V., L. A. T. Machado, E. R. Williams, and R. I. Albrecht. 2016. Polarimetric radar characteristics of storms with and without lightning activity. *Journal of Geophysical Research—Atmospheres* 121: 14201-14220. https://doi.org/10.1002/2016JD025142.

[43] McLaughlin, D., et al. 2009. Short-wavelength technology and the potential for distributed networks of small radar systems. *Bulletin of the American Meteorological Society* 90: 1797-1817. https://doi.org/10.1175/2009BAMS2507.1.

[44] Mecikalski,J. R. ,W. M. MacKenzie Jr. ,M. König,and S. Muller. 2010. Cloud top properties of growing cumulus prior to convective initiation as measured by Meteosat Second Generation. Part II:use of visible reflectance. *Journal of Applied Meteorology and Climatology* 49:2544-2558. https://doi. org/10. 1175/2010JAMC2480. 1.

[45] Melnikov,V. M. ,D. S. Zrni'c. 2017. Observations of convective thermals with weather radar. *Journal of Atmospheric and Oceanic Technology*. https://doi. org/10. 1175/JTECH-D-17-0068. 1.

[46] Melnikov,V. M. ,R. J. Doviak,D. S. Zrni'c,and D. J. Stensrud. 2011. Mapping Bragg scatter with a polarimetric WSR-88D. *Journal of Atmospheric and Oceanic Technology* 28: 1273-1285. https://doi. org/10. 1175/JTECH-D-10-05048. 1.

[47] Montopoli,M. 2016. Velocity profiles inside volcanic clouds from three-dimensional scanning micro-wave dual-polarization Doppler radars. *Journal of Geophysical Research—Atmospheres* 121: 7881-7900. https://doi. org/10. 1002/2015JD023464.

[48] Mosier,R. M. ,C. Schumacher,R. E. Orville,and L. D. Carey. 2011. Radarnowcasting of cloudto-ground lightning over Houston,Texas. *Weather Forecast* 26:199-212. https://doi. org/10. 1175/2010WAF 2222431. 1.

[49] Petersen,W. A. ,H. J. Christian Jr. ,and S. A. Rutledge. 2005. TRMM observations of the global relationship between ice water content and lightning. *Geophysical Research Letters* 32:L14819. https://doi. org/10. 1029/2005GL023236.

[50] Petersen,W. A. ,D. Marks,and D. Wolff. 2014. GPM ground validation NASA S-band dual polarimetric (NPOL)Doppler radar IFloodS. http://ghrc. nsstc. nasa. gov/ available from the NASA EOSDIS Global Hydrology Resource Center Distributed Active Archive Center,Huntsville,Alabama. http://dx. doi. org/10. 5067/GPMGV/IFLOODS/NPOL/DATA101.

[51] Rennie,S. J. ,A. J. Illingworth,S. L. Dance,and S. P. Ballard. 2010. The accuracy of Doppler radar wind retrievals using insects as targets. *Meteorological Applications* 17:419-432. https://doi. org/10. 1002/met. 174.

[52] Roberto,N. ,E. Adirosi,L. Baldini,D. Casella,S. Dietrich,P. Gatlin,G. Panegrossi,M. Petracca,P. San,and A. Tokay. 2016. Multi-sensor analysis of convective activity in central Italy during theHyMeX SOP 1. 1. *Atmospheric Measurement Techniques* 9:535-552. https://doi. org/10. 5194/amt-9-535-2016.

[53] Roberto,N. ,L. Baldini,E. Adirosi,L. Facheris,F. Cuccoli,A. Lupidi,and A. Garzelli. 2017. A support vector machine hydrometeor classification algorithm for dual-polarization radar. *Atmosphere*8: 134. https://doi. org/10. 3390/atmos8080134.

[54] Ruzanski,E. ,and V. Chandrasekar. 2015. Short-term predictability of weather radar quantities and lightning activity. In 2015 *IEEE International Geoscience and Remote Sensing Symposium*(*IGARSS*),Milan,2608-2611. https://doi. org/10. 1109/IGARSS. 2015. 7326346.

[55] Saunders,C. P. R. ,H. Bax-Norman,C. Emersic,E. E. Avila,and N. E. Castellano. 2006. Laboratory studies of the effect of cloud conditions ongraupel/crystal charge transfer in thunderstorm electrification. *Quarterly Journal of the Royal Meteorological Society* 132:2653-2673. https://doi. org/10. 1256/qj. 05. 218.

[56] Schultz,C. J. ,L. D. Carey,E. V. Schultz,and R. J. Blakeslee. 2015. Insight into the kinematic and microphysical processes that control lightning, jumps. *Weather Forecast* 30: 1591-1621. https://doi. org/10. 1175/WAF-D-14-00147. 1.

[57] Seliga,T. A. ,and V. N. Bringi. 1976. Potential use of radar differential reflectivitymeasurements at orthogonal polarizations for measuring precipitation. *Journal of Applied Meteorology* 15:69-76. https://doi. org/10. 1175/1520-0450(1976)015＜0069:PUORDR＞2. 0. CO;2.

[58] Srivastava, R., T. Matejka, and T. Lorello. 1986. Doppler radar study of the trailing anvil region associated with a squall line. *Journal of the Atmospheric Sciences* 43:356-377. https://doi.org/10.1175/1520-0469(1986)043<0356:DRSOTT>2.0.CO;2.

[59] Steiner, M., R. A. Houze Jr., S. E. Yuter. 1995. Climatological characterization of threedimensional storm structure from operational radar and rain gauge data. *Journal of Applied Meteorology*34: 1978-2007. https://doi.org/10.1175/1520-0450(1995)034<1978:CCOTDS>2.0.CO;2.

[60] Takahashi, T. 1978. Riming electrification as a charge generation mechanism in thunderstorms. *Journal of the Atmospheric Sciences*35: 1536-1548. https://doi.org/10.1175/1520-0469(1978)035<1536:REAACG>2.0.CO;2.

[61] Waldteufel, P., and H. Corbin. 1979. On the analysis of single-Doppler radar data. *Journal of Applied Meteorology* 18:532-542. https://doi.org/10.1175/1520-0450(1979)018<0532:OTAOSD>2.0.CO;2.

[62] Waldvogel, A., B. Federer, and P. Grimm. 1979. Criteria for the detection of hail cells. *Journal of Applied Meteorology* 18: 1521-1525. https://doi.org/10.1175/1520-0450(1979)018<1521:CFTDOH>2.0.CO;2.

[63] Wang, Y., V. Chandrasekar, and V. N. Bringi. 2006. Characterization and evaluation of hybrid polarization observation of precipitation. *Journal of Atmospheric and Oceanic Technology* 23: 552-572. https://doi.org/10.1175/JTECH1869.1.

[64] Weinheimer, A. J., and A. A. Few. 1987. The electric field alignment of ice particles in thunderstorms. *Journal of Geophysical Research* 92(D12):14833-14844. https://doi.org/10.1029/JD092iD12p14833.

[65] Wiens, K. C., S. A. Rutledge, and S. A. Tessendorf. 2005. The 29 June 2000 supercell observed during STEPS. Part II: lightning and charge structure. *Journal of the Atmospheric Sciences* 62: 4151-4177. https://doi.org/10.1175/JAS3615.1.

[66] Williams, E. R. 1989. The tripole structure of thunderstorm. *Journal of Geophysical Research* 94(D11): 13:151-167. https://doi.org/10.1029/JD094iD11p13151.

[67] Wilson, J. W. 1966. Movement and predictability of radar echoes. In *National Severe Storms Laboratory Technical Memorandum ERTM-NSSL*-28, 1966, 30.

[68] Wilson, J. W., R. D. Roberts, C. Kessinger, and J. McCarthy. 1984. Microburst wind structure and evaluation of Doppler radar for airport wind shear detection. *Journal of Climate and Applied Meteorology* 23: 898-915. https://doi.org/10.1175/1520-0450(1984)023<0898:MWSAEO>2.0.CO;2.

[69] Witt, A., M. D. Eilts, G. J. Stumpf, J. T. Johnson, E. D. Mitchell, and K. W. Thomas. 1998. An enhanced hail detection algorithm for the WSR-88D. *Weather Forecast* 13:286-303. https://doi.org/0.1175/1520-0434(1998)013<0286:AEHDAF>2.0.CO;2.

[70] Wood, V. T., and R. A. Brown. 1977. Effects of radar sampling on single-Doppler velocity signatures of mesocyclones and tornadoes. *Weather Forecast* 12: 928-938. https://doi.org/0.1175/1520-0434(1997)012<0928:EORSOS>2.0.CO;2.

[71] Xiao, Q., Y. Kuo, J. Sun, W. Lee, E. Lim, Y. Guo, and D. M. Barker. 2005. Assimilation of Doppler radar observations with a regional 3DVAR system: impact of Doppler velocities on forecasts of a heavy rainfall case. *Journal of Applied Meteorology* 44:768-788. https://doi.org/10.1175/JAM2248.1.

[72] Xu, Q., K. Nai, and L. Wei. 2010. Fitting VAD winds to aliased Doppler radial-velocity observations: a global minimization problem in the presence of multiple local minima. *Quarterly Journal of the Royal Meteorological Society*136:451-461. https://doi.org/10.1002/qj.589.

[73] Yoshida, S. , T. Adachi, K. Kusunoki, S. Hayashi, T. Wu, T. Ushio, and E. Yoshikawa. 2017. Relationship between thunderstorm electrification and storm kinetics revealed by phased array weather radar. *Journal of Geophysical Research—Atmospheres* 122:3821-3836. https://doi.org/10.1002/2016JD025947.

[74] Zhang, J. , and S. Wang. 2006. An automated 2Dmultipass Doppler radar velocity dealiasing scheme. *Journal of Atmospheric and Oceanic Technology* 23:1239-1248. https://doi.org/10.1175/JTECH1910.1.

[75] Zrni'c, D. S. , and P. R. Mahapatra. 1985. Two methods of ambiguity resolution in pulsed Doppler weather radars. *IEEE Transactions on Aerospace and Electronic Systems* 21:470-483. https://doi.org/10.1109/TAES.1985.310635.

[76] Zrni'c, D. S. , A. Ryzhkov, J. Straka, Y. Liu, and J. Vivekanandan. 2001. Testing a procedure for automatic classification of hydrometeor types. *Journal of Atmospheric and Oceanic Technology* 18: 892-913. https://doi.org/10.1175/1520-0426(2001)018<0892:TAPFAC>2.0.CO;2.

第5章　云雷达

Takeshi Maesaka

5.1　引言

到目前为止，世界各地已经部署了许多天气雷达，最近其中一些已经升级为双偏振雷达。业务雷达可提供实时观测结果，以预警即将由暴雨带来的灾害。历史上雷达气象学始于对军用雷达消除降水噪声的研究，而气象雷达的主要观测对象是降水。现在，大多数气象学者认识到气象雷达的回波来自雨、雪或雨夹雪，而不是来自云滴。

另一方面，云微物理的研究在天气和气候研究中尤为重要，云过程在数值模拟中至关重要，但云微物理的参数化仍然不够充分。云观测研究的难点之一是，除了雾外，不能在地面上进行原位测量。为了解决这个问题，必须使用飞机上的仪器或气球上的探空系统。然而，用这些方法，很难长时间观测云。因此，需要一种遥感仪器（主要是云雷达，就像实现了云探测功能的天气雷达），以便稳定和长期地观测云。

定量降水估算（Quantitative Precipitation Estimation，QPE）随着双偏振雷达的发展而日趋成熟，研究人员的兴趣也逐渐转向定量降水预报（Quantitative Precipitation Forecast，QPF），特别是局地强降水的预报是一个非常重要的问题。为了延长预报的提前时间，需要利用积雨云即积云的早期观测数据作为数值预报模式的初始条件。然而，大多数业务天气雷达只能捕捉降水，而不能探测到云。云雷达也被期望能够发布局地暴雨预警。

文献[4]中综合阐述了云雷达发展的历史概况。美国空军在二战后开始研制云雷达，并部署了垂直指向雷达（AN/TPQ-11）来观测机场附近的云。云雷达使用毫米波，比降水雷达使用的厘米波波长短，原因稍后会介绍。然而，这种无线电波的微波器件的可靠性还不足以保证其稳定运行。由于这个原因，空军的AN/TPQ-11已经退役，并且云雷达没有得到普遍使用。最近在毫米波器件和数据处理方面的进展使得稳定的云雷达系统得以发展。起初，云雷达是一种垂直指向雷达。后来雷达演变为机载、星载和扫描雷达。

本章描述了一些使用天气雷达观测云的实现和条件，重点介绍了灵敏度和波长的使用。然后，本文给出了一个最近使用双偏振扫描云雷达系统观测到的实例，并讨论了结果。

5.2　云雷达的灵敏度

气象雷达向云或降水发射无线电波，并接收由它们散射的无线电波。这些发射和接收的

无线电波功率(P_t和P_r)满足以下雷达方程(对于弥散目标):

$$P_r=\frac{G^2\lambda^2\theta^2 c\tau}{2^{10}\pi^2(\ln 2)}\cdot P_t\cdot\frac{1}{r^2}\cdot\eta \tag{5.1}$$

式中G为天线增益,λ为波长,θ为波束宽度,c为光速,τ为脉冲宽度,r为距雷达的距离,η为雷达反射率。注意G和θ几乎分别与λ^{-2}和λ成比例,因此式(5.1)中右边的第一个分数不是波长的函数。在瑞利近似下(当粒径远小于波长时),反射率因子η表示为:

$$\eta=\frac{\pi^2}{\lambda^4}|K|^2 Z \tag{5.2}$$

式中$|K|^2$为水凝物的折射指数因子,Z为雷达反射率因子。雷达反射率因子Z表示为:

$$Z=\sum_D N(D)D^6 \tag{5.3}$$

式中D为粒子直径,$N(D)$为粒径为D的粒子的数密度。通常,降雨时雷达反射率因子Z大于10 dBZ,毛毛雨时雷达反射率因子Z约为0 dBZ。

当水凝物被雷达探测到时,接收功率P_r必须大于雷达系统的噪声电平(或最低灵敏度,通常大约为-110 dBm),对应一个特定的雷达反射率因子(最小可测反射率因子)。最小可测反射率因子随距离变化而变化,但对于厘米波长的降水雷达,其典型值大于0 dBZ,这个值对于探测降水来说足够了。

由于云滴的典型直径为几十微米,根据式(5.3),云滴的反射率因子比降水的反射率因子要小得多。虽然云滴的数密度远大于降水的数密度,但在式(5.3)中,直径的六次方更为有效。云的反射率因子小于-10 dBZ,厘米波长的降水雷达很难探测到。为了使天气雷达能够进行云观测,需要根据式(5.1)和式(5.2)进行以下更改:

(1)使用更长的脉冲宽度

更长的脉冲宽度τ通过提高接收功率P_r来提高雷达灵敏度;然而,它也使距离分辨率变得粗糙。幸运的是,现在的脉冲压缩技术是通过调制发射脉冲(通常采用线性频率调制),然后计算接收和发射脉冲之间的相关性来实现的。该技术能够在保持灵敏度的同时压缩(提高)距离分辨率。

(2)使用高发射功率

式(5.1)表明,发射功率P_t越高,则接收功率P_r越高。这种方法的实施需要高功率微波器件,而它们比低功率微波器件更昂贵。

(3)使观测距离缩短

降水雷达的观测距离为几百千米。因为接收功率P_r与到雷达距离的二次方成反比,高灵敏度仅在接近雷达的距离内(约10 km)实现。

(4)使用较短的波长

在瑞利近似下,接收功率P_r与波长的四次方成反比(式(5.2))。波长越短,接收功率P_r越大。由于是四次方,这种方法的实施比上面描述的其他方法更有效。例如,如果波长从3 cm变化到3 mm,灵敏度就会提高10^4倍(40 dB)。这种灵敏度的提高就是为什么毫米波长通常用于云观测的原因。

5.3 布拉格散射

天气雷达有时会在非降水区域捕捉到晴空回波(也叫天使回波),这种回波来自于布拉格散射，是由大气折射指数不均匀引起的。非均匀性的实际情况是与湍流有关的水汽变化，这种回波虽然强度比降水回波弱，但有时会导致对降水的误判。另一方面，这种回波有时能够在非降水区域获得多普勒速度。

对于云观测，云回波受到布拉格散射的干扰。图 5.1 给出了将湍流变化造成的布拉格散射反射率等效为云的后向散射时，不同波长的雷达反射率因子[2]。布拉格散射的雷达反射率 η_B 是波长的函数，表示为：

$$\eta_B = 0.38\lambda^{-1/3} C_n^2 \tag{5.4}$$

式中 C_n^2 为表征折射率中湍流涨落的折射率结构参数。结合式(5.2)和(5.4)，推导出图 5.1 所示的直线。当来自云的后向散射的雷达反射率大于布拉格散射的反射率时(云的雷达反射率因子和对应的波长位于图 5.1 中线的右下方)，云信号被埋没在布拉格散射中。例如，在 $C_n^2 = 10^{-13}$ 条件下，雷达波长为 10 cm(S 波段无线电波)，云的雷达反射率因子为－20 dBZ，则雷达无法探测到云。图 5.1 显示，尽管通过前面所述的调整脉冲宽度、发射功率和观测距离，可以使厘米波长雷达灵敏度高到足以探测云信号，但是厘米波长的雷达不宜用于观测反射率因子小于－10 dBZ 的云，这也是为什么选择毫米波长观测云的原因。

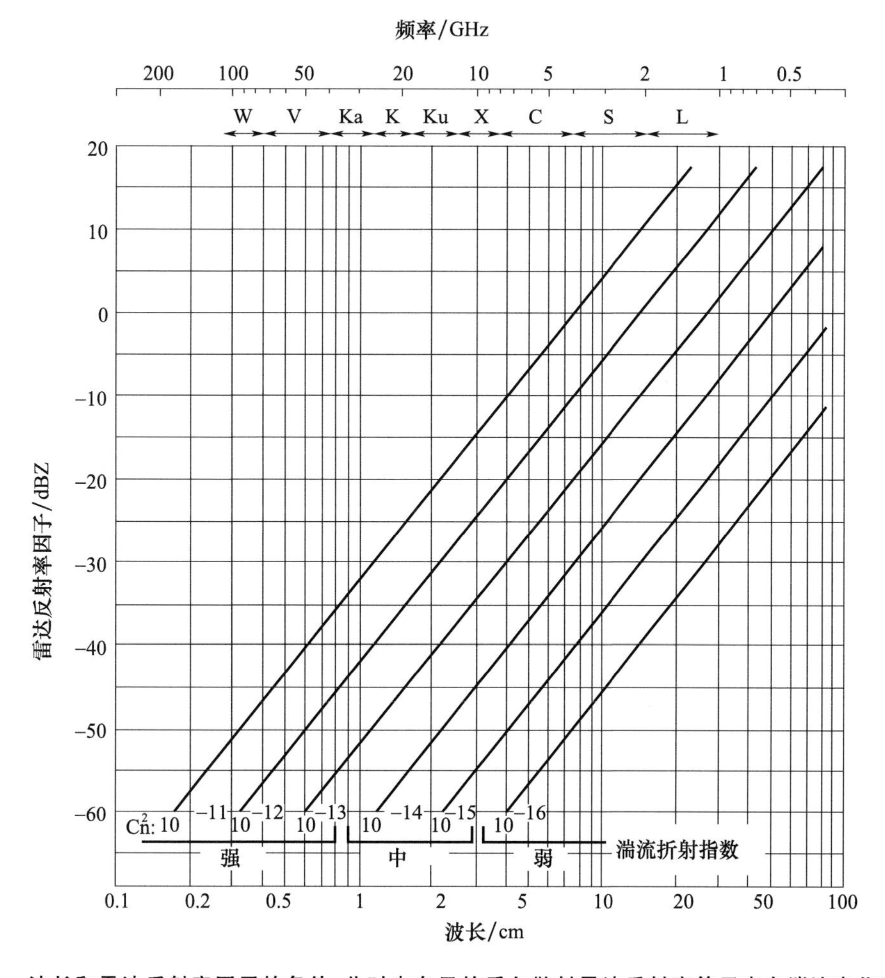

图 5.1　波长和雷达反射率因子的条件，此时来自云的后向散射雷达反射率等于来自湍流变化造成的布拉格散射的反射率[2](C_n^2 是表征折射指数中湍流波动的折射指数结构参数)

5.4 大气透射率

大气透射率对无线电波也很重要。如果无线电波在雷达和目标云体之间衰减，云信号就会变得微弱，难以与噪声区分。图 5.2 显示了干空气和水汽（大气中的主要成分）对各种无线电频率的比衰减率[5]。干空气的衰减是由氧气分子（O_2）的吸收引起的，在图 5.2 的频率范围内，在 60 GHz 和 120 GHz 附近有两个峰值，水汽衰减峰值在 22 GHz 左右。由于应该避免与这些峰值相关的频率，所以毫米波长的云雷达使用 Ka 波段和 W 波段（频率 27～40 GHz 和 75～110 GHz，波长 7.5～11 mm 和 2.7～4.0 mm）。这些波段处于大气无线电窗口（大气允许通过的无线电频率范围）；然而，这些波段的比衰减率（0.1～1 dB/km）要比厘米波长的无线电波段（通常为 0.01 dB/km）大得多。这就是为什么毫米波长的云雷达不能像厘米波长的降水雷达那样扩大观测范围的原因。

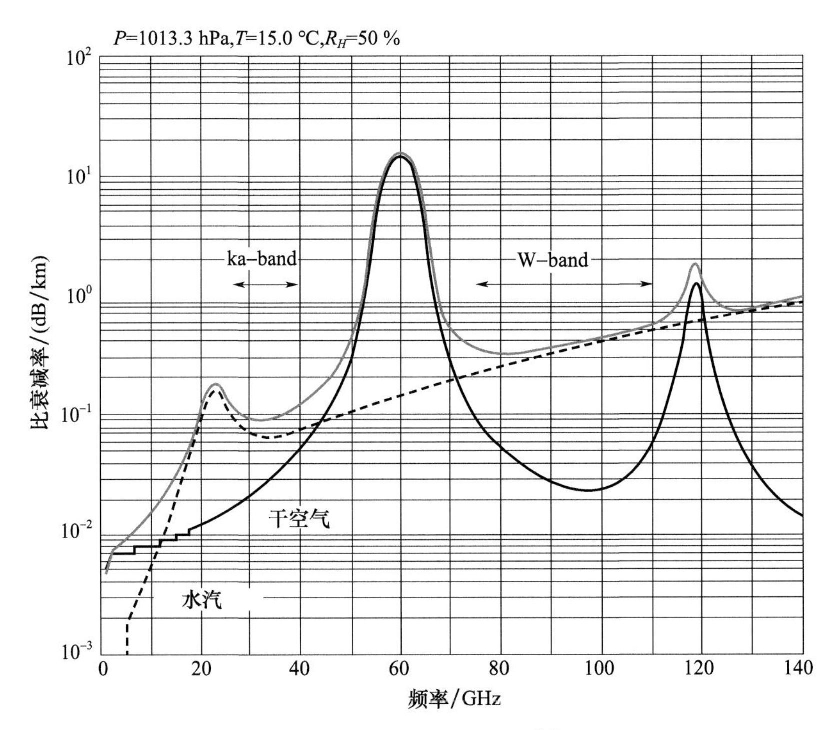

图 5.2 当气压、温度和相对湿度分别为 1013.6 hPa、15℃和 50%[5]时，大气对各种无线电频率的比衰减率。实线和虚线分别表示干燥空气（氧气 O_2）和水汽 H_2O 的衰减，粗灰色线表示总衰减

5.5 云雷达的类型

前面内容描述了使用雷达观测云所需的条件，并指出毫米波适用于云雷达观测。目前已研制出各种毫米波长的气象雷达来研究云微物理特征。这里从天线扫描和性能的角度对云雷

达的类型进行了分类，并对其优缺点进行了总结。

(1)垂直指向雷达

前面已经提到，云雷达的历史始于垂直指向雷达，由于机械装置非常简单，这种雷达适合长期监测云层，观测结果通常在时间-高图上显示和分析。

(2)扫描雷达

这种雷达有一个抛物面或卡塞格伦天线，其方位角和仰角可以通过轴向旋转独立调整。该雷达通常使用 PPI(平面位置显示)和 RHI(距离高度显示)扫描(分别以固定仰角水平旋转天线和固定方位角垂直旋转天线的方式)来捕捉云的三维结构。与垂直指向雷达相比，扫描雷达的积分时间(脉冲样本数)更短(更小)。这一特征使得扫描雷达对薄云的探测能力变差。为了恢复其探测能力，需要进一步提高灵敏度的操作，如使用脉冲压缩或高功率微波放大器。

(3)多普勒雷达

多普勒雷达通过测定发射和接收的无线电波的频率差(多普勒效应)，来确定目标物运动速度(多普勒速度)的一个径向(远离或朝向雷达的方向)分量，该分量大约是风和粒子下落速度的矢量相加。在脉冲雷达中，观测的多普勒速度受限于 Nyquist 速度，表示为：

$$V_N=\frac{\gamma \cdot \mathrm{PRF}}{4} \tag{5.5}$$

此处 PRF 为脉冲重复频率。超过 Nyquist 速度的速度被折叠。例如，如果 PRF 为 2000 Hz，则 3 cm 和 3 mm 波长对应的 Nyquist 速度分别为 15 m/s 和 1.5 m/s。这种差异表明，毫米波长云雷达观测的多普勒速度在风的探测中不如厘米波长降水雷达那么好用。

另一方面，垂直指向云雷达的多普勒性能良好。波长越短，Nyquist 速度越小；但它提高了多普勒速度的分辨率(雷达系统测量的是当前脉冲与前一个脉冲之间的相位差[－180°～＋180°]，相位差的范围对应的是$-V_N \sim +V_N$ 的速度范围)。垂直指向雷达观测的多普勒速度包含了大量的云粒子下落末速度信息。由于小冰晶和云滴的下落末速度非常小，云雷达观测有助于更好地理解云微物理。

(4)双偏振雷达

双偏振雷达通过同时或交替发射水平偏振波和垂直偏振波来探测目标物的形状信息。双偏振能力通过分析水平波和垂直波之间的差分相移，进行衰减订正并实现准确的 QPE。当雨滴变大时，由于空气动力学效应，雨滴的形状会改变得像汉堡包。这些 QPE 和衰减订正是基于轴比随液滴尺度发展的变化。然而，云滴的形状几乎是球形的，因为它的尺寸和下落末速度都很小。这种情况说明双偏振在估算云水含量和无线电波衰减方面是无用的。对于冰晶云和混合云，这种能力为分析粒子类型提供了有用的信息。请注意，双偏振极大地有助于数据质量控制，以消除非气象回波。

(5)双波长雷达

双波长雷达同时发射两种波长的无线电波。有时两种无线电波共用一根天线，有时两根天线并列以观测同一方向。尽管双偏振雷达对液态云观测毫无用处，但双波长雷达提供了双波长比(Dual-Wavelength Ratio，DWR)，即两个波长的雷达反射率因子之比，与云水含量和液态云对无线电波的衰减有很好的相关性。由于本节已经提到的原因，通常选择 Ka 波段和 W 波段的无线电波进行云观测。

5.6 双偏振扫描云雷达

市政工程师最近指出，城市区域容易受到局地、强降雨的影响。因此，降水监测和预报对水资源管理至关重要。天气雷达是实现这一目标的有力工具；然而，目前的业务雷达（S-、C-和X-波段）只能探测到降水粒子（水凝物，不包括云粒子）。这种情况意味着，当雷达捕捉到回波时，降雨已经开始了。同时，暴雨是由积云发展而来的成熟积雨云造成的。因此，对积云的探测有望延长强降水预报的提前时间。然而，目前的天气雷达无法观测到由云粒子组成的积云，日本国家地球科学与灾害恢复研究所（National Research Institute for Earth Science and Disaster Resilience，NIED），为预报暴雨，开始在东京首都区域周边部署云雷达网络[6]，该云雷达网络由 5 部云雷达组成，覆盖了东京周边的大部分人口密集区（图 5.3）。本节将介绍 NIED 云雷达的指标，并给出一些观测结果。

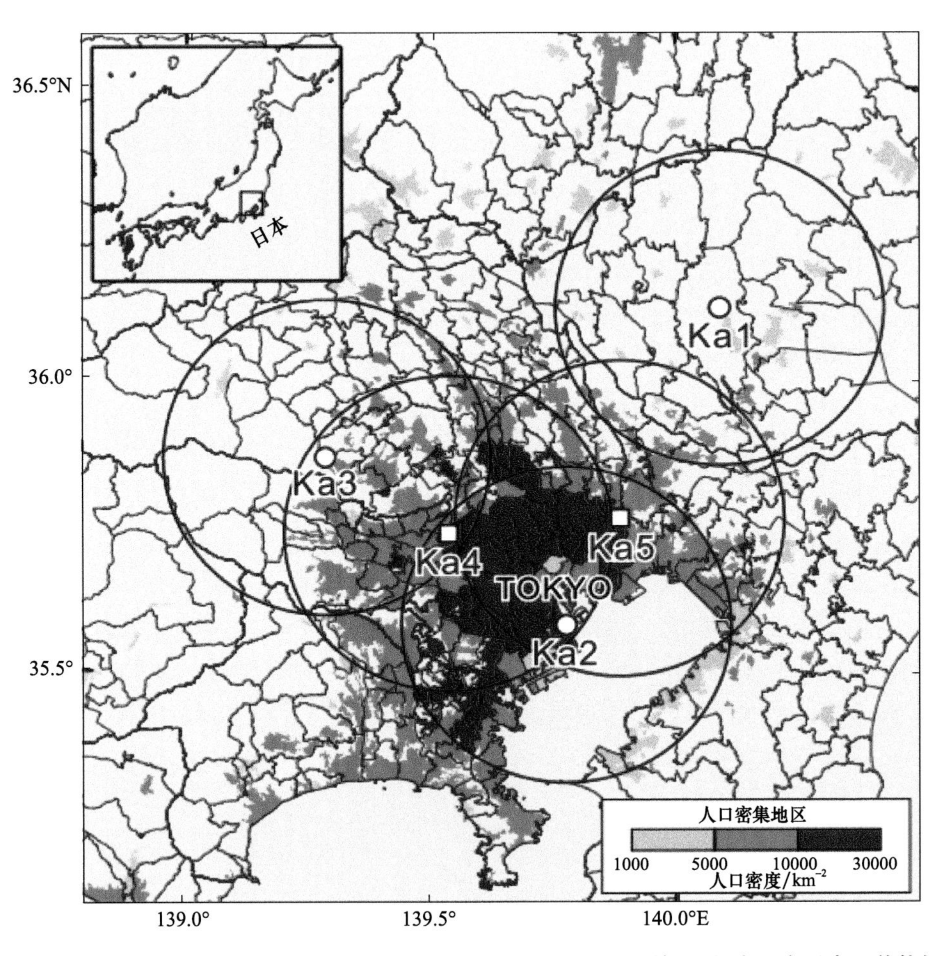

图 5.3 NIED 扫描云雷达的位置。空心圆表示双偏振雷达（Ka1～Ka3）的位置，空心方形表示单偏振雷达（Ka4 和 Ka5）的位置。蓝色实心圆圈表示每台雷达的观测范围（$r=30$ km）。彩色区域是由日本国家土地数字信息提供的人口密集地区

5.6.1 需求技术指标

从积云到积雨云演变的一个关键微物理过程是碰并过程。为了监测积雨云的形成，云雷达必须能够观测到云滴刚开始转变为雨滴的云。这一转变过程让人想起 Kessler 的暖雨参数化[3]，表达式：

$$\frac{\mathrm{d}q_r}{\mathrm{d}t}=\alpha(q_c-q_{c_0})H(q_c-q_{c_0}) \tag{5.6}$$

式中 q_r 和 q_c 分别是雨和云的混合比，α 是一个可调常数，$H(x)$ 是 Heaviside 阶跃函数。当云的混合比 q_c 超过阈值 q_{c_0} 时，转变被激活。深厚积云对流通常选用 α 约为 10^{-3}/s 和 q_{c_0} 约为 1.0 g/kg。这一参数化可导出云雷达的技术指标。

图 5.4 为 Atlas[1] 提出的云水含量与雷达反射率因子的关系。由于对流层低层空气密度接近 1 kg/m³，云的混合比和云水含量可相互替代。这一关系表明，1.0 g/kg 云混合比对应的雷达反射率因子为－13.2 dBZ。这意味着积云的雷达反射率因子小于－13.2 dBZ。考虑到 Kessler 的 q_{c0} 初始值为 0.5 g/kg(对应雷达反射率因子为－19.2 dBZ)，云雷达的需求技术指标确定为“单偏振雷达在 20 km 范围内可探测－20 dBZ 的雷达反射率因子”。对于双偏振雷达，反射率因子值为－17 dBZ，因为雷达发射功率平均分为水平和垂直偏振。通过将可探测接收功率定义为比噪声电平大 3 dB，根据式(5.1)和(5.2)导出了积云观测所需的雷达技术指标。大气的衰减(0.15 dB/km)也被考虑在内。

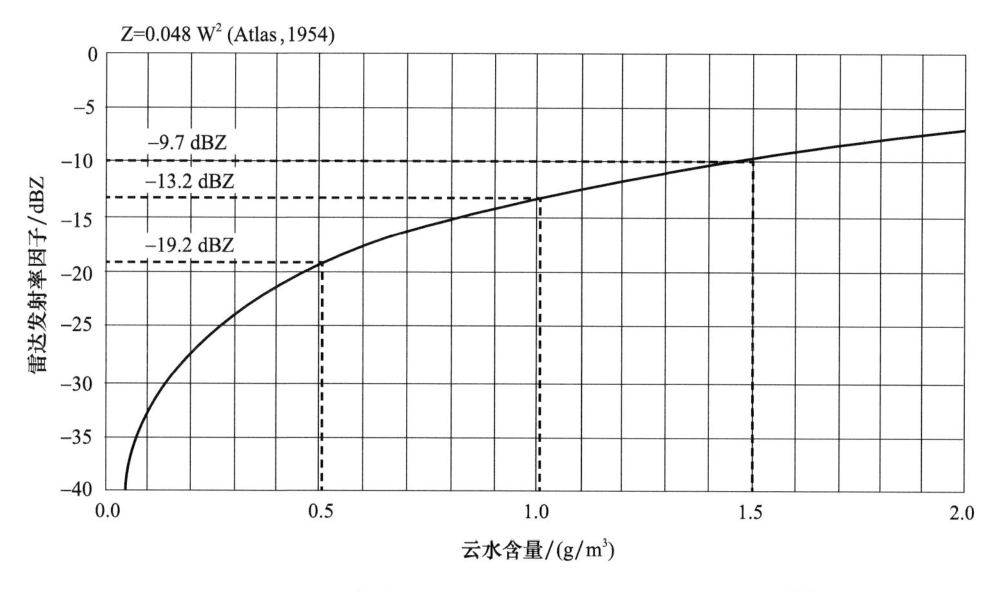

图 5.4 Atlas 提出的云水含量与雷达反射率因子之间的关系[1]

5.6.2 NIED 云雷达的技术指标

表 5.1 总结了满足上述条件的 NIED 云雷达的技术指标。雷达使用 Ka 波段无线电波，发射功率为 3 kW，由扩展交互速调管(EIK)放大。通常采用 55 μs(长脉冲)的脉冲宽度，配合脉冲压缩技术来提高距离分辨率和灵敏度。脉冲发射时，雷达系统无法接收到信号，这

意味着雷达无法观测到比(τc)/2(τ 为 55 μs 时，距离为 8.25 km)距离更近的回波，该范围被短脉冲(1.0 μs)观测覆盖。

云雷达安装了直径为 2.2 m 的卡塞格伦天线(图 5.5)。该天线产生了很窄的波束，波束宽度为 0.3°，因此和相同天线尺寸的 S、C 和 X 波段雷达相比，空间分辨率更精细。天线增益(54 dBi)也大于发射 S、C 和 X 波段无线电波的天线。

图 5.5　NIED 扫描云雷达的照片(Ka1)

如图 5.3 所示，云雷达中有 3 部为双偏振雷达，2 部为单偏振雷达，灵敏度比双偏振雷达高 3 dB。两种类型的雷达输出接收功率、雷达反射率因子、多普勒速度、多普勒谱宽、信号质量指数(signal quality index，SQI，又称归一化相干功率)和信噪比(SNR)。此外，双偏振雷达输出偏振参数，差分反射率(Z_{DR})、差分相位(Φ_{DP})、比差分相位(K_{DP})、相关系数(ρ_{HV})和线性退偏振比(LDR，可用于单偏振发射)。

表 5.1　NIED 扫描云雷达的技术指标

参数	技术指标
频率	34.815-34.905 GHz(Ka 波段)
使用带宽	13 MHz
微波放大器	扩展交互速调管(EIK)
发射功率	3 kW

续表

参数	技术指标
脉冲压缩	线性调频(2MHz)
脉冲宽度(短)	0.5 和 1.0 μs
脉冲宽度(长)	30、45、55、80 和 100 μs
脉冲重复频率	在脉冲宽度为 30 μs 时最大 2500 Hz
IF 数字转换器	16 bit,36 MHz
天线	卡塞格伦天线(Φ=2.2 m)
天线增益	54dBi(通常)
波束宽度	0.3°(通常)
天线旁瓣电平	−23 dB(通常)
偏振	H/V 同时(3 部雷达)
	单 H 或 V(2 部雷达)
观测距离	30 km
输出数据分辨率	75 或 150 m
Nyquist 速度	PRF=2500 Hz 时为 5.38 m/s
双 PRF	3:2、4:3 和 5:4 用于退速度模糊
杂波滤波器	IIR 或频谱插值
输出数据	接收功率(P_r)
	雷达反射率因子(Z)
	多普勒速度(V)
	多普勒谱宽(W)
	信号质量指数(SQI)
	信噪比(SNR)
输出数据(双偏振)	差分反射率(Z_{DR})
	差分相位(Φ_{DP})
	比差分相位(K_{DP})
	相关系数(ρ_{HV})
	线性退偏振比(L_{DR},可用于单偏振发射)

5.6.3 云雷达观测

图 5.6 为 2015 年 9 月 7 日 1925 JST(日本标准时间),NIED 和国土建设交通和旅游部(MLIT)的 X 波段双偏振雷达观测到的降雨强度。静止锋位于东京南部,已经停滞了几个小时。分析了在锋面附近的中雨区和在锋面以北的弱雨区(<1 mm/h)。此时,Ka1 雷达进行距离高度显示(Range Height Indicator,RHI)扫描,来观察图 5.6 中所示的垂直截面。

图 5.7 为 Ka1 雷达在 2015 年 9 月 7 日 1921 JST 观测到的 RHI 垂直截面。在雷达反射率因子场(图 5.7a)中,也对噪声进行了说明,在 8.25 km 处的异常对应于短脉冲和长脉冲观测边界。距离雷达 20 km 处观测到 15 dBZ_e 回波强度的浅层弱降水,与图 5.6 刻度线上的弱

降水相对应。反射率因子小于－20 dBZ_e 的浅薄层状云(雾)也在距地面 2 km 以内被雷达捕捉到。Ka 波段云雷达成功地观测到了 X 波段、C 波段和 S 波段气象雷达无法探测到的层云。云雷达较好地捕捉到了距地面 5.5～9.0 km 高度的卷层云。由于当时 Ka1 雷达附近的 0 ℃层高度为 4650 m,卷层云中主要是冰粒子。ka 波段云雷达对冰云观测非常有用。

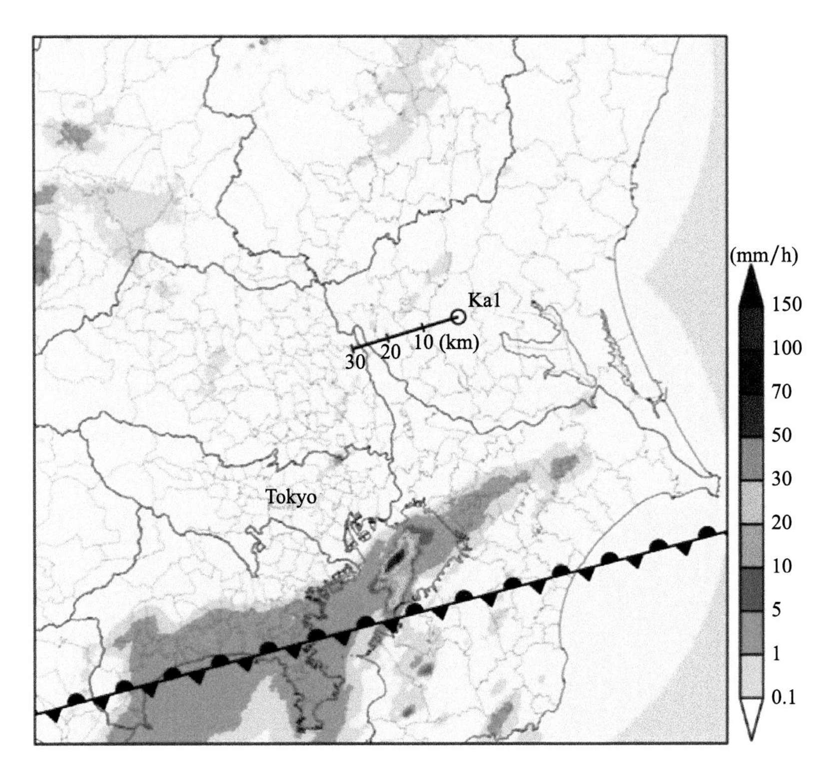

图 5.6　2015 年 9 月 7 日 1925 JST(日本标准时间),NIED 以及国土建设交通和旅游部(MLIT)的 X 波段双偏振雷达观测到的降雨强度。空心圆表示 NIED Ka1 雷达的位置,刻度线表示图 5.7 中所示的距离高度显示扫描(RHI)的垂直横截面位置。静止锋是根据日本气象厅地面观测资料主观绘制的

图 5.7b 为 Ka1 雷达观测到的退模糊后的多普勒速度。该 RHI 扫描是由 PRF 分别为 1980 和 1584 Hz(5:4)的双 PRF(脉冲重复频率)雷达观测的。这些 PRF 对应的 Nyquist 速度分别为 4.257 和 3.4056 m/s(来自公式(5.5),且 λ =8.6 mm)。采用双 PRF 退模糊方法,能将最大不模糊多普勒速度提高到 Nyquist 速度的最小公倍数(17.028 m/s)。但部分距地面 7 km 以上的多普勒速度超过该最大值,出现模糊(负多普勒速度嵌入正速度区域)。由于退模糊失败,多普勒速度出现了部分不连续。由于 Ka 波段雷达的 Nyquist 速度小于 X、C 和 S 波段,所以在风的空间变化较大的区域有时会退模糊失败。图 5.7c 显示了 Ka1 雷达观测到的差分反射率(Z_{DR})。Z_{DR}是水平和垂直雷达反射率因子的差值,通常用分贝表示。其定义是:

$$10\ \log Z_{DR}=10\log\frac{Z_H}{Z_V}=10\log Z_H-10\log Z_V \tag{5.7}$$

式中,下标 H 和 V 分别表示水平和垂直极化。Z_{DR}为正时表明存在水平轴比垂直轴长的粒子,在弱降雨(毛毛雨)和层云区域,毛毛雨和云滴的形状几乎为球形,Z_{DR}值较小。另一方面,在卷层云(距地高度约 7 km)中可见较大的正 Z_{DR}值,这表明了扁平冰粒子的存在,比如板状或枝状冰晶。

图 5.7d 为 Ka1 雷达观测的差分相位(Φ_{DP})。当无线电波在云和降水中传播时,传播速度略有延迟,如果云和降水粒子的横轴比纵轴长,则水平偏振比垂直偏振延迟更大。延迟的差异被探测为水平和垂直偏振之间的差分相位,一个波长的距离延迟对应的 Φ_{DP} 为 360°。注意,在由水平轴较长的粒子组成的降水和云中,Φ_{DP} 随着与雷达距离的增加而单调增加,这是因为延迟的影响随着无线电波传播(前向散射相移)而累积。在图 5.7d 的低层,随着到雷达距离的增大,Φ_{DP} 略有增加。这表明存在略微扁平的雨滴,但这种雨滴的数量可能没有那么多。在卷层云中,在高 Z_{DR} 区域(图 5.7 中虚线表示)出现了 Φ_{DP} 的峰值,峰值附近的 Φ_{DP} 并不是单调增加的,这个峰可能是由后向散射相移产生的,这种相移有时在大而扁平的粒子存在的区域被观测到。在图 5.7 虚线所示的 Z_{DR} 和 Φ_{DP} 的大值下方,在图 5.7a 中可以看到雷达反射率因子的峰值。这可能是板状和枝状冰晶丛集增长的结果。

图 5.7e 为 Ka1 雷达观测的水平与垂直偏振的相关系数(ρ_{HV}),该值的范围是 0~1。当粒子的横/纵轴比在不同脉冲之间波动时(如非球形粒子的翻滚),ρ_{HV} 值较小,纯降雨的 ρ_{HV} 几乎为 1。观测到的 ρ_{HV} 也取决于信噪比(SNR),当弱回波区域 SNR 很小时,ρ_{HV} 值较小。这就是为什么在较低的层云中 ρ_{HV} 较小的原因。

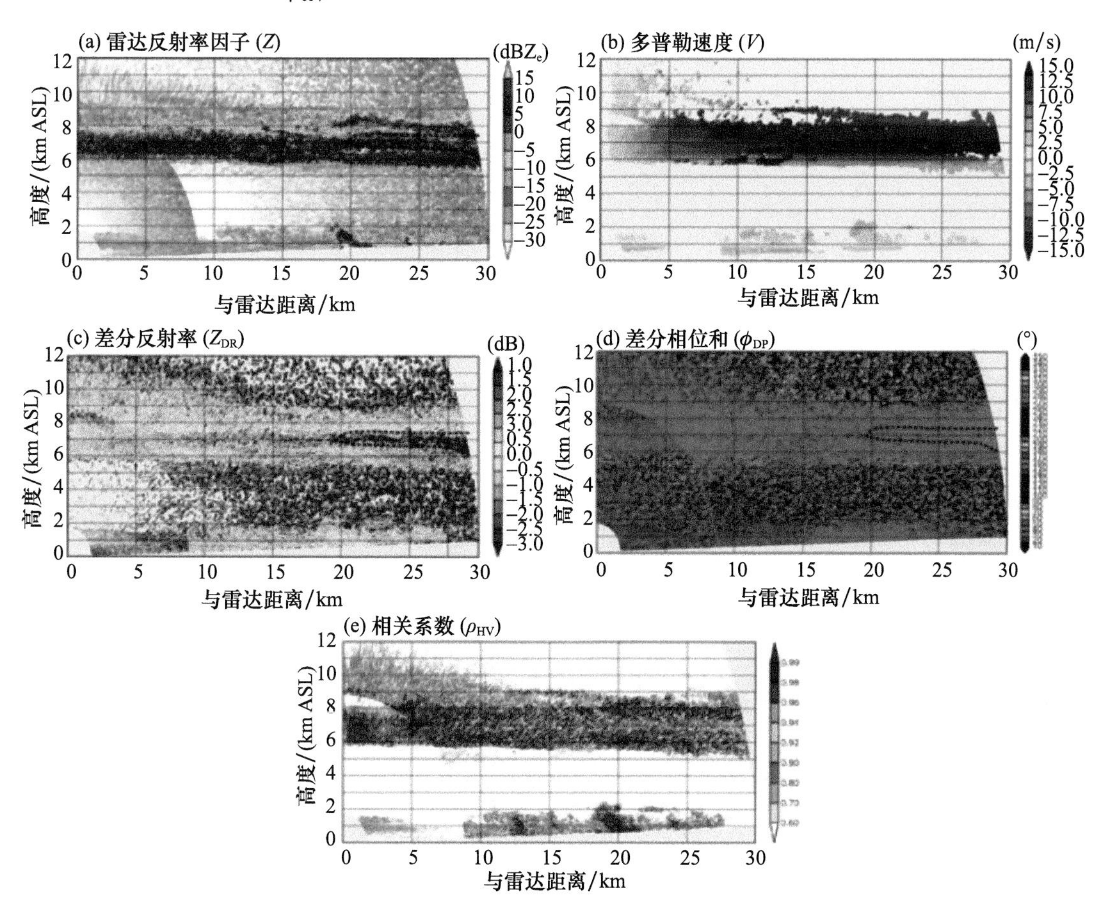

图 5.7 NIED Ka1 雷达在 2015 年 9 月 7 日 1921 JST 观测的 RHI(距离高度显示)扫描,RHI 扫描方位角为 256.6°。(a)反射率因子,(b)多普勒速度(退模糊),(c)差分反射率,(d)差分相位和(e)相关系数。(a)、(c)、(d)中的虚线表示高差分反射率区域

5.7 小结

由于云的雷达反射率相对于降水的反射率要小得多，用于云观测的气象雷达需要进行各种调整以提高灵敏度。本章描述了双偏振云雷达如何提供云微物理方面有价值的信息。云雷达的观测结果可用于强降雨事件的预警。近年来雷达技术的进步使我们能够生产出稳定可靠的双偏振云雷达，并将逐渐在世界范围内普及。这种类型的雷达有望促进云微物理研究的进展，并为更好地表征云提供所需数据。

致谢 我非常感谢 Constantin Andronache 博士给我这个机会为这一章做出贡献以及给我鼓励。我还要感谢国家地球科学与灾害复原研究所的所有同事，感谢他们安装、运行和维护了云雷达网。本章使用的一些 X 波段雷达数据由日本国土建设交通和旅游部通过日本数据集成与分析系统(DIAS)提供。

参考文献*

[1] Atlas, D. 1954. The estimation of cloud parameters by radar. *Journal of Meteorology* 11:309-317.

[2] Gossard, E. E. 1988. Measuring drop-size distributions in clouds with a clear-air-sensing Doppler radar. *Journal of Atmospheric and Oceanic Technology* 5:640-649.

[3] Kessler, E. 1969. On distribution and continuity of water substance in atmospheric circulations. *Meteorological Monographs* 10:1-84.

[4] Kollias, P., E. E. Clothiaux, M. A. Miller, B. A. Albrecht, G. I. Stephens, and T. P. Ackerman. 2007. Millimeter-wavelength radars. *Bulletin of the American Meteorological Society* 88:1608-1624.

[5] Liebe, H. J., G. A. Hufford, and M. G. Cotton. 1993. Propagation modeling of moist air and suspended water/ice particles at frequencies below 1000 GHz. In *AGARD Conference Proceedings*. Vol. 542, 3/1-10.

[6] Maesaka, T., K. Iwanami, S.-I. Suzuki, Y. Shusse, and N. Sakurai. 2015. Cloud radar network in Tokyo metropolitan area for early detection of cumulonimbus Generation. Paper Presented at 37th Conference on Radar Meteorology. Norman: American Meteorological Society. https://ams.confex.com/ams/37RADAR/webprogram/Paper275910.html.

* 参考文献沿用原版书中内容，未改动

第二卷 星基遥感

02

第6章　热带降雨观测任务

Mircea Grecu and David T. Bolvin

6.1 引言

热带降雨测量任务(TRMM)的目标是提供热带降水及其对大气能量平衡的非绝热贡献的准确估计,它对卫星遥感降水科学和技术的贡献远超最初设定的目标。TRMM 取得显著成功的一个主要因素无疑是其仪器的质量和寿命,它们提供了跨越 17 年(即 1997 年 12 月 1 日至 2015 年 4 月 15 日)的高质量红外和微波辐射计以及降水雷达观测的连续记录。本章的目标是描述 TRMM 仪器和相关观测、TRMM 观测的主要产品以及 TRMM 产品和观测在地球物理中的应用。

6.2 观测和主要产品

值得注意的是,与之前的卫星平台相比,TRMM 搭载了第一个卫星降水雷达(PR)。TRMM PR 旨在提供使 TRMM 成为"飞行雨量计"[60]所需的信息,即一个可准确估计降水值并用于校验其他卫星降水产品的卫星平台。TRMM PR 被证明是用于星基降水估计算法开发和检验的关键工具。PR 的主要特征总结在表 6.1 中(改编自文献[33])。虽然星基雷达能够提供关于降水的更高分辨率和更详细的资料,但从来没有打算取代卫星微波辐射计作为提供关于降水全球资料的首选仪器。这是因为,虽然成本高得多,但星基雷达提供的覆盖范围却小得多,因为它们的高分辨率本质上与明显较窄的扫描带相关,即使是 TRMM PR 这样的三维雷达。值得一提的是,在逐月和更长的时间尺度上,PR 采样不再是一种限制。为了缓解短时间尺度上的采样限制,除了 PR 之外,TRMM 还配备了一个微波成像仪(TRMM 微波成像仪或 TMI),旨在联合 TMI/PR 共同的扫描带内提供与 PR 观测信息互补的信息,并促进推导出与基于 PR 观测得到的产品一致但受时间采样误差影响较小的降水产品。表 6.2(改编自文献[33])总结了 TMI 的主要特征。从表 6.2 可以明显看出,TMI 仪器类似于以前的微波辐射计,即自 1987 年以来安装在美国国防气象卫星计划(DMSP)卫星上的特殊微波传感器/成像仪 SSM/I 仪器,但包括一个额外的 10 GHz 频率通道。此外,由于 TRMM 运行在比 DMSP 卫星低得多的轨道高度(2001 年 8 月前为 350 km,之后为

402.5 km),TMI的有效视场(EOV),即视场在采样间隔内的积分[29],几乎比典型SSM/I EOV小1倍。文献[33]中显示了PR和TMI的扫描几何图形的示意图。从扫描角度来看,TMI和PR仪器的主要区别在于TMI是一种圆锥扫描仪器,即其特征是仪器视线与星下点(定义为从仪器到地心的垂直线)之间的角度恒定;PR是一种跨轨道扫描仪器,仪器视线与星下点之间的角度从−17°到17°不等。基于跨轨道扫描仪器收集的观测资料进行反演通常比根据锥形扫描仪器的观测资料反演更具挑战性,因为随着仪器指向远离星下点,有效视场的分辨率下降。然而,PR对指向角的有效视场依赖性很小,实际上可以应用相同的反演算法,而不考虑指向角。然而,有效视场对指向角的依赖性体现在TRMM PR产品中,因为众多因素中地面杂波的范围是亮带强度的函数。除了PR和TMI外,TRMM还配备了一个可见光和红外辐射计系统(VIRS)和一个闪电成像传感器(LIS)。虽然不像PR和TMI观测那样广泛用于卫星降水估计,但VIRS和LIS仪器在种类繁多的研究中被广泛使用。

表6.1　PR的主要特征参数

特征	技术参数
频率	13.8 GHz
灵敏度	17dBZ(0.7 mm/h)
垂直分辨率	250 m
水平分辨率	4.3 km
刈幅	215 km
射线数	49

表6.2　TMI的主要特征参数

通道	1	2	3	4	5	6	7	8	9
频率/GHz	10.65	10.65	19.35	19.35	21.3	37	37	85.5	85.5
偏振	*V*	*H*	*V*	*H*	*V*	*V*	*H*	*V*	*H*
EFOV-CT/km	9.1	9.1	9.1	9.1	9.1	9.1	9.1	4.6	4.6
EFOV-DT/km	63.2	63.2	30.4	30.4	22.6	16	16	7.2	7.2
每次扫描的EFOV	104	104	104	104	104	104	104	208	208

注:CT和DT分别表示交叉轨道和下行轨道。

6.2.1　观测结果

虽然TRMM PR和TMI收集到的观测结果与降水有很强的相关性,但它们之间的关系并不是完全一一对应的。也就是说,TRMM观测结果与实际降水变量之间不存在唯一的、准确的数学关系。但是,已经发展了完善的理论和方法,从非唯一可解释的观测结果得出最优估计,并应用于包括降水遥感在内的许多学科中出现的问题。但在深入研究这些理论之前,对TRMM PR和TMI观测结果进行定性分析,了解它们的优势和局限性是有益的。

图6.1的上部所示为2001年9月16日发生在大西洋上空的飓风Felix期间,TRMM PR提供的三维(3D)反射率场的横截面。如图6.1所示,TRMM PR在其观测范围内提供了关于风暴垂直结构的详细而清晰的信息。在图6.1中,融化层以上降水的垂直范围和强度是显而易见的。此外,亮带结构在图中很明显(由于冻结水凝物的融化反射率增强[2])。在一些扫描

中(扫描 4125 左右)也很明显的是,在低层(1 km 以下)的反射率值相当低,而在 3.0 km 以上高度的反射率值相当大。虽然云物理过程如蒸发和风切变可以解释低层的反射率明显小于其上部,但由雨水和云水造成的接收信号衰减可能是这种现象的成因。这是因为 TRMM PR 的工作频率为 13.8 GHz,在这个频率下,信号受明显的雨和云水衰减影响。因此,TRMM PR 观测结果只有经过衰减订正后才能用于定量分析。有些信息,比如观测的降水结构深度及其类型(例如对流相对层状降水[22])能够在衰减订正步骤之前获得,但从 TRMM PR 的观测资料推导降水估计值涉及衰减订正的步骤。这将在下一节中进行描述。图 6.1 的下部显示的是衰减订正后的反射率。从图中可以看出,观测的反射率与衰减订正后的反射率存在显著差异,这说明在理解观测到的 TRMM PR 反射率时必须考虑衰减。

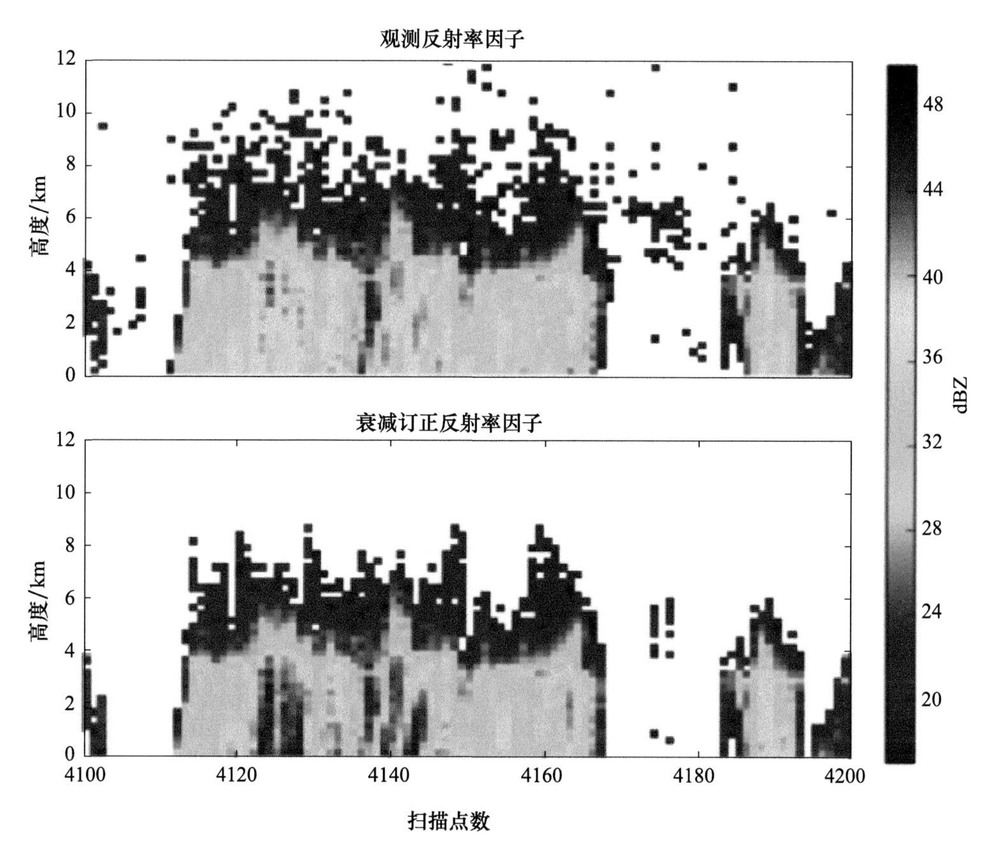

图 6.1　2001 年 9 月 16 日轨道 21899 观测和衰减订正反射率场的天底点截面

(图中使用的是版本 7 的 TRMM PR 产品)

与 TRMM PR 不同,TRMM TMI 提供的观测结果范围明显较大。具体来说,TRMM TMI 测量在广阔观测体积内(水平延伸数十千米,从地球表面到大气顶部)太空、地球表面和大气发射、被大气散射或吸收的辐射;而 TRMM PR 测量的是在相当小的观测体积内(水平方向延伸数公里,垂直方向延伸 250 m)降水粒子对雷达发射辐射的后向散射。为了便于观测目标的三维重建,TMI 依赖于多频率、双偏振的多通道运行。早期的星载微波辐射计主要是为了提供对推断海洋表面温度、海洋表面风速、海洋上空积分大气液态水和水汽等地球物理变量估计有价值的观测结果[12,53],但人们很快意识到,星载辐射计的观测结果可以成功地用于探测强降水，甚至量化其强度[64]。随着星载辐射计的改进，很明显，多个辐射计通道可以提供关于海洋上空降水系统总体垂直结构的互补信息[65]。具体而言，低频通道(10 和 19 GHz)提供的信息主要是融化层之下的降水量，而 85 GHz 通道提供的信息是冻结层之上的固态降水量。36 GHz 通道对融化层以下的液态降水发射的辐射和融化层以上固态降水粒子散射的辐射都很敏感。22 GHz 对大气水汽量也很敏感，在将水汽量的信息纳入估计过程中很有用。偏振信息可能有助于减少地表发射率变化对辐射计信号解析的影响。

图 6.2 显示的是 2001 年 9 月 16 日在飓风 Felix 上空 TRMM 21899 轨道的一部分观测到的 19-GHz V 和 85-GHz V 亮度温度。飓风的涡旋结构在两个通道上都很明显。然而，正如各种研究的理论计算所预测的[1,63,65]，19-GHz V 通道的亮温主要表现为发射辐射，即在融化层之下的雨滴发射辐射，辐射计探测到的辐射量增加，而 85-GHz V 亮温主要表现为散射辐射，即辐射计探测到的辐射量降低，因为散射过程使大气顶部的入射辐射比水凝物和地球表面的发射辐射更重要。飓风左侧狭窄的雨带在两个通道中都表现出发射特征，这是弱的冰相过程迹象。此外，在图 6.2 中可以明显看出，85-GHz 频段的分辨率要比 19-GHz 频段的分辨率高得多。

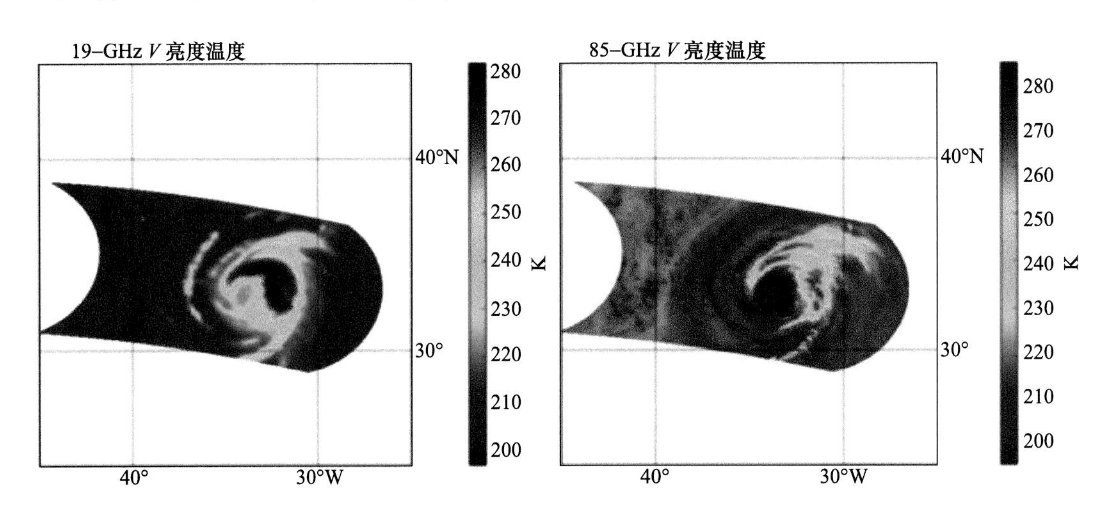

图 6.2　2001 年 9 月 16 日在飓风 Felix 上空的 TRMM 21899 轨道的一部分观测到 19-GHz V 和 85-GHz V 亮度温度

6.2.2　主要算法和产品

6.2.2.1　雷达算法和降水产品

如上所述及图 6.1 所示，PR 的反射率观测结果可能会受到明显的衰减影响。因此，任何定量使用 PR 反射率观测的方法都必须包含衰减订正步骤。衰减订正并不是一个简单的过程，尽管已有成熟的方法[18]，因为衰减是雷达观测体积内粒子谱分布的多元函数，这使得衰减订正过程具有不确定性。在官方 TRMM PR 算法[27]中使用的订正方法通常称为 Hitschfeld 和 Bordan[18]方法。具体地说，从距离 r 处测得的反射率 $Z_m(r)$ 与距离 r 处真实的等效反射率 $Z_e(r)$ 之间的关系开始，即

$$Z_m(r) = Z_e(r)\exp\left(-0.2\ln(10)\int_0^r k(s)\,\mathrm{d}s\right) \tag{6.1}$$

式中 $k(s)$ 为比衰减（dB/km），假设比衰减与真实反射率通过 $k=\alpha Z^{\beta}$ 幂律形式相关联，可以推导出具有解析解的微分方程，实际推导过程读者可以参考[26]的研究。将 $Z_e(r)$ 和 $Z_m(r)$ 联系起来的解析表达式为

$$Z_e(r) = Z_m(r)/(1-\in q\cdot\beta)\exp(-0.2\ln(10)\int_0^r k(s)\mathrm{d}s) \tag{6.2}$$

式中 $q=0.2\ln(10)$，$\in$ 是影响路径积分衰减（PIA）的因子，即 $\int_0^{r_s} k(s)\mathrm{d}s$，其中 r_s 是离地表的距离，与根据地表参考技术（SRT）估计的 PIA 一致。当 $\in=1$ 时，得到 Hitschfeld 和 Bordan[18] 的原始公式。插值技术可利用无雨时从地表返回的观测值 $P_m(r_s$；无雨$)$ 来估计降雨时的 $P_m(r_s$；有雨$)$。那么 PIA 可以由下式估算[26,48]：

$$\mathrm{PIA_{SRT}} = 10\log_{10}(P_m(rs;\text{无雨})/P_m(rs;\text{有雨})) \tag{6.3}$$

公式（6.2）和公式（6.3）提供的 PIA 估计之间的差异可能表明衰减订正过程中存在许多潜在缺陷，这些缺陷通常是由不准确的衰减-反射率关系造成的。为了将降水引起的衰减与雷达观测体积内的相关反射率联系起来（由雷达脉冲宽度、脉冲的传播时间和天线波束宽度决定），需要了解降水粒子在观测体积内的分布与其粒径的函数关系[47]。其中，已知每单位直径内直径为 D 的降水粒子数为 $N(D)$，可根据下式计算比衰减和等效反射率[47]：

$$k = 0.434\int\sigma_t(D)N(D)\mathrm{d}D \tag{6.4}$$

$$Z_e = \frac{\lambda^4}{\pi^5\left|K_w\right|^2}\int\sigma_b(D)N(D)\mathrm{d}D \tag{6.5}$$

式中 σ_t 和 σ_b 为总消光和后向散射截面，λ 为雷达波长，K_w 为水的介电常数。总消光和后向散射截面可以用麦克斯韦电磁方程的 Mie 解来推导[5]，通常在降雨阶段在反射率衰减关系中不是一个很大的不确定来源，而谱分布函数 $N(D)$ 则是相当多的不确定因素的来源。为了描述谱分布的变化，两个或多个变量的函数是必要的，即计算 k 和 Z_e 需要两个或两个以上的变量，而单频 TRMM PR 系统每个观测体积只提供一个测量值 Z_m。因此，函数 $N(D)$ 不能从 Z_m 唯一确定。此外，没有唯一的系数存在，以使用 $k=\alpha Z^{\beta}$ 幂律关系准确描述作为有效反射率函数的比衰减。因此，由解析衰减过程导出的 PIA 不一定与由 SRT 导出的 PIA 一致。Iguchi 和 Meneghini[26] 研究表明，基于 SRT 的 PIA 估算结果的衰减订正调整解析可提高降雨估计精度。结果显示[10]，调整因子与一个 $N(D)$ 参数有关，即归一化函数的截距 a[69]。由于 SRT PIA 估计不是没有不确定性的，因此使分析订正 $[1-\in q\beta\int_0^{r_s}\alpha(s)Z_m^{\beta}(s)\mathrm{d}s]^{1/\beta}$ 完全匹配 SRT PIA 并不是最优的。取而代之的是，应用概率程序来优化利用关于反演降水和“观测”降水的“先验”信息[27]。也就是说，对于强降水，SRT PIA 估计是可靠的。在这种情况下，可以将 $\in$ 设为 $\dfrac{1-10^{\beta\mathrm{PIA_{SRT}}/10}}{\beta\int_0^{r_s}\alpha(s)Z_m^{\beta}(s)\mathrm{d}s}$。然而，对于弱降水，SRT PIA 估计被认为是不确定的，这种情况需要将 $\in$ 设置为 1.0。对于这两个极端之间的情况，基于概率将 $\in$ 设置为 1.0 到 $\dfrac{1-10^{\beta\mathrm{PIA_{SRT}}/10}}{\beta\int_0^{r_s}\alpha(s)Z_m^{\beta}(s)\mathrm{d}s}$ 之间的值。

当推导出真实等效反射率 Z_e 后，利用指数关系 $R=aZ_e^b$ 得到降水率，与公式（6.4）和（6.5）

中的计算结果一致。如果衰减调节因子∈不为 1.0，则系数 a 和 b 被调节。a、b 的实际值及调节步骤请参见官方 TRMM PR 算法手册。虽然这种调节在程序上不同于 Ferreira 等[10]提出的调节，但应该认识到这两种类型的调节都是同一假设的结果，即分析 PIA 和 SRT 估计之间的差异表明粒径分布与 k-Z 和 R-Z 关系中默认的假设不同。

官方的 TRMM PR 降水估计是三维的。水平方向为扫描数和射线数，垂直方向为距离库数。雷达距离库由接收的能量脉冲宽度定义，其距离分辨率为 250 m。由于扫描中每条射线的海平面位置、地理位置以及射线倾角都在输出文件中指定，因此可以很容易地从 TRMM PR 算法输出文件中保存的信息导出三维降水场。降水场的状态可以从输出文件中为每条射线指定的五节点结构中推断。将 5 个节点定义为[27]：(1)A 为 Ku 波段可探测回波的最高距离库；(2)B 为混合相态层的顶部；(3)C 为层状降水亮带峰值以及估算的对流降水冻结高度对应的距离库；(4)D 为混合相态层底部；(5)E 为地表杂波以上的最低距离库。TRMM PR 观测和产品文件由 NASA 降水处理系统(PPS)托管在一个公共服务器上，即 ftp://arthurhou.pps.eosdis.nasa.gov。注册用户可以随时使用它们。注册过程很简单，并向拥有有效电子邮件地址的任何人授予对数据的即时访问权。用户可能感兴趣的其他变量，如降水类型，公式(6.2)中∈的值，PIA 的最终值，也保存在各种 PR 产品的输出文件中。NASA PPS 在网站(https://pps.gsfc.nasa.gov/ppsdocuments.html)上提供了这些产品和相关文件结构的详细描述。

6.2.2.2 TMI 算法和降水产品

TMI 不如雷达观测量详细，且更难以定量和明确地解析。这是因为与 PR 观测相比，TMI 观测是更复杂的辐射过程的结果，这些辐射过程发生在更大的观测体积中。虽然专门的参数化和模型可以用于将一组 TMI 观测与降水结构独特地联系起来，但贝叶斯估计理论[61]等统计方法更可取，因为它们提供了严格的方法来量化和减轻与估计过程相关的不确定性。辐射计降水算法从专门参数化到贝叶斯算法的演变并不完全清楚，但到 20 世纪 90 年代中期，至少存在一种严格的贝叶斯反演公式[9]，随后不久出现了业务适用的公式[32,56]。业务公式相对于一般公式的一个明显特点是，它们依赖于物理上一致的降水廓线和相关亮温的“先验”数据库。在反演过程中，对“先验”数据库中每个观测的亮温多光谱集，估算程序给数据库中每个降水廓线赋概率值，即为“观测的”廓线。最终的估计是数据库中降水廓线的概率组合，数学上，这可以写成：

$$P_r(\boldsymbol{R}|\boldsymbol{T}_b) \propto P_r(\boldsymbol{R}) \cdot P_r(\boldsymbol{T}_b|\boldsymbol{R}) \tag{6.6}$$

其中 $P_r(\boldsymbol{R}|\boldsymbol{T}_b)$ 为给定亮温观测 $\boldsymbol{T}_b$ 的降水廓线 $\boldsymbol{R}$ 的条件概率，$P_r(\boldsymbol{R})$ 为观测某一廓线 $\boldsymbol{R}$ 的概率，$P_r(\boldsymbol{T}_b|\boldsymbol{R})$ 为给定某一特定降雨廓线 $\boldsymbol{R}$ 的亮温矢量 $\boldsymbol{T}_b$ 的概率[34]，以及

$$\boldsymbol{R} = \sum_{i=1}^{n} \boldsymbol{R}_i P_r(\boldsymbol{T}_b|\boldsymbol{R}_i) \Big/ \sum_{i=1}^{n} P_r(\boldsymbol{T}_b|\boldsymbol{R}_i) \tag{6.7}$$

其中 $\boldsymbol{R}_i$ 为“先验”数据库中的降水廓线，$P_r(\boldsymbol{T}_b|\boldsymbol{R}_i)$ 为给定 $\boldsymbol{R}_i$ 时 $\boldsymbol{T}_b$ 的条件概率。虽然数据库中的降水廓线与通过前向辐射传输计算(例如[34])或直接观测得到的唯一确定性值相关联，但概率 $P_r(\boldsymbol{T}_b|\boldsymbol{R}_i)$ 不是 delta 函数(即 $\boldsymbol{T}_b=\boldsymbol{T}_b(\boldsymbol{R}_i)$ 时概率为 1，其中 $\boldsymbol{T}_b(\boldsymbol{R}_i)$ 在“先验”数据库中是唯一与降水廓线 $\boldsymbol{R}_i$ 相关的确定性值，在其他数据库中为 0)，而是高斯函数[34]。这是因为，无论是观测数据还是用于导出与数据库中给定廓线相关的亮温的辐射传输模型是不确定的、随机

的，而忽略这些不确定性将导致公式(6.7)产生的估计出现较大的随机误差(甚至可能是偏差)。上面描述的 TRMM TMI 降水算法的早期版本，通常被称为 Goddard 廓线算法(GPROF)，采用了从云解析模型(CRM)模拟中导出的"先验"亮温降水数据库。然而，随着任务的进行，重点转向基于"先验"数据库的观测结果推导，因为 CRMs 在再现降水廓线的整体分布以及降水廓线和 TMI 探测的亮温的联合分布方面被认为有些不足[35]。特别地，利用随高度变化的概率分布图(CFADs)对 CRM 模拟的反射率数据与雷达观测的反射率数据进行了系统比较，结果表明 CRM 模拟的雪量偏大[36]。这与文献[3]中的观测结果一致，文中发现模拟的 85 GHz 亮温与 TMI 观测到的不同。此外，CRMs 是使用相对较小的一组大尺度环境条件进行初始化的，这些条件在统计上可能不能代表自然界大尺度环境的分布。尽管 CRM 模拟水凝物的垂直分布不断改善(例如[37])，但 CRM 推导的"先验"数据库仍不如从观测结果中推导的数据库稳定，这促使在最新版本的 GPROF 算法中使用基于观测的"先验"数据库[35]。图 6.3 是 TMI 月降水产品的一个示例。

1998年1月TMI月降水量/(mm/d)

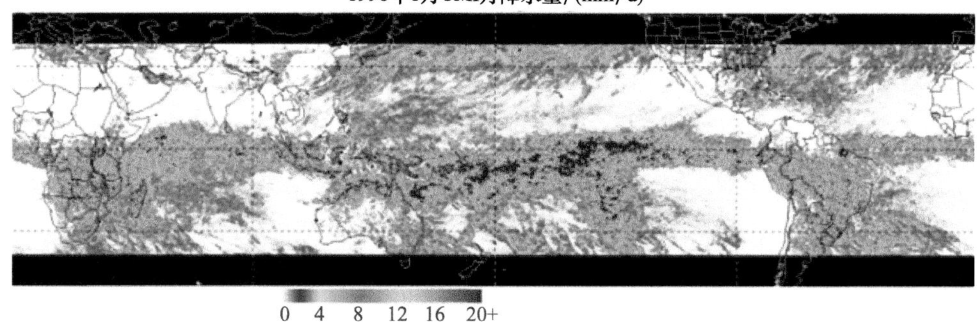

1998年1月PR与TMI融合的月降水量/(mm/d)

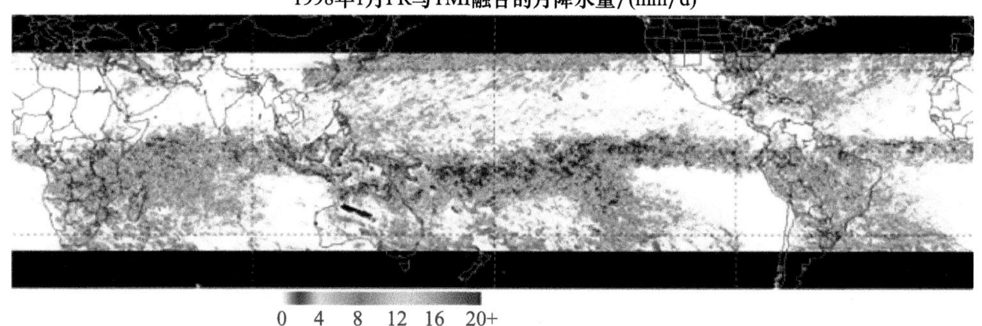

1998年1月3B43月降水量/(mm/d)

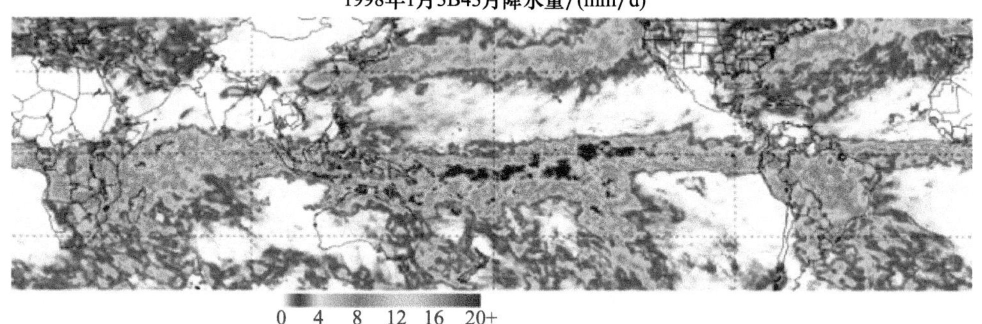

图 6.3　结合 TRMM TMI 和多卫星分析 1998 年 1 月的月降水量估计

6.2.2.3 雷达辐射计联合算法和降水产品

虽然雷达和辐射计的算法看起来明显不同，但是同一通用算法的实例。即，式(6.6)的通用形式可写成：

$$P_r(\boldsymbol{R}|\boldsymbol{Y}) \propto P_r(\boldsymbol{R}) \cdot P_r(\boldsymbol{Y}|\boldsymbol{R}) \tag{6.8}$$

其中$\boldsymbol{Y}$是一组观测数据，可以由任何通用传感器或传感器组合提供。但是，我们可以通过最大化条件概率$P_r(\boldsymbol{R}|\boldsymbol{Y})$来估计$\boldsymbol{R}$，而不是通过对条件概率$P_r(\boldsymbol{R}|\boldsymbol{Y})$积分来计算$\boldsymbol{R}$的期望值。需要指出的是，$P_r(\boldsymbol{R}|\boldsymbol{Y})$的最大化并不是导出广义雷达-辐射计算法的唯一过程，但为了简单起见，这里只讨论这个过程。

设$\boldsymbol{Y}$为天基雷达观测到的反射率廓线，即$\boldsymbol{Y}=(Z_1,Z_2,\dots,Z_n)$，其中下标表示雷达距离库，为了便于计算将式(6.8)近似为：

$$P_r(\boldsymbol{R}|\boldsymbol{Y} \propto P_r(\boldsymbol{R}) \cdot P_r(\boldsymbol{Y}|\boldsymbol{R}) = P_r(\boldsymbol{R}) \cdot P_r(Z_1|R_1) \cdot P_r(Z_2|R_1,R_2) \cdots \cdot P_r(Z_n|R_1,R_2,\dots,R_n) \tag{6.9}$$

其中降水廓线$\boldsymbol{R}$由一组降水值组成，每个降水值的位置由雷达距离库确定。假设似然函数$P_r(\boldsymbol{R}|\boldsymbol{Y})$为分段函数，通过将$P_r(Z_1|R_1)$优化为$R_1$的函数、$P_r(Z_2|R_1,R_2)$优化为$R_2$的函数等，可对其进行依次优化。这相当于对衰减进行依次订正，并根据距离库i处衰减订正后的反射率估算降水R_i，在相同的衰减与反射率、反射率与降水的关系假设下，这正是Hitschfeld和Bordan解。但显然，式(6.8)和(6.9)形式更一般，因为它们可以处理衰减-反射率和反射率-降水关系的不确定性。此外，它们提供了一个严格的框架对来自多个来源的观测进行融合(如PR和TMI)。雷达和辐射计联合观测的好处是，如果联合考虑它们观测的信息，可以减少雷达和辐射计估计固有的不确定性。第一个结合PR和TMI观测的TRMM反演降水算法是官方的1日(day-1)TRMM组合算法[15]。与辐射计算法相似，组合算法整合了分布在条件$P_r(\boldsymbol{Z},\boldsymbol{T_b}|\boldsymbol{R})$上的可能的降水值，其中$\boldsymbol{Z}$为PR雷达观测值，$\boldsymbol{T_b}$为TMI观测值。图6.3中部展示了TRMM联合月降水反演的一个例子。如图所示，TMI和联合反演之间有很好的一致性。其他基于最大似然法的PR/TMI联合反演公式也在卫星发射后不久制定[13,50]。这些公式基于显式优化程序，最大化条件概率$P_r(\boldsymbol{Z},\boldsymbol{T_b}|\boldsymbol{R})$的对数。在数学上，这等价于一个包含两项的二次函数的最小化。第一项最小化TRMM实际观测与从反演降水廓线得到的模拟结果之间的不一致，而第二项最小化反演的降水廓线与其预期气候学的偏离。这种类型公式的优点是不需要对条件概率进行显式积分，但它的计算量非常大，而且容易由于前向模拟模型的偏差而产生偏差。文献[45]中开发了一种组合算法，不需要对观测数据进行模拟，而是依赖于云解析模型模拟和迭代反演程序，以确定与实际观测最一致的预计算观测数据。

6.3 TRMM多卫星降水分析

在应用中使用卫星近地轨道(LEO)估算降水的一个主要挑战是其有限的空间和时间采样。此外，TRMM只覆盖热带和中低纬度地区，这排除了需要中高纬度降水估算的应用。为了减轻这些限制并与其他活动相协调[24,28]，开发了TRMM多卫星降水分析(TMPA)[25]。

TMPA 结合了其他与降水相关的卫星的估计结果，以 TRMM 为基础，产生了覆盖 50°N 到 50°S 完整区域的 3 h 降水场。在与 TRMM 合并之前，所有其他卫星降水估计结果都根据 PR/TMI 联合产品进行校准，因为结果表明，与实地观测相比，组合产品的估算结果最一致。这种相互校准过程分三步：(1)所有其他卫星降水估算均根据 TMI 使用基于匹配直方图的固定季节气候校准来进行校准；(2)使用基于直方图匹配的月度动态校准，根据 PR/TMI 联合产品校准 TMI；(3)使用(2)中的校准方法，根据 PR/TMI 联合产品校准(1)中其他卫星的产品。这个多步骤的过程是必要的，由于采样的限制，其他卫星估计产品无法直接按 PR/TMI 联合估计产品校准。随后，所有卫星估计值被合并，产生单个“高质量”降水估计值。虽然这种高质量的降水场是由许多 LEO 卫星贡献的，但在 3 h 场方面仍然存在空白。半小时的 NOAA CPC 4 km 红外(IR)估测被用来填补高质量降水场的这些空白。红外亮温直方图直接匹配高质量降水估计，以创建月度动态校准。然后将此校准应用于红外亮温，并用于完成 TMPA 降水场。与提供降水“直接”观测的高质量微波估计不同，校准后的红外估计是从冷云推断出来的，不那么准确，因此质量较低。对于每个日历月，将 TMPA 3 h 的估计值汇总到月度，并根据逆误差加权与全球降水气候中心的月度测量分析合并，从而计算出 TMPA 的月度估计值。为了保持 3 小时和月度产品之间的一致性，3 h 的估计值被汇总以匹配月度估计值，因此 3 h 的估计值具有度量“影响”。TMPA 包括实时监测和后实时研究产品。这一记录始于 1998 年 1 月，一直持续到现在。在此期间，输入的卫星数据已经演变了，但相互校准的过程力求保持记录的同质性。此外，考虑到气候校准的需要，虽然 TRMM 卫星不再运行，TMPA 产品仍在继续。

6.4 应用

TRMM 观测结果及其产品的应用范围比发射前设想的要广泛得多，但本节将主要讨论大气科学方面的应用。即使在这个有限的应用领域内，也很难根据其关注的领域对应用进行客观和明确的分类，因为大量应用是由多个目标驱动的。然而，在 TRMM 研究中经常制订若干通用目标，这些目标可以单独使用，也可以与其他目标结合使用，以达到分类目的。这些目标包括：

(1)由 TRMM 观测结果和产品导出的降水和相关变量的统计描述。

(2)使用统计分析来制定、完善和验证各种大气过程的概念模型。

(3)使用 TRMM 产品验证数值天气预报(NWP)和气候模式。

(4)开发适用于 TRMM 或其他任务/仪器的新的遥感技术和反演方法。

(5)业务和准业务应用。

大量使用 TRMM 观测数据的同行评议期刊文章多达数千篇，这使得我们几乎不可能对 TRMM 观测数据在科学应用中的使用进行全面的综述。尽管如此，值得一提的是，使用文献检索及分析软件(PoP)[17]表明，除了几篇关于 TRMM 任务及其算法的论文外，针对目标(1)的研究是被引用最多的 TRMM 研究[51,52,54,72]。这些研究主要关注热带地区降水的时空分布。这并不奇怪，因为 TRMM 提供了关于热带地区降水时空结构的前所未有的信息，以及一个合乎逻辑的方式来评估我们通过包括建立一个参考的模型理解和再现大气过程的能力。涉及目标(2)的研究通常在一定程度上也涉及目标(1)。例如文献[58]研究了热带地区层状云降雨的分布并将其与大尺度大气环流联系起来,文献[55]的研究将亚马逊地区降水的垂直分布与大尺度系统联系起来,文献[23]分析了TRMM所描述的喜马拉雅地区夏季季风对流的三维结构与气象条件的关系。其他的例子包括使用TRMM观测来研究大气季节内振荡(MJO)并制定新的可能的传播机制[46],以及使用TRMM来研究垂直风切变和风暴运动对热带气旋降雨不对称的影响[6]。

甚至在TRMM发射之前,一些研究[11,66]就表明,观测的雷达反射率分布可能与CRMs模拟得出的结果存在显著差异。然而,考虑到CRM的性能依赖于天气条件,且不容易获得代表多种天气条件的高质量雷达观测,对CRM和现有云微物理方案的系统评估并不容易实现。随着TRMM的发射,雷达观测的可用性不再是一个考虑的因素,已经进行了许多研究(属于上述第3类)来评价利用雷达和辐射计的观测资料在各种天气条件下CRMs的表现。这类研究的示例包括很多文献[16,36,39,71,37]。这些研究的结果使得云微物理方案中的几个缺陷已经被发现并在很大程度上得到了解决。但很明显,TRMM观测和产品的效用作为评估模式的手段,并不仅局限于CRMs,还延伸到全球环流和气候模式。具体而言,TRMM的观测结果和产品具有并已被用于评估全球环流模式再现观测降水的空间分布[7]及其日循环[8,57]的能力,以及评估新参数化相对于旧参数化的优点[67,68]。多项研究[30,40]表明,全球环流模式难以再现大气季节内振荡(MJO)[42]。从这个角度来看,第6.3节描述的TMPA降水产品是一个极其重要的数据集,因为其时空分辨率高,能够严格评估全球环流模式重现MJO现象的能力[31]。

从第6.2.1.1节对辐射计和雷达-辐射计联合算法的简要描述可以明显看出,反演方法在TRMM的整个生命周期中不断发展。被动辐射计反演从CRM衍生数据库发展到观测衍生数据库[35],联合反演以多种方式发展[13,46,50]。虽然这些工作的主要目标是根据观测得到更准确的降水估计,但其结果超出了TRMM的范畴,因为在这些过程中发展的技术被用于发展同样适用于其他任务的反演方法。例如,全球降水测量任务(GPM)的雷达-辐射计算法[14],虽然深深植根于TRMM时代发展起来的联合反演方法,但它是专门为组合仪器设计的程序,在许多方面与TRMM的仪器有显著不同[62],是上述第4类TRMM应用的一个例子。其他工作包括发展适用于TRMM观测的变分反演方法,但可扩展到其他平台的观测[49,59]。

还应提及的是,虽然TRMM不是一个业务任务,但它使开发可用于业务的工具和方法成为可能。数据同化研究属于这一类的应用,即上述目标(5)。数值天气预报(NWP)模式的卫星资料同化在TRMM之前就已经是一个非常活跃的研究领域,而TMI观测的高分辨率和现成可用性使其成为数据同化研究的热点。其中一项研究[44]表明,由于额外的低频通道和TMI仪器分辨率的提高,TRMM TMI观测比SSMI观测能够更好地进行分析,从而能够更好地估计热带条件下的地表参数和水汽。其他研究[20,21,43]表明,对TMI和SSM/I降水率的同化提高了对热带气旋以及其他热带和中纬度降水系统的预报和分析。同样,TRMM PR观测虽然比TMI观测的覆盖范围窄得多,但对热带气旋[4]的预报分析有积极影响,而将TRMM TMPA降水产品同化进入全球模式可改善5日预报[41]。除了提高NWP模式数据同化的知识和能力外,TRMM观测和降水产品还被用于近实时热带气旋监测[38]、实时全球洪水估计[70]和实时滑坡灾害评估[19]。

6.5 总结

本章介绍了热带降雨测量任务。TRMM 的观测寿命长、观测质量高、独特的仪器组合(包括首个星载降水雷达)以及其领导层和科学界的热情和奉献,使我们在理解和量化卫星观测降水相关的基本大气过程方面取得了显著进展。本章简要地描述了这方面的一些进展和使之成为可能的因素。

本章简要介绍了 TRMM 雷达和辐射计及其性能,然后介绍根据实际卫星观测结果估算降水的主要算法。首先介绍了 PR 降水估计算法及其在雷达观测体积中与 PSDs 的多变量变化相关的挑战。接着,给出了官方辐射计算法。与雷达估计算法不同的是,因为辐射计观测比雷达观测更不明确,辐射计算法在本质上更具有统计学意义(尽管包含了复杂的物理模型)。本章还介绍了官方 TRMM 雷达-辐射计联合算法背后的动机及其概念公式。除了利用单一或组合 TRMM 仪器观测进行单个降水估算外,TRMM 任务还采用了多卫星降水分析产品,该产品统一了来自 TRMM 仪器和非 TRMM 仪器的降水产品,以减少单个降水估算中的时间采样误差。本章还介绍了该产品及其意义。

本章包括对 TRMM 观测结果的使用以及在科学和社会应用中的产品的讨论。TRMM 产生了相当多的科学和同行评审的出版物,几乎不可能以非常简洁的方式完整地描述它们。相反,这里的重点是数据应用的方向和示例,虽然在很多方面不完整,但希望这样的介绍包含足够的信息,以鼓励读者自己探索 TRMM 任务所提供的一切。

致谢 作者感谢 Ramesh Kakar 博士(NASA 总部)和 TRMM 项目科学家博士 Christian Kummerow(科罗拉多州立大学)、Robert Adler(马里兰大学)和 Scott Braun(NASA 戈达德太空飞行中心)对所有 TRMM 相关科学项目的支持。作者也要感谢名古屋大学的 Hirohiko Masunaga 博士,他审阅了本书的章节并提供了建设性的意见。

参考文献*

[1] Adler, R., H. Yeh, N. Prasad, W. Tao, and J. Simpson. 1991. Microwave simulations of a tropical rainfall system with a three-dimensional cloud model. *Journal of Applied Meteorology* 30:924-953. https://doi.org/10.1175/1520-0450-30.7.924.

[2] Battan, L. J. 1973. *Radar Observation of the Atmosphere*, 324 pp. Chicago: University of Chicago Press.

[3] Bauer, P. 2001. Over-ocean rainfall retrieval from multisensor data of the Tropical Rainfall Measuring Mission. Part I: design and evaluation of inversion databases. *Journal of Atmospheric and Oceanic Technology* 18:1315-1330.

* 参考文献沿用原版书中内容,未改动

[4] Benedetti, A., P. Lopez, P. Bauer, and E. Moreau. 2005. Experimental use of TRMM precipitation radar observations in 1D+4D-Var assimilation. *Quarterly Journal of the Royal Meteorological Society* 131:2473-2495.

[5] Bohren, C. F., and D. R. Huffman. 1983. *Absorption and Scattering of Light by Small Particles*. New York: Wiley.

[6] Chen, S. S., J. A. Knaff, and F. D. Marks. 2006. Effects of vertical wind shear and storm motion on tropical cyclone rainfall asymmetries deduced from TRMM. *Monthly Weather Review* 134:3190-3208. https://doi.org/10.1175/MWR3245.1.

[7] Deng, Y., K. P. Bowman, and C. Jackson. 2007. Differences in rain rate intensities between TRMM observations and community atmosphere model simulations. *Geophysical Research Letters* 34:L01808. https://doi.org/10.1029/2006GL027246.

[8] Dirmeyer, P. A., B. A. Cash, J. L. Kinter, T. Jung, L. Marx, M. Satoh, C. Stan, H. Tomita, P. Towers, N. Wedi, and D. Achuthavarier. 2012. Simulating the diurnal cycle of rainfall in global climate models: resolution versus parameterization. *Climate Dynamics* 39(1-2):399-418.

[9] Evans, K., J. Turk, T. Wong, and G. Stephens. 1995. A Bayesian approach to microwave precipitation profile retrieval. *Journal of Applied Meteorology* 34:260-279. https://doi.org/10.1175/1520-0450-34.1.260.

[10] Ferreira, F., et al. 2001. Study and tests of improved rain estimates from the TRMM precipitation radar. *Journal of Applied Meteorology* 40(11):1878-1899.

[11] Ferrier, B. S., W. Tao, and J. Simpson. 1995. A double-moment multiple-phase four-class bulk ice scheme. Part II: simulations of convective storms in different large-scale environments and comparisons with other bulk parameterizations. *Journal of the Atmospheric Sciences* 52:1001-1033. https://doi.org/10.1175/1520-0469(1995)052<1001:ADMMPF>2.0.CO;2 $.

[12] Gloersen, P., and F. T. Barath. 1977. A scanning multichannel microwave radiometer for Nimbus-G and SeaSat-A. *IEEE Journal of Oceanic Engineering* 2:172-178.

[13] Grecu, M., W. S. Olson, and E. N. Anagnostou. 2004. Retrieval of precipitation profiles from multiresolution, multifrequency active and passive microwave observations. *Journal of Applied Meteorology* 43:562-575.

[14] Grecu, M., W. S. Olson, S. J. Munchak, S. Ringerud, L. Liao, Z. Haddad, B. L. Kelley, and S. F. McLaughlin. 2016. The GPM combined algorithm. *Journal of Atmospheric and Oceanic Technology* 33:2225-2245. https://doi.org/10.1175/JTECH-D-16-0019.1.

[15] Haddad, Z. S., et al. 1997. The TRMM 'day-1' radar/radiometer combined rain-profiling algorithm. *Journal of the Meteorological Society of Japan. Series II* 75(4):799-809.

[16] Han, M., S. A. Braun, W. S. Olson, P. O. Persson, and J. Bao. 2010. Application of TRMM PR and TMI measurements to assess cloud microphysical schemes in the MM5 for a winter storm. *Journal of Applied Meteorology and Climatology* 49:1129-1148. https://doi.org/10.1175/2010JAMC2327.1.

[17] Harzing, A. W. 2007. Publish or Perish. Available from http://www.harzing.com/pop.html.

[18] Hitschfeld, W., and J. Bordan. 1954. Errors inherent in the radar measurement of rainfall at attenuating wavelengths. *Journal of Meteorology* 11:58-67.

[19] Hong, Y., R. Adler, and G. Huffman. 2006. Evaluation of the potential of NASA multi-satellite precipitation analysis in global landslide hazard assessment. *Geophysical Research Letters* 33(22). http://dx.doi.org/10.1029/2006GL028010.

[20] Hou, A. Y., S. Q. Zhang, A. M. da Silva, and W. S. Olson. 2000. Improving assimilated global datasets using TMI rainfall and columnar moisture observations. *Journal of Climate* 13:4180-4195. https://doi.

org/10. 1175/1520-0442(2000)013<4180:IAGDUT>2. 0. CO;2 $.

[21] Hou, A. Y. , S. Q. Zhang, and O. Reale. 2004. Variational continuous assimilation of TMI and SSM/I rain rates: impact on GEOS-3 hurricane analyses and forecasts. *Monthly Weather Review*132: 2094-2109. https://doi. org/10. 1175/1520-0493(2004)132<2094:VCAOTA>2. 0. CO;2.

[22] Houze, R. A. Jr. 2014. *Cloud Dynamics*. Vol. 104. New York: Academic.

[23] Houze, R. A. , Wilton, D. C. , and B. F. Smull. 2007. Monsoon convection in the Himalayan region as seen by the TRMM precipitation radar. *Quarterly Journal of the Royal Meteorological Society*133(627): 1389-1411.

[24] Huffman, G. J. , R. F. Adler, M. Morrissey, D. T. Bolvin, S. Curtis, R. Joyce, B. McGavock, and J. Susskind. 2001. Global precipitation at one-degree daily resolution frommultisatellite observations. *Journal of Hydrometeorology* 2:36-50.

[25] Huffman, G. J. , D. T. Bolvin, E. J. Nelkin, D. B. Wolff, R. F. Adler, G. Gu, Y. Hong, K. P. Bowman, and E. F. Stocker. 2007. The TRMMmultisatellite precipitation analysis(TMPA): quasi-global, multiyear, combined-sensor precipitation estimates at fine scales. *Journal of Hydrometeorology*8: 38-55. https://doi. org/10. 1175/JHM560. 1.

[26] Iguchi, T. , and R. Meneghini. 1994. Intercomparison of single-frequency methods for retrieving a vertical rain profile from airborne or spaceborne radar data. *Journal of Atmospheric and Oceanic Technology*11: 1507-1516. https://doi. org/10. 1175/1520-0426(1994)011.

[27] Iguchi, T. , T. Kozu, R. Meneghini, J. Awaka, and K. Okamoto. 2000. Rain-profiling algorithm for the TRMM precipitation radar. *Journal of Applied Meteorology* 39:2038-2052.

[28] Joyce, R. J. , J. E. Janowiak, P. A. Arkin, and P. Xie. 2004. CMORPH: a method that produces global precipitation estimates from passive microwave and infrared data at high spatial and temporal resolution. *Journal of Hydrometeorology* 5:487-503.

[29] Kidder, S. Q. , and T. H. von derHaar. 1995. *Satellite Meteorology-An Introduction*, 446. New York: Academic.

[30] Kim, D. , Sperber, K. , Stern, W. , Waliser, D. , Kang, I. S. , Maloney, E. , Wang, W. , Weickmann, K. , Benedict, J. , Khairoutdinov, M. , Lee, M. I. , Neale, R. , Suarez, M. , Thayer-Calder, K. , and G. Zhang. 2009. Application of MJO simulation diagnostics to climate models. *Journal of Climate*22:6413-6436.

[31] Klingaman, N. P. , and S. J. Woolnough. 2014. Using a case-study approach to improve the MaddenJulian oscillation in the Hadley Centre model. *Quarterly Journal of the Royal Meteorological Society*140:2491-2505. https://doi. org/10. 1002/qj. 2314.

[32] Kummerow, C. , W. S. Olson, and L. Giglio. 1996. A simplified scheme for obtaining precipitation and vertical hydrometeor profiles from passive microwave sensors. *IEEE Transactions on Geoscience and Remote Sensing*34(5):1213-1232.

[33] Kummerow, C. , W. Barnes, T. Kozu, J. Shiue, and J. Simpson. 1998. The tropical rainfall measuring mission(TRMM)sensor package. *Journal of Atmospheric and Oceanic Technology* 15:809-817.

[34] Kummerow, C. , Y. Y. Hong, W. S. Olson, S. S. Yang, R. F. Adler, J. J. McCollum, R. R. Ferraro, G. G. Petty, D. B. Shin, and T. T. Wilheit. 2001. The evolution of the Goddard profiling algorithm(GPROF) for rainfall estimation from passive microwave sensors. *Journal of Applied Meteorology*40:1801-1820.

[35] Kummerow, C. D. , S. Ringerud, J. Crook, D. Randel, and W. Berg. 2011. An observationally generated a priori database for microwave rainfall retrievals. *Journal of Atmospheric and Oceanic Technology*28:113-

130.

[36] Lang, S. E., W.-K. Tao, R. Cifelli, W. Olson, J. Halverson, S. Rutledge, and J. Simpson. 2007. Improving simulations of convective systems from TRMM LBA: easterly and westerly regimes. *Journal of the Atmospheric Sciences* 64: 1141-1164.

[37] Lang, S. E., W. Tao, J. Chern, D. Wu, and X. Li. 2014. Benefits of a fourth ice class in the simulated radar reflectivities of convective systems using a bulk microphysics scheme. *Journal of the Atmospheric Sciences* 71: 3583-3612. https://doi.org/10.1175/JAS-D-13-0330.1.

[38] Lee, T. F., F. J. Turk, J. Hawkins, and K. Richardson. 2002. Interpretation of TRMM TMI images of tropical cyclones. *Earth Interactions* 6: 1-17. https://doi.org/10.1175/1087-3562(2002)006<0001:IOTTIO>2.0.CO;2.

[39] Li, X., W.-K. Tao, T. Matsui, C. Liu, H. Masunaga. 2010. Improving a spectral bin microphysical scheme using TRMM satellite observations. *Quarterly Journal of the Royal Meteorological Society* 136: 382-399. https://doi.org/10.1002/qj.569.

[40] Lin, J. L., G. N. Kiladis, B. E. Mapes, K. M. Weickmann, K. R. Sperber, W. Lin, M. Wheeler, S. D. Shubert, A. Del Genio, L. J. Donner, S. Emori, J. F. Gueremy, F. Hourdain, P. J. Rasch, E. Roeckner, and J. F. Scinocca. 2006. Tropical intraseasonal variability in 14 IPCC AR4 climate models. Part I: convective signals. *Journal of Climate* 19: 2665-2690.

[41] Lien, G., T. Miyoshi, and E. Kalnay. 2016. Assimilation of TRMM Multisatellite precipitation analysis with a low-resolution NCEP global forecast system. *Monthly Weather Review* 144: 643-661. https://doi.org/10.1175/MWR-D-15-0149.1.

[42] Madden, R. A., and P. R. Julian. 1971. Detection of a 40-50 day oscillation in the zonal wind in the tropical Pacific. *Journal of the Atmospheric Sciences* 28: 702-708.

[43] Mahfouf, J. F., P. Bauer, and V. Marécal. 2005. The assimilation of SSM/I and TMI rainfall rates in the ECMWF 4DVar system. *Quarterly Journal of the Royal Meteorological Society* 131(606): 437-458.

[44] Marcal, V., É. Gérard, J.-F. Mahfouf, and P. Bauer. 2001. The comparative impact of the assimilation of SSM/I and TMI brightness temperatures in the ECMWF 4D-Var system. *Quarterly Journal of the Royal Meteorological Society* 127: 1123-1142. https://doi.org/10.1002/qj.49712757322.

[45] Masunaga, H., and C. D. Kummerow. 2005. Combined radar and radiometer analysis of precipitation profiles for a parametric retrieval algorithm. *Journal of Atmospheric and Oceanic Technology* 22: 909-929.

[46] Masunaga, H., T. S. L'Ecuyer, and C. D. Kummerow. 2006. The Madden-Julian oscillation recorded in early observations from the tropical rainfall measuring mission (TRMM). *Journal of the Atmospheric Sciences* 63: 2777-2794. https://doi.org/10.1175/JAS3783.1.

[47] Meneghini, R., and T. Kozu. 1990. *Spaceborne Weather Radar*, 199 pp. London: Artech House.

[48] Meneghini, R., T. Iguchi, T. Kozu, L. Liao, K. Okamoto, J. A. Jones, and J. Kwiatkowski. 2000. Use of the surface reference technique for path attenuation estimates from the TRMM Precipitation radar. *Journal of Applied Meteorology* 39: 2053-2070.

[49] Moreau, E. 2003. Variational retrieval of rain profiles from spaceborne passive microwave radiance observations. *Journal of Geophysical Research* 108: D16.

[50] Munchak, S. J., and C. D. Kummerow. 2011. A modular optimal estimation method for combined radar-radiometer precipitation profiling. *Journal of Applied Meteorology* 50: 433-448.

[51] Nesbitt, S. W., and E. J. Zipser. 2003. The diurnal cycle of rainfall and convective intensity according to

three years of TRMM measurements. *Journal of Climate* 16(10):1456-1475.

[52] Nesbitt, S. W., E. J. Zipser, and D. J. Cecil. 2000. A census of precipitation features in the tropics using TRMM: radar, ice scattering, and lightning observations. *Journal of Climate* 13(23):4087-4106.

[53] Njoku, E. G., J. M. Stacey, and F. T. Barath. 1980. The seasat scanning multichannel microwave radiometer(SMMR): instrument description and performance. *IEEE Journal of Oceanic Engineering* 5(2): 100-115.

[54] Petersen, W. A., and S. A. Rutledge. 2001. Regional variability in tropical convection: observations from TRMM. *Journal of Climate* 14:3566-3586.

[55] Petersen, W. A., S. W. Nesbitt, R. J. Blakeslee, R. Cifelli, P. Hein, and S. A. Rutledge. 2002. TRMM observations ofintraseasonal variability in convective regimes over the amazon. *Journal of Climate* 15:1278-1294. 10. 1175/1520-0442(2002)015<1278:TOOIVI>2. 0. CO;2

[56] Pierdicca, N., F. S. Marzano, G. d'Auria, P. Basili, P. Ciotti, and A. Mugnai. 1996. Precipitation retrieval from spaceborne microwave radiometers using maximum a posteriori probability estimation. *IEEE Transactions on Geoscience and Remote Sensing* 34:831-846.

[57] Sato, T., H. Miura, M. Satoh, Y. N. Takayabu, and Y. Wang. 2009. Diurnal cycle of precipitation in the tropics simulated in a global cloud-resolving model. *Journal of Climate* 22:4809-4826. https://doi. org/10. 1175/2009JCLI2890. 1.

[58] Schumacher, C., and R. A. Houze. 2003. Stratiform rain in the tropics as seen by the TRMM precipitation radar. *Journal of Climate* 16:1739-1756. https://doi. org/10. 1175/1520-0442(2003)016<1739:SRITTA>2. 0. CO;2.

[59] Seo, E., S. Yang, M. Grecu, G. Ryu, G. Liu, S. Hristova-Veleva, Y. Noh, Z. Haddad, and J. Shin. 2016. Optimization of cloud-radiation databases for passive microwave precipitation retrievals over ocean. *Journal of Atmospheric and Oceanic Technology* 33: 1649-1671. https://doi. org/10. 1175/JTECH-D-15-0198. 1.

[60] Simpson, J., R. Adler, and G. North. 1988. A proposed tropical rainfall measuring mission (TRMM) satellite. *Bulletin of the American Meteorological Society* 69:278-295.

[61] Sivia. D. S. 1996. *Data Analysis-A Bayesian Tutorial*. Oxford: Oxford University Press.

[62] Skofronick-Jackson, G., W. A. Petersen, W. Berg, C. Kidd, E. F. Stocker, D. B. Kirschbaum, R. Kakar, S. A. Braun, G. J. Huffman, T. Iguchi, P. E. Kirstetter, C. Kummerow, R. Meneghini, R. Oki, W. S. Olson, Y. N. Takayabu, K. Furukawa, and T. Wilheit. 2017. The global precipitation measurement(GPM) mission for science and society. *Bulletin of the American Meteorological Society*. https://doi. org/10. 1175/BAMS-D-15-00306. 1.

[63] Smith, E., H. Cooper, X. Xiang, A. Mugnai, and G. Tripoli. 1992. Foundations for statistical-physical precipitation retrieval from passive microwave satellite measurements. Part I: brightness-temperature properties of a time-dependent cloud-radiation model. *Journal of Applied Meteorology* 31:506-531.

[64] Spencer, R., W. Olson, W. Rongzhang, D. Martin, J. Weinman, and D. Santek. 1983. Heavy thunderstorms observed over land by the Nimbus 7 scanning multichannel microwave radiometer. *Journal of Climate and Applied Meteorology* 22:1041-1046.

[65] Spencer, R., H. Goodman, and R. Hood. 1989. Precipitation retrieval over land and ocean with the SSM/I: identification and characteristics of the scattering signal. *Journal of Atmospheric and Oceanic Technology* 6:254-273.

[66] Swann, H. 1998. Sensitivity to the representation of precipitating ice in CRM simulations of deep convection. *J. Atmos. Oceanic. Technol.* 47-48:415-435.

[67] Tao, W.-K., and M. W. Moncrieff. 2009. Multiscale cloud system modeling. *Reviews of Geophysics* 47: RG4002. https://doi.org/10.1029/2008RG000276.

[68] Tao, W.-K., and J.-D. Chern. 2017. The impact of simulated mesoscale convective systems on global precipitation: a multiscale modeling study. *Journal of Advances in Modeling Earth Systems* 9. https://doi.org/10.1002/2016MS000836.

[69] Testud, J., S. Oury, X. Dou, P. Amayenc, and R. Black. 2001. The concept of normalized distribution to describe raindrop spectra: a tool for cloud physics and cloud remote sensing. *Journal of Applied Meteorology* 40:1118-1140.

[70] Wu, H., R. F. Adler, Y. Tian, G. J. Huffman, H. Li, and J. Wang. 2014. Real-time global flood estimation using satellite-based precipitation and a coupled land surface and routing model. *Water Resources Research* 50:2693.2717. https://doi.org/10.1002/2013WR014710.

[71] Zhou, Y. P., W. Tao, A. Y. Hou, W. S. Olson, C. Shie, K. Lau, M. Chou, X. Lin, and M. Grecu. 2007. Use of High-resolution satellite observations to evaluate cloud and precipitation statistics from cloud-resolving model simulations. Part I: South China sea monsoon experiment. *Journal of the Atmospheric Sciences* 64: 4309-4329. https://doi.org/10.1175/2007JAS2281.1.

[72] Zipser, E. J., C. Liu, D. J. Cecil, S. W. Nesbitt, and D. P. Yorty. 2006. Where are the most intense thunderstorms on Earth? *Bulletin of the American Meteorological Society* 87(8):1057-1071.

第7章　全球降水测量(GPM):星基联合降水估计

Gail Skofronick-Jackson,Wesley Berg,Chris Kidd,
Dalia B. Kirschbaum,Walter A. Petersen,
George J. Huffman,Yukari N. Takayabu

7.1 引言

全球降水测量(GPM)任务是由美国国家航空航天局(NASA)和日本宇宙航空研究开发机构(JAXA)发起的一项国际卫星任务,旨在联合和推进星基全球降水测量[13]。该任务主要是部署一个GPM核心观测台(GPM-CO,于2014年2月发射),携带Ka/ku波段双频降水雷达(DPR)和一个多频(10-183 GHz)微波辐射计,以及GPM微波成像仪(GMI)。GPM-CO作为一个降水系统的物理观测台,也是一个由国际合作伙伴联盟运作的微波辐射计群估算降水的校准参考[13]。该任务旨在为科学研究和社会应用提供新一代基于卫星星座的全球降水产品[47]。

作为一个具有综合应用目标的科学探索任务,GPM任务的目标是:(1)利用主动和被动遥感技术建立新的星基降水测量参考标准;(2)通过增强对全球降水时空分布的测量,提高对降水系统、水循环变率和淡水可用性的认识;(3)通过更好地测量潜热、降水微物理和地表水通量,改进气候模式和预报能力;(4)通过更准确地估计瞬时降水信息和误差特征,提高天气预报技术;(5)通过改进时间采样和高分辨率模式降尺度降水产品,改进对高影响自然灾害事件(如洪水、干旱、滑坡和飓风)的水文模拟和预测。通过向传统科学界以外的业务机构和受益相关者提供近实时(NRT)数据,GPM使星基降水观测能够在各种实际应用中使用,从而直接造福社会[19]。

GPM是NASA和JAXA合作的产物,建立在成功的美日热带降雨测量任务(TRMM)[28]基础上。GPM-CO由NASA和JAXA提供,于2014年2月28日在日本种子岛发射。GPM-CO已经完成了3年的主要任务寿命,现在正在延长运行。如果电池和仪器仍能正常工作,预计剩下的燃料将支撑其运行到21世纪30年代。

本章第7.2节描述GPM任务;第7.3节介绍核心观测台传感器性能;第7.4节介绍卫星星座覆盖和采样;第7.5节介绍卫星星座辐射计相互校准;第7.6节介绍降水反演方法;第7.7节介绍地面校准;第7.8节介绍应用;第7.9节是本章总结。

7.2 任务描述

GPM核心观测台在一个非太阳同步轨道上运行，对赤道的倾角为65°，高度为407 km，可对降水的日变化进行观测并保持与卫星星座辐射计的同步测量，作为星座成员辐射测量和降水反演的相互校准的参考标准。在降水反演中，GPM-CO数据对于减少假设数量是必要的。这是因为卫星的降雨量估计不能用简单的数学反演来表示。降水云包含更多的自由参数(包括降雨和非降雨液态和冰粒子的大小、形状以及内部分布)，比从有限的卫星探测数据中实际得到的还要多，因此需要对云的组成进行重要的假设[48]。为了解决这个问题，GPM首先设计了高度精密的仪器，其次，利用GPM-CO使用最先进的主动和被动降雨传感器提供的详细信息，创建了一个自然出现的云状态的通用清单。然后，这些云的清单可以用作卫星星座传感器反演的贝叶斯或最优估计框架中的先验状态[29]。虽然这项技术并没有完全消除对假设的需要，但它确实使用了高度精确的GPM-CO数据，而不是性能较差的辐射计或模式来独立地做出假设。GPM-CO选择65°倾角以提供广泛的纬度覆盖，而不被锁定在太阳同步近地轨道(LEO)，同时保持相对较短的进动周期，以采样降水的日变化。倾角为35°的TRMM观测表明，由LEO卫星在固定当地时间提供的额外资料对近实时的飓风监测和预报很重要。GPM-CO首次提供了追踪中纬度和温带风暴的天气观测。非太阳同步轨道还提供了多个同步的通道，其中大部分是LEO卫星星座传感器，以便在广泛的纬度范围内进行传感器间校准。

GPM任务的地面支撑包括：

(1)NASA提供的用于GPM核心运行的任务操作系统。

(2)地面验证(GV)系统，由NASA、JAXA和一系列GPM地面验证伙伴提供的专用和合作地面验证站点组成。

(3)NASA降水处理系统(PPS)和JAXA GPM任务操作系统(GPM-MOS)与其他GPM合作数据处理站点协调，提供NRT和标准全球降水产品。

在卫星间校准、反演算法开发、地面验证和科学应用方面，NASA降水测量任务(PMM)科学团队和JAXA PMM科学团队在支持GPM。GPM对于国际社会研究和了解地球水文和能量循环是必要的。GPM任务是地球观测卫星委员会(CEOS)正在开发的监测全球降水的多国卫星星座的科学基石，并提供关键探测数据，以推进一系列国际科学计划和活动的目标，其中包括全球能源和水循环实验(GEWEX)、全球水循环综合观测(IGWCO)，以及国际降水工作组(IPWG)。

7.3 GPM核心观测台传感器性能

TRMM主要关注热带海洋的中到大雨。由于小雨(<0.5 mm/h)和降雪量是热带以外地区总降水事件的重要组成[37]，因此需要改进这些量的测量，以便更好地了解全球水通量和降水特征(如频率、强度、分布等)。GPM-CO携带了首个在Ku和Ka波段的星载双频降水雷达

(DPR),以及一个包括微波成像仪和湿度探空仪的典型频率范围的多光谱(10～183 GHz)GMI,能够在云中逐层进行详细的微物理探测。GPM-CO GMI 和 DPR 的功能规格和在轨性能见表 7.1。GPM-CO 搭载的主动(DPR)和被动(GMI)传感器提供互补信息,用于提高基于物理的降水反演的准确性。DPR 相对于 TRMM 降水雷达(PR)和高频 GMI 通道的灵敏度增加,使 GPM 能够探测小雨(低至 0.2 mm/h)和降雪。

表 7.1 GPM-CO GMI 和 DPR 的(功能)规格和在轨性能

通道/GHz	空间分辨率/km	NEDT/K(规定的)	NEDT/K(在轨)	波束宽度/°(在轨)	波束效率/%(在轨)
GMI 10.65 V-pol	19.4×32.1	0.96	0.77	1.72	91
GMI 10.65H-pol	19.4×32.1	0.96	0.78	1.72	91
GMI18.7 V-pol	10.9×18.1	0.84	0.63	0.98	91
GMI 18.7 H-pol	10.9×18.1	0.84	0.60	0.98	91
GMI 23.8 V-pol	9.7×16.0	1.05	0.51	0.85	93
GMI 36.64 V-pol	9.4×15.6	0.65	0.41	0.82	98
GMI 36.64H-pol	9.4×15.6	0.65	0.42	0.82	98
GMI89 V-pol	4.4×7.2	0.57	0.32	0.38	97
GMI89 H-pol	4.4×7.2	0.57	0.31	0.38	97
GMI166 V-pol	4.1×6.3	1.5	0.70	0.38	97
GMI 166 H-pol	4.1×6.3	1.5	0.66	0.37	97
GMI 183.31 ± 3 V-pol	3.8×5.8	1.5	0.56	0.37	95
GMI 183.31 ± 7 V-pol	3.8×5.8	1.5	0.47	0.37	95
DPR 13.6(Ku)[a]	5 km 水平				
	250 m 垂直[b]				
DPR 35.5(Ka)[c]	5 km 水平				
	250 m 垂直				
	500 m 垂直[d]				

注:a. 最小可探测反射率约 12～13 dBZ。
b. 过采样到 125 km。
c. 最小可探测反射率约 17～18 dBZ(高灵敏度模式为 12～13 dBZ)。
d. 在高灵敏度模式(过采样分别至 125 km 和 250 km)。

7.3.1 双频率降水雷达(DPR)

DPR 仪器提供三维降水结构的探测,由 35.5 GHz 的 Ka 波段降水雷达(KaPR)和 13.6 GHz 的 Ku 波段降水雷达(KuPR)组成。其中,Ku 波段雷达是在 TRMM 的升级版本[27]。JAXA 和日本国家信息通信技术研究所(NICT)共同研制了 DPR 仪器。KuPR 和 KaPR 是共同放置的,它们的 5 km 信号覆盖区在地球表面重合,交叉轨道的刈幅分别为 245 km 和 125 km(见图 7.1),垂直距离分辨率为 250 m。当 KaPR 在高灵敏度的空间过采样模式下工作时,距离分辨率为 500 m。Ka 和 Ku 均采用垂直过采样技术,使得采样距离减半。DPR 使用的可变脉冲重复频率技术增加了每个瞬时视场(IFOV)的样本数量，以实现 12 dBZ 或 0.2 mm/h 的探测阈值。DPR 的校准系统包括系统内部定标和由 JAXA 提供的一系列地面主动雷达进行定标。

利用 DPR 的资料，通过 Ka 对小粒子的敏感性、Ku 对大粒子的敏感性以及基于两个频率观测值之间差分衰减的反演技术，降水探测可以更准确地反演体粒径分布的两个参数。双频返回信号让我们能深入了解微物理过程(蒸发、碰撞/合并、聚合)，可以区分液相、冰相和混合相降水区域。此外也提供降水的整体属性，如水通量和测量柱中的含水量。双波长雷达提升了的准确性和详细的微物理信息被用来约束 GMI 的降水反演，并开发一个先验贝叶斯云模式数据库，用于基于辐射计的反演。该数据库被卫星星座辐射计用于统一全球降水反演。

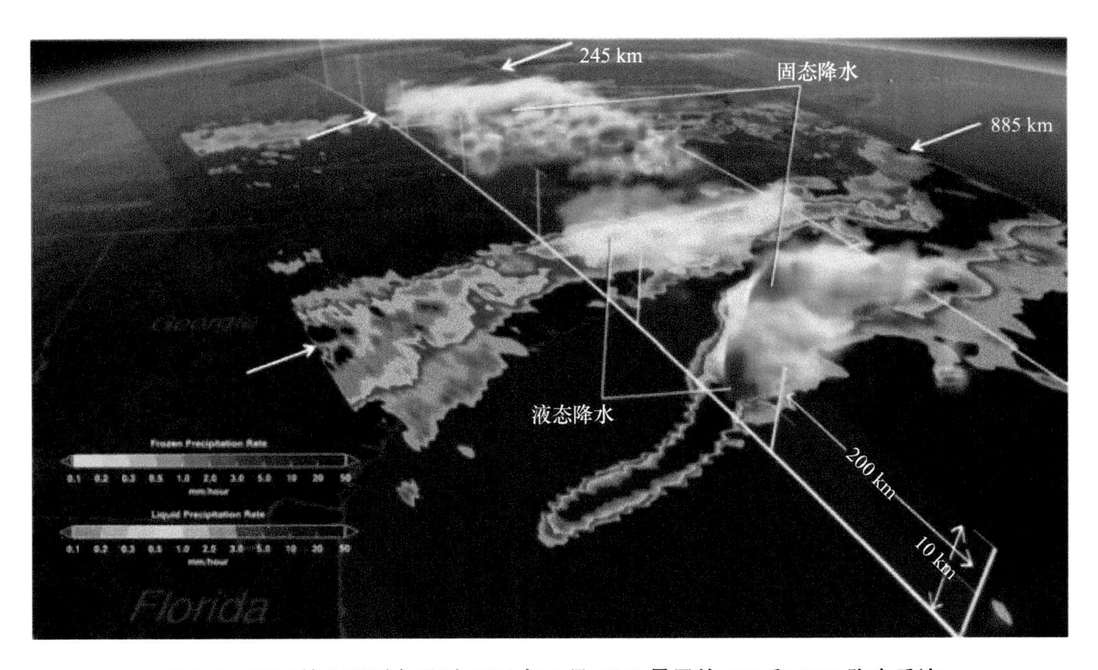

图 7.1 GPM 核心观测台观测 2014 年 3 月 17 日暴雪的 GMI 和 DPR 降水反演

7.3.2 GPM 微波成像仪(GMI)

GMI 是一种多通道锥形扫描微波辐射计，其特点是有 13 个通道、频率范围从 10～183 GHz(表 7.1)，以及一个直径为 1.2 m 的天线，提供比 TMI 更好的空间分辨率。GMI 的偏移天顶角为 48.5°，对应地球入射角为 52.8°，因此保持了与其前任 TRMM 辐射计相似的几何结构。GMI 主反射器以每分钟 32 转的速度旋转，在以航天器地面轨道为中心的 140°扇形区域上收集微波辐射亮温(T_b)的测量数据，从 GPM-CO 407 km 的高度在地球表面形成 885 km 的交叉轨道带(见图 7.1)。Ball 航空航天技术公司与 NASA 戈达德太空飞行中心签订合同并建造 GMI。GMI 扫描带的中心部分与雷达扫描带重叠，GMI 和雷达观测时间之间大约有 67 s 的差异(由于几何构造和航天器运动)。DPR 和 GMI 重叠区域内的探测结果对于改进和交叉验证降水反演结果，特别是基于辐射计的反演结果具有重要意义。

考虑到参考辐射计的作用，GMI 的设计采用了几种创新的工程特性，以实现更高的仪器精度和稳定性。在主反射器和冷校准副反射器上使用了高质量的反射涂层。在 10.7～36.64 GHz

通道的定标方案中加入了噪声二极管,从而可以显示解决辐射计响应函数中的非线性问题[7]。热负载中嵌入了 14 个温度传感器,以更好地表征其物理温度和任何热梯度。冷定标副反射器尺寸有些过大,以减少溢出问题;且暖定标负载被遮盖/遮蔽,以减少太阳光入侵[6]。

发射后,进行了一系列航天器姿态调转,以检查和改进定标。其中一项调转涉及将 GPM 航天器仰起,这样主波束和冷校准镜都能观察到冷的宇宙空间。辐射计的电压被确定为与主反射器和副反射器的测量不确定度相同,表明任一反射器都没有显著的发射率问题。还进行了俯仰调转,以验证双极化通道在天底点的垂直和水平极化之间的一致性,并对辐射计对磁场的敏感性造成的定标误差进行了修正,以及对发射前推导出的溢出修正进行了更新。仪器设计和详细的在轨定标检查和改进相结合,使得定标非常稳定和准确,误差在 0.5 K 以内[50]。

7.4　覆盖和采样

鉴于目前微波遥感技术的能力,GPM 任务采取了一种基于卫星星座的方法,将近地轨道上的一系列微波传感器的探测结果结合起来,以实现全球覆盖,并为广泛的科学和社会应用提供所需的高时间分辨率采样(图 7.2)。这是几十年来从星基设备估算降雨的主要方法。与圆锥扫描成像仪相比,微波探测仪器的观测结果与降水的直接关系较小,因此在最初的任务设计中没有包括在内。然而,近年来,随着微波探测仪反演算法的不断改进,150～183 GHz 高频波段的微波探测仪对当前降雨的反演质量已被证明与地基锥形扫描辐射计相当[33]。2006 年重新配置了基准 GPM 卫星星座,加入了微波探测仪器,以增加地基锥形扫描辐射计的采样,在 GPM 任务寿命期间,重访时间在 1～2 h[13]。GMI 上的高频探空通道也将为使用 GPM-CO 传感器探测结果来进一步改善 GPM 卫星星座的反演提供纽带/桥梁。

GPM 卫星星座的建立遵循具有灵活架构的滚动波策略,利用卫星(观测)时机信号与合作伙伴提供的资源,提供尽可能最好的采样和覆盖。2017 年的基准卫星星座搭档配置有 10 个传感器,包括这些锥形扫描成像仪:日本全球变化观测任务-水 1 号(GCOM-W1)上的先进微波扫描辐射计-2(AMSR-2)[35],美国国防气象卫星计划(DMSP)卫星上的特殊传感器微波成像仪/探空仪(SSMIS)仪器[30]。2017 年卫星星座内交叉扫描温湿度探测仪包括:法国-印度超级热带卫星上的用辐射法测定热带湿气廓线的大气探测仪[42]、国家极地轨道业务环境卫星系统(NPOESS)的预备项目(NPP)上的先进技术微波探空仪(ATMS)[9]、NOAA 和欧洲 MetOp 卫星上的四个微波湿度探测仪(MHS)[22]。表 2 提供了基准配置中卫星星座传感器的基本传感器和频率的详细信息。每个卫星星座搭档卫星辐射计都有自己的科学或业务目标,但将微波 T_b 观测结果贡献给 GPM 任务,以获得统一的基于卫星星座的全球降水产品。在一致的框架内使用 GPM-CO 传感器测量值作为参考标准进行卫星间校准(在辐射和反演层面)是 GPM 任务的一个关键目标。

有了这样一组多样的降水传感器,卫星间校准是 GPM 任务的一个重要内容。事实上,GPM 的多卫星综合多卫星反演(IMERG[15])几乎每 30 min 提供一次全球降水估计,通过红外(IR)观测改变卫星星座传感器的降水估计。IMERG 产品可在降水事件发生 5 h 后在 NRT 以 0.1°×0.1°(或约 10 km×10 km)的地表分辨率获得。所有 GPM-CO 产品都可以在 NRT

中为应用用户提供，以后也可以作为研究产品提供，其中包括所有辅助数据，以获得更好的性能和更低的不确定性。与 IMERG 对应的日本版是全球卫星降水制图（GSMaP），其水平分辨率为 0.1°×0.1°×0.1°，且有不同的延迟，临近预报、4 h 预报等[25]这些产品分别可以在网站 gpm.nasa.gov 和 sharaku.eorc.jaxa.jp/GSMaP 上找到。

7.5 卫星星座传感器的交叉辐射定标

气候质量数据集需要跨越多个观测平台在长时间内保持准确和一致。不同 GPM 微波卫星星座仪器之间的任何 T_b 偏差都可能妨碍全球降水估计的一致性。尽管这些卫星星座传感器有相似之处，但在中心频率、观测角度和空间分辨率（表 7.2）方面存在差异，必须调整以形成统一的全球降水估计。交叉校准工作的一个重要副产品是能够监测 GPM 卫星星座中每个仪器的突然和/或逐步校准变化以及其他退化。

表 7.2 2017 年 GPM-CO 和卫星星座搭档被动微波辐射计的传感器性能

	通道[a]/GHz								
	6～7	10	19	23	31～37	50～60	89～91	150～167	183～190
GMI		10.65	18.70	23.80	36.50		89.0	165.6	183.31±3V
分辨率[b]		V/H 26 km	V/H 15 km	V 12 km	V/H 11 km		V/H 6 km	V/H 5 km	183.31±8V 5 km
AMSR 2	6.925/7.3	10.65	18.70	23.80	36.50		89.0		
分辨率[b]	V/H 62/58 km	V/H 42 km	V/H 22 km	V/H 26 km	V/H 12 km		V/H 5 km		
SSMIS			19.35	22.235	37.0	50.3-63.28	91.65	150	183.31±1H
分辨率[b]			V/H 59 km	V 59 km	V/H 36 km	V/H 22 km	V/H 14 km	H 14 km	183.31±3H 14 km
SAPHIR									183.31±0.2H_Q
									183.31±1.1H_Q
									183.31±2.8H_Q
									183.31±4.2H_Q
									183.31±6.8H_Q
分辨率[b]									183.31±11H_Q 10 km
MHS							89V_Q	157V_Q	183.31±1H_Q
									183.31±3H_Q
分辨率[b]							17 km	17 km	190.311V_Q 17 km
ATMS				23.8	31.4	50.3～57.29	87～91	165.5H	183.31±1H_Q
				V_Q	V_Q	H_Q	V_Q	H_Q	183.31±1.8H_Q

续表

	通道[a]/GHz								
	6～7	10	19	23	31～37	50～60	89～91	150～167	183～190
									183.31±3H_Q
									183.31±4.5H_Q
									183.31±7H_Q
分辨率[b]				74 km	74 km	32 km	16 km	16 km	16 km

注:a. 通道中心频率(GHz):V 为垂直极化,H 为水平极化,V_Q 为准垂直,H_Q 为准水平(V_Q 和 H_Q 表示交叉轨道扫描仪的混合极化)。

b. 平均空间分辨率(km)。

通过卫星间校准实现一致性需要几个步骤。每个仪器供应商都有责任提供符合既定标准的经过良好定标的 T_b 产品。基于合作伙伴提供的这些标准 T_b 产品,GPM 检查数据,以发现和消除由俯仰和转动误差、航天器干扰和更微妙的原因导致的像素对像素的偏差。对于锥形扫描仪,探测偏差在概念上很简单;每个波束位置的平均 T_b 应该是相同的。由纬度和小入射角异常引起的偏差很容易消除[10]。由于对入射角的依赖关系需要非常精确地建模/准确地模拟,交叉轨迹扫描器探测和订正偏差以实现自洽要困难得多。

如果仪器在扫描中显示出自一致,那么它在轨道上也必须自一致。为此,利用全球数据同化系统(GDAS)等环境数据来计算整个轨道上非降雨海洋像素的 T_b。虽然这些计算可能不足以进行绝对校准,但它们适用于识别 T_b 的异常。当探测到一致的异常模式时,开发了一种方法来订正它们。

如果单个仪器 T_b 是自洽的,卫星星座内的多个辐射计将相互校准,以在一致的框架内产生统一的全球 T_b 产品。为此,GMI 观测在时空上与其他卫星星座传感器相匹配。对于匹配的数据集,算法开发遵循两种方法:(1)将系统中同一位置的 T_{bs} 转换为共同要素并进行比较[1,12];(2)分别推导每个辐射计的极限值并通过模型进行替代性比较[3,43]。这些匹配比较所指出的订正随后应用于互校准数据文件中的卫星星座 T_{bs},这些数据文件被用作辐射计反演算法的输入[1]。

7.6 降水反演和产品

从不同种类的卫星星座辐射传感器进行统一的降水估算,不仅需要 T_b 的稳健互校,还需要一个根据共同标准推导降水的物理框架。GPM 通过使用 GPM-CO 来生成由不同卫星星座辐射计观测到的所有通道降水廓线及其对应 T_b 的最佳先验统计数据来实现这一目标。GPM 卫星星座辐射计参考他们自己观测的共同先验数据库,使其与任务设想的理念在物理上一致。GPM 任务生成了三种类型的降水产品,分别基于 DPR 反演、DPR/GMI 联合反演以及基于 DPR/GMI 联合反演生成的先验数据库的辐射计反演。DPR 反演已被证明可以提供最好的结果,当直接与地基雷达[18]和以往从多参数地基雷达获得的微物理数据比较时,DPR/GMI 联合反演设计为从 DPR 解决方案开始,并根据需要对该方案进行修改,以确保不仅与 DPR 反射率一致，而且与 GMI 辐射率也一致。这种方法改善了整体结果——特别是在 DPR 无法完全描述水滴粒径分布、云液态含水量或降雨量低于 Ka 波段雷达探测阈值的情况下。

雷达算法利用了 Ka 和 Ku 波段对降水粒子的敏感性。DPR 的两个反射率在每个距离库上互为独立，用于确定垂直采样廓线中的液滴粒径分布[32]。这是向前迈出的重要一步，因为 TRMM 和地基雷达依赖于单一频率，因此需要对液滴粒径分布进行假设。雷达反演估计降雨率、液态含水量和雨雪粒径分布参数的廓线，以及相关的误差统计。特别强调了对极弱降雨、极强降雨和降雪个例（TRMM 轨道没有对这些情况进行反演）的反演方法。利用两个频率的测量值，DPR 反演也得益于差分衰减订正技术，而不是更严格的地面参考技术[36]。

雷达-辐射计联合反演在概念上通过结合 DPR 和 GMI 信息提供星载平台对降水结构的最佳估计[11]，该反演自然发挥了用于卫星星座辐射计反演降水的云/降水数据库的先验作用。虽然 TRMM 已经在热带海洋上空使用了这种方法，但 GPM 联合算法开发的一个挑战是创建更新的物理参数，用于高纬度和增强的 GMI 和 DPR 通道集。由于这些纬度地区的零度层高度较低，因此 DPR 需要更好地识别对流和层状降水区域从冰相到混合相到液相转变发生的高度。此外，还改进了适用于 GMI 10-183 GHz 频率范围的冰相和混合相降水的消光和散射的描述[31,39]。对于 GPM 算法的发展，改进的物理参数得到了一系列地面验证外场试验的支持（见 7.7 节）。

DPR、GMI 和联合算法都利用更高频率的通道（GMI 的 166、183 和 DPR 的 Ka）以感知冻结粒子的特征，进而估计降雪量。2006 年，在一系列论文表明它们对探测和估计降雪有用之后，这些更高频率的通道被添加到设计中[5,23,34,45]。自发射以来，在文献[4,49]中对 GPM 反演降雪的性能进行了评价。

用于全球统一降水估计的辐射计算法必须在多个频率和所有卫星星座成员之间物理上保持一致。贝叶斯或最优估计算法非常适合于反演降水及其垂直结构，方法是将任一多频辐射计提供的信息内容与降水云垂直廓线的常见先验数据库提供的信息内容以及每条廓线和传感器的模拟 T_b 进行比较。通过将联合算法反演的降水与 DPR 反射率廓线和 GMI 的 T_b 相关联，构建了一个健全的由 DPR 和 GMI 探测数据约束的全球先验数据库，消除了对仅依赖于云解析模式的先验数据库的需求。当 DPR 未检测到降水，但 GMI 的 166 GHz 和 183 GHz 通道中存在降水信号时，将 CloudSat 数据纳入先验数据库，这种方法被称为“雷达增强”辐射计算法，提高了辐射计反演的准确性和物理一致性。虽然成像仪和探空仪通常有不同的通道集和扫描方式（锥形和交叉追踪），但所有被动传感器的算法开发在概念上是相同的，贝叶斯数据库反映了仪器的差异。T_b 数据库由 DPR/GMI 联合反演得到的地球物理参数计算而来，充分考虑了传感器通道、入射角和空间分辨率的差异。由于地表特征的变化及其对降水的影响，贝叶斯数据库被分为大约 14 种地表类型，这有助于通过对地表发射率的影响降到最低来改善反演。由于同样的方法可以用于以前和未来的传感器，因此有可能使用 SSM/I 观测将 GPM 降雨气候学追溯至 1987 年，并有可能超越 GPM-CO 卫星本身的时代。更长的时间序列产品将于 2018 年向公众发布。

虽然卫星星座多个微波辐射计提供的频繁扫描很重要，但水文等应用需要更高的时间和空间分辨率。为此，已经发展了一些多卫星技术[8]，确保均匀、频繁覆盖的一种方法是使用被动微波遥感观测进行降雨估测，这种估算是经过 IMERG 和 GSMap 等地球静止卫星红外观测增强的[24,25]。GPM 保证了在这些合并的多卫星微波/红外方法中用于偏差校正的微波产品彼此一致。GPM 还计划探索如何通过统计降尺度或利用某些水文应用的云解析模式进行数据同化,来开发非常高的空间和时间分辨率(1～2 km 和 5～10 min)的降水估计。

为了满足 GPM 任务的目标(3),我们还生成了三维潜热数据。有两种具有代表性的潜热算法,分别是对流-层云加热(CSH)算法和频谱潜热(SLH)算法,前者基本使用 GMI 数据,利用对流/层云比和区域性生成的潜热表,后者使用 GPM DPR 产品以及对流和层云降水各自的频谱表。潜热数据可以作为 GPM 标准产品得到。

7.7 地面验证

NASA GPM 项目建立了强有力的地面验证(GV)计划,利用地基观测(包括飞机测量)支持卫星发射前和卫星发射后降水产品评估。GPM GV 基础设施的三个关键组成部分包括大量的外场试验[14,16,38,46]和现场测量、NOAA 多雷达多传感器(MRMS)产品的开发[51]以及下文所述的验证网络(VN)架构的开发。外场试验充分获取了地表降水、云中的降水物理样本,并利用地面、原位探测和遥感飞机以及卫星测量进行了云中主动/被动微波遥感。这些外场试验数据有助于通过减少假设来发展卫星反演算法,并有助于测量用于水文气象应用的综合数量。MRMS 数据对美国和邻近地区(130°－60°W,20°－55°N)的降水率和类型(雨/雪)的雷达估计进行了雨量计偏差调整。GV MRMS 数据集来源于 NOAA MRMS 产品,但经过进一步处理[20,21]来产生雨量计校正的参考降水率数据集,其中增加了本地 2 min/0.01°×0.01°分辨率的降水类型和数据质量信息。验证网络(VN)雷达数据包括美国网络、海洋和/或其他国家/国际研究特定地点的约 80 部双偏振(dual-pol)雷达及其推导的降水率、DSD 和水凝物类型估计,其覆盖区和几何体积与 GPM-CO 的过境覆盖区匹配[2,40,44]。

GPM GV 降水测量设备目录庞大[47]。随着 GPM 继续参与全球范围内与降水相关的外场试验,大多数设备设计为便携式,便于移动到外场试验地点。当不用于 GPM 相关的外场试验时,这些仪器在 NASA Wallops 飞行设施(WFF)网络(或伙伴站点)内运行,特别强调在 GPM-CO 过境期间有针对性和用户自适应的数据收集,以满足 GPM 算法和 GV 团队的需求和要求。全球卫星观测的所有数据都可以通过 GPM 网页上的链接免费获取(gpm. nasa. gov)。

7.8 应用

通过改进雨雪测量,GPM 产品系列有助于广泛的社会应用,例如:热带和温带气旋定位和降雨监测、饥荒预警、干旱监测、水资源管理、农业预报、数值天气预报、陆地表面模拟、全球气候模拟、疾病追踪、经济研究和动物迁徙;其中许多最初是用 TRMM 数据开发的。其中一些应用需要 NRT 数据以及长期、校准良好的融合卫星降水信息[41];GPM 任务支持这些产品延迟。在世界其他区域,特别是在缺乏充分的地基探测信息的区域,IMERG 正被用作预报模式的输入量(例如洪水、滑坡、农业/饥荒、降水引起的疾病暴发)[17]。选定的应用在文献[19,26]中报告。GPM 的遥感降水数据为各机构、科研院所和全球社会提供了多种应用。

7.9 总结

GPM 是一项国际卫星任务，旨在统一和推进由专用和业务化的卫星星座微波传感器进行的降水测量，为研究和应用提供下一代统一的全球降水产品。GPM 是 NASA 和 JAXA 合作的结晶，也得到了在其他国内和国际合作伙伴的支持。2014 年 2 月，NASA 和 JAXA 在 65°倾角的轨道上发射了 GPM-CO，携带了首个双频雷达和先进的微波辐射计，作为降水物理观测站和所有卫星星座辐射测量及反演的校准参考。

GPM 是一项具有综合应用目标的科学任务，旨在提高对全球水/能量循环变化和淡水可用性的认识，并改善天气、气候和水文预测能力。在 GEO 和 GEOSS 框架内，GPM 是发展 CEOS 降水卫星星座（PC）的基石。在其任务期间，GPM 实现了 CEOS-PC 的目标：为科学研究和社会应用提供最先进的全球降水产品。

致谢 感谢 GPM 卫星间校准工作组、地面验证工作组、GPM 算法团队和 PMM 科学团队的贡献。非常感谢来自 NASA 科学可视化工作室的图像帮助。这些活动的资金由 NASA 总部提供，特别是 Ramesh Kakar。

参考文献*

[1] Berg, W., S. Bilanow, R. Chen, S. Datta, D. Draper, H. Ebrahimi, S. Farrar, W. L. Jones, R. Kroodsma, D. McKague, V. Payne, J. Wang, T. Wilheit, and J. X. Yang. 2016. Intercalibration of the GPM Radiometer constellation. *Journal of Atmospheric and Oceanic Technology* 33:2639-2654. https://doi.org/10.1175/JTECH-D-16-0100.

[2] Bolen, S., and V. Chandrasekar. 2003. Methodology for aligning and comparing spaceborne radar and ground-based radar observations. *Journal of Atmospheric and Oceanic Technology* 20:647-659. https://doi.org/10.1175/1520-0426(2003)20<647:MFAACS>2.0.CO;2$.

[3] Brown, S. T., and C. S. Ruf. 2005. Determination of a hot blackbody reference target over the amazon rainforest for the on-orbit calibration of microwave radiometers. *AMS Journal of Oceanic and Atmospheric Technology* 22(9):1340-1352.

[4] Casella, D., G. Panegrossi, P. Sanò, A. C. Marra, S. Dietrich, B. T. Johnson, and M. S. Kulie. 2017. Evaluation of the GPM-DPR snowfall detection capability: comparison withCloudSat-CPR. *Atmospheric Research*. https://doi.org/10.1016/j.atmosres.2017.06.018.

[5] Chen, F. W., and D. H. Staelin. 2003. AIRS/AMSU/HSB precipitation estimates. *IEEE Transactions on Geoscience and Remote Sensing* 41(2):410-417. https://doi.org/10.1109/TGRS.2002.808322.

* 参考文献沿用原版书中内容，未改动

[6] Draper, D. W., D. A. Newell, D. S. McKague, and J. R. Piepmeier. 2015. Assessing calibration stability using the global precipitation measurement(GPM) microwave imager(GMI) noise diodes. *IEEE Journal of Selected Topics in Applied Earth Observations and Remote Sensing* 8: 4239-4247. https://doi.org/10.1109/JSTARS.2015.2406661.

[7] Draper, D. W., D. A. Newell, F. J. Wentz, S. Krimchansky, and G. Skofronick-Jackson. 2015. The global precipitation measurement(GPM) microwave imager(GMI): instrument overview and early on-orbit performance. *IEEE Journal of Selected Topics in Applied Earth Observations and Remote Sensing* 8:3452-3462. https://doi.org/10.1109/JSTARS.2015.2403303.

[8] Ebert, E., J. Janowiak, and C. Kidd. 2007. Comparison of near real time precipitation estimates from satellite observations and numerical models. *Atmospheric Research* 88:47-64.

[9] Goldberg, M. D., H. Kilcoyne, H. Cikanek, and A. Mehta. 2013. Joint polar satellite system: the United States next generation civilian polar-orbiting environmental satellite system. *Journal of Geophysical Research—Atmospheres*118:13463-13475. https://doi.org/10.1002/2013JD020389.

[10] Gopalan, K., W. L. Jones, S. Biswas, S. Bilanow, T. Wilheit, and T. Kasparis. 2009. A timevarying radiometric bias correction for the TRMM microwave imager. *IEEE Transactions on Geoscience and Remote Sensing*47(11):3722-3730.

[11] Grecu, M., W. S. Olson, S. J. Munchak, S. Ringerud, L. Liao, Z. Haddad, B. L. Kelley, and S. F. McLaughlin. 2016. The GPM combined algorithm. *Journal of Atmospheric and Oceanic Technology*33:2225-2245. https://doi.org/10.1175/JTECH-D-16-0019.1.

[12] Hong, L., W. L. Jones, T. T. Wilheit, and T. Kasparis. 2009. Two approaches for inter-satellite radiometer calibrations between TMI and WindSat. *Journal of the Meteorological Society of Japan. Series II*87: 223-235.

[13] Hou, A. Y., R. K. Kakar, S. Neeck, A. A. Azarbarzin, C. D. Kummerow, M. Kojima, R. Oki, K. Nakamura, and T. Iguchi. 2014. The global precipitation measurements mission. *Bulletin of the American Meteorological Society*95:701-722. https://doi.org/10.1175/BAMS-D-13-00164.1.

[14] Houze, R. A., L. A. McMurdie, W. A. Petersen, M. R. Schwaller, W. Baccus, J. Lundquist, C. Mass, B. Nijssen, S. A. Rutledge, D. Hudak, S. Tanelli, G. G. Mace, M. Poellot, D. Lettenmaier, J. Zagrodnik, A. Rowe, J. DeHart, L. Madaus, and H. Barnes. 2017. The olympic mountains experiment(OLYMPEX). *Bulletin of the American Meteorological Society*. https://doi.org/10.1175/BAMS-D-16-0182.1.

[15] Huffman, G. J., D. T. Bolvin, D. Braithwaite, K. Hsu, R. Joyce, C. Kidd, E. J. Nelkin, S. Sorooshian, J. Tan, and P. Xie. 2017. Algorithm theoretical basis document(ATBD) version 4.6 for the NASA global precipitation measurement(GPM) Integrated Multi-satellitE Retrievals for GPM(I-MERG), 32 pp. GPM Project: Greenbelt, MD. Available online at https://pmm.nasa.gov/sites/default/files/document_files/IMERG_ATBD_V4.6.pdf.

[16] Jensen, M. P., W. A. Petersen, A. Bansemer, N. Bharadwaj, L. D. Carey, D. J. Cecil, S. M. Collis, A. D. DelGenio, B. Dolan, J. Gerlach, S. E. Giangrande, A. Heymsfield, G. Heymsfield, P. Kollias, T. J. Lang, S. W. Nesbitt, A. Neumann, M. Poellot, S. A. Rutledge, M. Schwaller, A. Tokay, C. R. Williams, D. B. Wolff, S. Xie, and E. J. Zipser. 2016. The midlatitude continental convective clouds experiment(MC3E). *Bulletin of the American Meteorological Society* 97:1667-1686. https://doi.org/10.1175/BAMS-D-14-00228.1.

[17] Kidd, C., G. J. Huffman, A. Becker, G. Skofronick-Jackson, D. Kirschbaum, P. Joe, and C. Muller. 2017. So, how much of the Earth's surface is covered by rain gauges? *Bulletin of the American Meteorological*

*Society*98:69-78. https://doi.org/10.1175/BAMS-D-14-00283.1.

[18] Kidd, C., J. Tan, P. Kirstetter, and W. A. Petersen. 2017. Validation of the version 05 level 2 precipitation products from the GPM core observatory and constellation satellite sensors. *Quarterly Journal of the Royal Meteorological Society*. First Published:4 December 2017. http://dx.doi.org/10.1002/qj.3175.

[19] Kirschbaum, D. B., G. J. Huffman, R. F. Adler, S. Braun, K. Garrett, E. Jones, A. McNally, G. Skofronick-Jackson, E. Stocker, H. Wu, and B. F. Zaitchik. 2017. NASA's remotely-sensed precipitation: a reservoir for applications users. *Bulletin of the American Meteorological Society*98:1169-1184. https://doi.org/10.1175/BAMS-D-15-00296.1.

[20] Kirstetter, P., Y. Hong, J. J. Gourley, S. Chen, Z. L. Flamig, J. Zhang, M. Schwaller, W. Petersen, and E. Amitai. 2012. Toward a framework for systematic error modeling of spaceborne precipitation radar with NOAA/NSSL ground radar-based National Mosaic QPE. *Journal of Hydrometeorology*13:1285-1300. https://doi.org/10.1175/JHM-D-11-0139.1.

[21] Kirstetter, P. E., Y. Hong, J. J. Gourley M. Schwaller, W. Petersen, and Q. Cao. 2015. Impact of sub-pixel rainfall variability on spaceborne precipitation estimation: evaluating the TRMM 2A25 product. *Quarterly Journal of the Royal Meteorological Society* 141:953-966. https://doi.org/10.1002/qj.2416.

[22] Klaes, K. D., M. Cohen, Y. Buhler, P. Schlüssel, R. Munro, A. Engeln, E. Clérigh, H. Bonekamp, J. Ackermann, J. Schmetz, and J. Luntama. 2007. An introduction to the EUMETSAT polar system. *Bulletin of the American Meteorological Society* 88:1085-1096. https://doi.org/10.1175/BAMS-88-7-1085.

[23] Kongoli, C., P. Pellegrino, R. R. Ferraro, N. C. Grody, and H. Meng. 2003. A new snowfall detection algorithm over land using measurements from the advanced microwave sounding unit (AMSU). *Geophysical Research Letters* 30:1756. https://doi.org/10.1029/2003GL017177, 14.

[24] Kubota, T., S. Shige, H. Hashizume, K. Aonashi, N. Takahashi, S. Seto, M. Hirose, Y. N. Takayabu, K. Nakagawa, K. Iwanami, T. Ushio, M. Kachi, and K. Okamoto. 2007. Global precipitation map using satelliteborne microwave radiometers by the GSMaP project: production and validation. *IEEE Transactions on Geoscience and Remote Sensing* 45:2259-2275.

[25] Kubota, T., K. Aonashi, T. Ushio, S. Shige, Y. N. Takayabu, Y. Arai, T. Tashima, M. Kachi, R. Oki. 2017. Recent progress in global satellite mapping of precipitation (GSMaP) product. In *Proceedings of IGARSS*2017, Fort Worth.

[26] Kucera, P. A., E. E. Ebert, F. J. Turk, V. Levizzani, D. Kirschbaum, F. J. Tapiador, A. Loew, M. Borsche. 2013. Precipitation from space: advancing earth system science. *Bulletin of the American Meteorological Society*94:365-375. https://doi.org/10.1175/BAMS-D-11-00171.1.

[27] Kummerow, C., W. Barnes, T. Kozu, J. Shiue, and J. Simpson. 1998. The tropical rainfall measuring mission (TRMM) sensor package. *Journal of Atmospheric and Oceanic Technology* 15:809-817.

[28] Kummerow, C., J. Simpson, O. Thiele, W. Barnes, A. T. C. Chang, E. Stocker, R. F. Adler, A. Hou, R. Kakar, F. Wentz, P. Ashcroft, T. Kozu, Y. Hong, K. Okamoto, T. Iguchi, K. Kuriowa, E. Im, Z. Haddad, G. Huffman, B. Ferrier, W. S. Olson, E. Zipser, E. A. Smith, T. T. Wilheit, G. North, T. Krishnamurti, and K. Nakamura. 2000. The status of the tropical rainfall measuring mission (TRMM) after two years in orbit. *Journal of Applied Meteorology* 39(12, Part 1):1965-1982.

[29] Kummerow, C. D., D. L. Randel, M. Kulie, N. Y. Wang, R. Ferraro, S. J., Munchak, and V. Petkovic. 2015. The evolution of the Goddard profiling algorithm to a fully parametric scheme. *Journal of Atmospheric and Oceanic Technology*32:2265-2280. https://doi.org/10.1175/JTECH-D-15-0039.1.

[30] Kunkee,D. B. ,G. Poe,D. Boucher,S. Swadley,Y. Hong,J. Wessl,and E. Uliana. 2008. Design and evaluation of the first special sensor microwave imager/sounder. *IEEE Transactions on Geoscience and Remote Sensing* 46:863-883.

[31] Kuo,K. ,W. S. Olson,B. T. Johnson,M. Grecu,L. Tian,T. L. Clune,B. H. van Aartsen,A. J. Heymsfield,L. Liao,and R. Meneghini. 2016. The microwave radiative properties of falling snow derived from nonspherical ice particle models. Part I:an extensive database of simulated pristine crystals and aggregate particles,and their scattering properties. *Journal of Applied Meteorology and Climatology* 55:691-708. https://doi.org/10.1175/JAMC-D-15-0130.1.

[32] Liao,L. ,and R. Meneghini. 2005. A study of air/space-borne dual-wavelength radar for estimation of rain profiles. *Advances in Atmospheric Sciences* 22(6):841-851.

[33] Lin,X. ,and A. Y. Hou. 2008. Evaluation of coincident passive microwave rainfall estimates using TRMM PR and ground-based measurements as references. *Journal of Applied Meteorology and Climatology* 47:3170-3187.

[34] Liu,G. 2004. Approximation of single scattering properties of ice and snow particles for high microwave frequencies. *Journal of the Atmospheric Sciences* 61:2441-2456. https://doi.org/10.1175/1520-0469(2004)061<2441:AOSSPO>2.0.CO;2.

[35] Maeda,T. ,Y. Taniguchi,and K. Imaoka. 2016. GCOM-W1 AMSR2 Level 1R product:dataset of brightness temperature modified using the antenna pattern matching technique. *IEEE Transactions on Geoscience and Remote Sensing* 54(2):770-782. https://doi.org/10.1109/TGRS.2015.2465170.

[36] Meneghini,R. ,T. Iguchi,T. Kozu,L. Liao,K. Okamoto,J. A. Jones,and J. Kwiatkowski. 2000. Use of the surface reference technique for path attenuation estimates from the TRMM precipitation radar. *Journal of Applied Meteorology* 39:2053-2070.

[37] Mugnai,A. ,et al. 2007. Snowfall measurements by the proposed European GPM Mission. In *Measuring Precipitation from Space:EURAINSAT and the Future*,ed. by V. Levizzani,P. Bauer,and J. Turk,750 pp. Berlin:Springer.

[38] Nayak,M. A. ,G. Villarini,and A. A. Bradle. 2016. Atmospheric rivers and rainfall during NASA'siowa flood studies(IFloodS)campaign. *Journal of Hydrometeorology* 17:257-271. https://doi.org/10.1175/JHM-D-14-0185.1.

[39] Olson,W. S. ,L. Tian,M. Grecu,K. -S. Kuo,B. T. Johnson,A. J. Heymsfield,A. Bansemer,G. M. Heymsfield,J. R. Wang,and R. Meneghini. 2017. The microwave radiative properties of falling snow derived from nonspherical ice particle models. Part II:initial testing using radar,radiometer and in situ observations. *Journal of Applied Meteorology and Climatology* 55:709-722. https://doi.org/10.1175/JAMC-D-15-0131.1.

[40] Pippitt,J. ,D. B. Wolff,W. A. Petersen,and D. Marks. 2015. Data and operational processing for NASA's GPM ground validation program. In *37th Conference on Radar Meteorology*,Norman,OK. Boston:American Meteorological Society.

[41] Reed,P. M. ,N. W. Chaney,J. D. Herman,M. P. Ferringer,and E. F. Wood. 2015. Internationally coordinated multi-mission planning is now critical to sustain the space-based rainfall observations needed for managing floods globally. *Environmental Research Letters* 10:024010. https://doi.org/10.1088/1748-9326/10/2/024010.

[42] Roca,R. ,H. Brogniez,P. Chambon,O. Chomette,S. Cloché,M. E. Gosset,J. -F. Mahfouf,P. Raberanto,and N.

Viltard. 2015. The Megha-tropiques mission: a review after three years in orbit. *Frontiers in Earth Science* 3: https://doi.org/10.3389/feart.2015.00017.

[43] Ruf, C. S., Y. Hu, and S. T. Brown. 2006. Calibration of windsat polarimetric channels with a vicarious cold reference. *IEEE Transactions on Geoscience and Remote Sensing* 44(3): 470-475.

[44] Schwaller, M. R., and K. R. Morris. 2011. A ground validation network for the global precipitation measurement mission. *Journal of Atmospheric and Oceanic Technology* 28: 301-319. https://doi.org/10.1175/2010JTECHA1403.1.

[45] Skofronick-Jackson, G. M., M. J. Kim, J. A. Weinman, and D. E. Chang. 2004. A physical model to determine snowfall over land by microwave radiometry. *IEEE Transactions on Geoscience and Remote Sensing* 42(5): 1047-1058.

[46] Skofronick-Jackson, G., D. Hudak, W. Petersen, S. W. Nesbitt, V. Chandrasekar, S. Durden, K. J. Gleicher, G. Huang, P. Joe, P. Kollias, K. A. Reed, M. R. Schwaller, R. Stewart, S. Tanelli, A. Tokay, J. R. Wang, and M. Wolde. 2015. Global precipitation measurement cold season precipitation experiment (GCPEX): for measurement's sake, let it snow. *Bulletin of the American Meteorological Society* 96: 1719-1741. https://doi.org/10.1175/BAMS-D-13-00262.1.

[47] Skofronick-Jackson, G., W. A. Petersen, W. Berg, C. Kidd, E. F. Stocker, D. B. Kirschbaum, R. Kakar, S. A. Braun, G. J. Huffman, T. Iguchi, P. E. Kirstetter, C. Kummerow, R. Meneghini, R. Oki, W. S. Olson, Y. N. Takayabu, K. Furukawa, and T. Wilheit. 2017. The global precipitation measurement (GPM) mission for science and society. *Bulletin of the American Meteorological Society*. https://doi.org/10.1175/BAMS-D-15-00306.1.

[48] Stephens, G. L., and C. D. Kummerow. 2007. The remote sensing of clouds and precipitation from space: a review. *Journal of the Atmospheric Sciences* 64: 3742-3765.

[49] Tang, G., Y. Wen, J. Gao, D. Long, Y. Ma, W. Wan, and Y. Hong. 2017. Similarities and differences between three coexisting spaceborne radars in global rainfall and snowfall estimation. *Water Resources Research* 53. https://doi.org/10.1002/2016WR019961.

[50] Wentz, F., and D. Draper. 2016. On-orbit absolute calibration of the global precipitation mission microwave imager. *Journal of Atmospheric and Oceanic Technology* 33: 1393-1412. https://doi.org/10.1175/JTECH-D-15-0212.1.

[51] Zhang, J., K. Howard, C. Langston, B. Kaney, Y. Qi, L. Tang, H. Grams, Y. Wang, S. Cocks, S. Martinaitis, A. Arthur, K. Cooper, and J. Brogden. 2016. Multi-radar multi-sensor (MRMS) quantitative precipitation estimation: initial operating capabilities. *Bulletin of the American Meteorological Society* 97: 621-638. https://doi.org/10.1175/BAMS-D-14-00174.1.

第 8 章　云的主动传感器:来自 CloudSat、CALIPSO 和 EarthCARE 的新视角

Hajime Okamoto and Kaori Sato

8.1 引言

云是大气-海洋耦合模式预测的气候变化演进的最重要贡献者之一[8]。各大环流模式对云的描述差异很大。20 个模式对液态水路径的模拟相差 10 倍以上[18],对冰水路径的模拟差异更大(20 倍)[45]。这可能是由于对云的生成和消散机制认识不足所致。

关于云的卫星遥感,一般有两种类型的传感器:被动和主动。前一种传感器探测来自云层的天然来源,如反射的太阳光或发射的热辐射,并沿着视线观察上述描述的积分能量。因此,原则上很难获得云的高分辨率垂直结构。被动传感器可以提供广阔的水平覆盖范围。云的被动遥感有着悠久的历史,国际卫星云气候学项目(ISCCP)是一个国际卫星项目,使用自 1983 年 7 月 1 日以来提供云特性观测数据的气象卫星上的几个被动传感器[32]。

主动传感器是指发射电磁波并探测目标返回信号的云廓线雷达和激光雷达。与被动遥感相比,主动遥感技术具有优势,能够识别云顶、云底高度和云的多层结构。主动仪器的一个缺点是扫描覆盖面积小,即与被动仪器相比,它们的扫描带更窄。同一目标如云在云雷达和激光雷达波长处的散射特性存在差异。正是由于这些特性,云雷达和激光雷达对粒子的敏感性形成了区分,导致了云的可探测性的差异。

94 GHz(W 波段雷达)的云廓线雷达已经由 Lhermitte 首先发展[17]。选择大气气体在该波长处吸收小的频率,3.16 mm 的波长远短于降水雷达,如 13.8 GHz(2.17 cm)。当考虑与波长相比较小的粒子时,后向散射强度与 λ^{-4} 成正比。即在其他因素相同的情况下,较短的波长(如 3.16 mm)比降水雷达的灵敏度更高。由于除强降水外,衰减一般较小,因此 94 GHz 云雷达能够获得水成物的多层结构信息。

用于云研究的地基偏振激光雷达有着较长的历史,它始于 20 世纪 70 年代的那 10 年。Scotland 等人[40]提出,获得退极化比的能力对于云相态的识别是不可或缺的,自那时以来,利用偏振激光雷达进行了大量的观测,以用于云研究[34]。一般来说,在粒子总体质量相同的情况下,激光雷达探测有效粒径小粒子比大粒子更加敏感,当目标与激光雷达之间的衰减较小

时，激光雷达可以探测到气溶胶和云。

CloudSat 和 CALIPSO 卫星于 2006 年 4 月 28 日发射，自 2006 年 6 月以来一直运行。虽然 CloudSat 最初的寿命是 22 个月，CALIPSO 在轨道上的寿命是 3 年，但它们在 11 年后仍在提供数据。他们位于 NASA 的下午卫星星座，名为 A-Train，这样 CloudSat 雷达和 CALIPSO 激光雷达就能与其他卫星在同一时间观测到同一个地方[41]。它们的轨道与太阳同步，位于赤道，海拔约 705 km，跨越赤道的时间为下午 1:30，周期为 16 天。CloudSat 在太空中携带了第一部 94 GHz 雷达，它测量云层和降水中的雷达反射率因子(Z_e)[42]。云-气溶胶激光雷达和红外探路者卫星观测(CALIPSO)搭载了一种后向散射型激光雷达——正交偏振云-气溶胶激光雷达(CALIOP)，这是空基第一部偏振激光雷达。CALIOP 可以测量 32 nm 和 1064 nm 处的衰减后向散射系数(β_{att})和 532 nm 处的退偏振比[46]。CALIPSO 还携带红外成像辐射计(IIR)和广角相机(WFC)。CloudSat 和 CALIPSO 提供了来自太空的云和降水的内部结构数据。

为了通过云辐射效应和对流的相互作用来评价云对气候系统的作用，需要知道云生成的三维结构和多层结构、云粒子相态、云微物理以及温度、压力和水汽含量等大气条件。为此，引入这些算法来提供上述物理量。在第 8.2 节中，介绍了 CloudSat 和 CALIPSO 的观测数据以及用于云分析的辅助数据。在第 8.3 节中，介绍了用于云研究的一套算法，云掩码、云粒子类型和云微物理分别在 8.3.1、8.3.2 和 8.3.3 节中讨论；介绍了两种散射理论：离散偶极子近似方法(DDA)和物理光学方法，这些都与 CloudSat 雷达和 CALIPSO 激光雷达的观测资料理解有关，用于云微物理反演。第 8.4 节是云掩码、云类型和冰相微物理的分析。总结、其余的问题分析和可能的未来方向都在第 8.5 节的讨论中给出。

8.2 云产品的输入数据

本研究使用了来自 2B GEOPROF 产品(R04 版本)的 CloudSat 数据和来自 CALIPSO 的 1B 级激光雷达(版本 3)数据，以及来自欧洲中期天气预报中心(ECMWF)的大气廓线数据。

自 2006 年 8 月 15 日以来，CloudSat 的星下点偏角已经设置为 0.16°。垂直分辨率为 480 m，实测后向散射返回信号过采样，即在 240 m 处采样。返回脉冲沿天底点轨迹平均约 0.16 s，覆盖范围 1.4 km×1.7 km[21]。CPR 沿轨道每 1.1 km 报告一次云和降水的 Z_e 廓线。CloudSat 的最小可检测信号约为−30 dBZ。

CALIOP 使用掺钕的钇铝石榴石(Nd:YAG)固体激光器，发射 532 和 1064 nm 的脉冲。合成波束直径在地球表面约为 70 m，接收器覆盖区直径在地面约为 90 m[47]。对于 532 nm 和 1064 nm，基本水平分辨率在 8.2 km 以下为 1/3 km，在 8.2 km 以上为 1.0 km。对于 532 nm，垂直分辨率在 8.2 km 上下分别为 60 m 和 30 m。1064 nm 的垂直分辨率在 20.2 km 之下为 60 m。

我们创建了融合数据集，其中来自 CloudSat CPR 和 CALIPSO 激光雷达的观测数据具有相同的水平和垂直分辨率。CALIPSO 在 CloudSat 和 CALIPSO 覆盖区之间沿轨和跨轨距离分别在 0.55 和 0.7 km 以内时进行采样[9]，然后对平行和垂直通道衰减的激光雷达后向散射系数和总后向散射系数求平均，使其水平分辨率和垂直分辨率与 CloudSat 相同。这些数据被称为 KU 融合数据集，其中 KU 表示九州大学。

8.3 研究云宏观和微物理特性的算法

8.3.1 针对 CloudSat 和 CALIPSO 的云掩码方案

有四种云掩码方案。该方案分别针对 CloudSat(C1)、CALIPSO 激光雷达(C2)、CloudSat 和 CALIPSO 同时探测到的云以及 CloudSat 或 CALIPSO 探测到的云。它们最初是由在日本附近的西太平洋和热带西太平洋的舰载观测发展起来的[27,28]。根据定义,C3-云和 C4-云的占比分别为最小和最大。

我们首先描述了考虑信噪比(SNR)和特殊连续性测试的 C1 云检测方案。我们遵循 CloudSat 的标准云掩码算法。初始检测依赖于噪声的平均功率(P_n)及其标准差(σ_n)。每个记录的 P_n 和 σ_n 值是通过平均平流层中的回波功率来估计的[21]。当目标距离库上的回波功率超过 P_n 和 σ_n 之和时,目标库被指定为候选有云库,然后采用一种特殊的连续性检验方法来降低云检测的不确定性。通过检测算法,我们可以推断出水凝物的出现,即云和降水。为了进一步区分云和降水,有必要引入云的经验阈值如 Z_e 应小于−15 dBZ$_e$。由于降水的干扰,单靠雷达信号确定云底边界通常是困难的。另一个难点是 C1 低空云的检测,这是由于约 720 m 以下的地表杂波的影响。

CALIPSO C2 方案[9]与 CALIPSO 激光雷达 2 级垂直特征掩码(VFM)中提供的标准掩码不同[44]。C2 掩码需要两个条件,第一个判据是云的后向散射功率应该超过公式(8.1)中的阈值:

$$P_{\mathrm{obs}}(R,r) > P_{\mathrm{th}}(R) \tag{8.1}$$

式中 $P_{\mathrm{th}}(R)$ 由平均噪声信号及其标准差估计,得到原始 CALIPSO 激光雷达分辨率,由式(8.2)表示:

$$\begin{aligned} P_{th}(R,r) = &\frac{1}{2}(P_{\mathrm{aerosol}}(r)+(P_m(R,r)+P_n+\sigma_n)) - \\ &\frac{1}{2}(P_{\mathrm{aerosol}}(r)+(P_m(R,r)+P_n+\sigma_n))\tanh(R-5) \end{aligned} \tag{8.2}$$

式中 R(km)为云层距地面的高度。P_m 和 β_{mol} 表示分子信号和用 ECMWF 数据估计的分子衰减后向散射系数。这里 P_m 由式(8.3)估算:

$$P_m(R,r) = \beta_{mol}(R)/r^2 \tag{8.3}$$

气溶胶的阈值由星载激光雷达数据确定(式(8.4)):

$$P_{\mathrm{aerosol}}(r) = \beta_{\mathrm{aerosol}}/r^2, \beta_{aerosol} = 10^{-5.25} \tag{8.4}$$

P_{th} 对应的后向散射系数估计为 $\beta_{\mathrm{th}}(R,r) = P_{\mathrm{th}}(R)r^2$。$R$ 小于 4 km 时,系数 β_{th} 接近 β_{aerosol},R 大于 8 km 时,接近 $\beta_{\mathrm{mol}}(R)+(P_n+\sigma_n)r^2$。图 8.1 展现了 β_{th}、β_{mol} 和 $\beta_{\mathrm{att,obs}}$ 的垂直廓线示例。在 7~12 km 处有高云,在 1 km 处有薄的低空云层。图 8.1 中约 17 km 处有一些杂散信号。为了去除噪声引起的杂散信号,我们引入了第二个判据——特殊连续性测试。对于关注的目标像素,我们在水平和垂直方向考虑 5X5 维像素。如果超过 13 个像素满足云掩码的第一个条件,则认为该像素是有云的。事实证明,这对消除虚假信号是有效的。

Hagihara 等[9]确定了 CALIPSO 标准云掩码产品——垂直特征掩码(VFM)存在一个问题，即使用 CALIPSO 版本 2 数据时会产生过多的低层云。Rossow 和 Zhang[33]比较了国际卫星云气候学项目(ISCCP)、CloudSat 和 CALIPSO 探测的两个融合数据集的结果，一个来自 2BGeoprof 激光雷达产品[19]，一个来自 KU-掩码。他们发现，ISCCP 和 2BGeoprof 激光雷达产品(由 CloudSat 2B-Geoprof 产品和 VFM 产品融合而成的数据集)在低层云量上的不一致主要是由于 CALIPSO 的 2BGeoprof 激光雷达产品对低层云量的过高估计造成的。KU-掩码显著降低了这种不一致。Hagihara 等[10]也对 CALIPSO level1 版本 3 数据进行了类似的比较。他们也发现了相同的高估低层云的趋势。

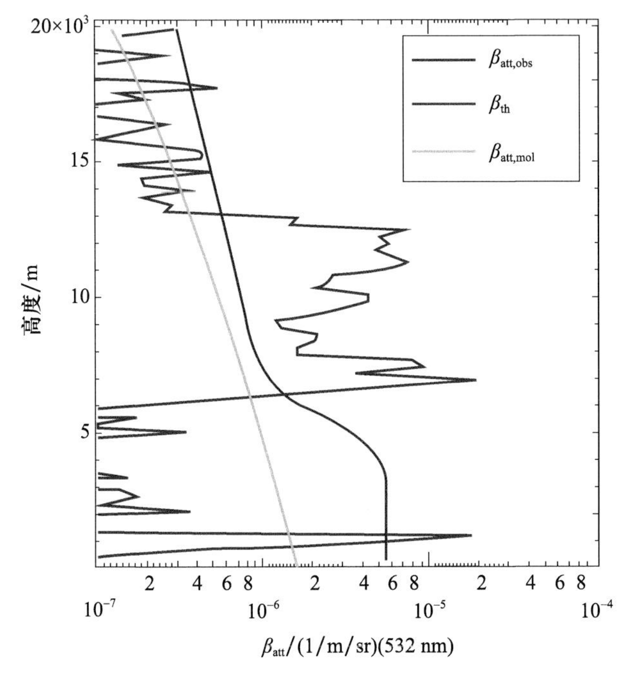

图 8.1 CALIPSO 的 532nm 通道衰减后向散射系数 β_{att} 的廓线(蓝色)，用于 CALIPSO 激光雷达掩码的阈值廓线(红色)和分子造成的衰减后向散射系数的廓线(绿色)

8.3.2 云粒子类型方案

地基激光雷达观测结果显示，水云的退偏振比大多小于 10%左右。这个小值表明多次散射的贡献有限，并且用米氏理论的单次散射解释的退偏振比不会偏离球形粒子的值。相比之下，冰云产生较大的退偏振比(约 40%)。在地基激光雷达观测中，根据观测数据和理论估计，可以用 δ 进行云相态识别[34]。

然而，这种基于 δ 的简单判别在星载激光雷达观测的情况下不起作用。星载激光雷达搭载的仪器在地球表面观测的覆盖范围比传统的地基激光雷达要大得多。由于卫星高度为 700 km，前者的覆盖面直径约为 90 m，视场为 0.13 mrad，后者在 1 km 的高度上的覆盖直径约为 1 m。由于 CALIOP 的覆盖面积大，对于相同的消光廓线，多次散射对星载激光雷达信号的贡献比对地基雷达信号的贡献大得多。由于多次散射的影响，随着光学厚度的增加，CALIOP 测量的水云亮度与冰云亮度相当。结果表明，单凭退偏振比分辨冰云和水云存在很大的困难。

Yoshida 等[49]引入二维图来区分云粒子类型。他们对两个垂直连续的层使用 δ 和衰减后向散射系数(β_{att})之比。这里我们使用普通的对数形式 $\chi(R_i)$如下所示：

$$\chi(R_i)=\log_{10}\left[\frac{\beta_{\mathrm{att}}(R_i)}{\beta_{\mathrm{att}}(R_{i+1})}\right] \tag{8.5}$$

式中 R_i为第 i 层的中心高度。当云微物理参数垂直均匀时，χ 与云层消光成正比。

温度>0℃时，预计水云为主导，根据 CALIPSO 观测结果，此温度范围内的云衰减大，退偏振比中等(约 30%)。当温度低于－40℃时，冰粒子占主导。对于这些低温条件下的云，δ 大于 20%，衰减很小。将这种冰云归为三维冰晶类型，即随机指向的冰粒子。当温度在－5 至－20℃之间时，除了 CALIPSO 观测揭示的三维冰晶特征外，还有一些衰减小 δ 很低的云(<3%)。这种云中的粒子被分类为二维平板状冰晶，即水平指向的平板状冰晶[35]。在相同的温度范围内，可以存在水和冰粒子，类似的特征出现在 $T>0$℃时，液滴同样存在[12,49]。

云粒子类别的精确边界是用三条直线和两条曲线作为 χ 和 δ[%]的函数来确定的(图 8.2)。(1)$\chi\leqslant0.2$ 时 $\delta=10$(图 8.2 中的蓝线)；(2)$0\leqslant\chi\leqslant\sqrt{1.5}$ 时 $\delta=60\chi^2+10$(红色曲线)；(3)$0.2\leqslant\chi\leqslant2$ 时 $\delta=7.5\exp(-4(\chi-0.2)^2)+2.5$(绿色曲线)；(4)$\chi\leqslant1.0228$ 时 $\delta=3$(品红线)；(5)$\chi=0.5$ 时 $7.7326\leqslant\delta\leqslant25$(紫色线)。

三维冰晶的面积由蓝线 1 和红曲线 2 确定。三维冰晶 δ 的数值大于 10%，三维冰与水粒子的分离采用红曲线 2。这种分离保证了大的退极化和小的三维冰晶衰减。

液态粒子类别由红曲线 2、绿曲线 3 和紫线 5 确定。红色和绿色曲线是在假设 CALIOP 观测条件下，进行水云的蒙特卡洛模拟确定[49]。与三维冰晶相反，红色曲线的分离反映星载激光雷达观测水云时相对较大的 δ 和较大的衰减，这个已经在前文说明，根据二维图确定液态类别后，根据温度将水云分为暖水云和过冷水云。

二维平板状冰晶类别在 $\chi\leqslant1.0228$ 时存在于品红线 4 之下，在 $\chi\geqslant1.0228$ 时在绿线 3 之下，值 1.0228 是品红线 4 和绿色曲线 3 的交叉点对应的 χ，即由 $3=7.5\exp(-4(\chi-0.2)^2)+2.5$ 给出。这些条件保证了二维平板状冰晶的较小 δ。虽然实际上没有 χ 的条件，但从 CALIOP 观测到的二维平板状冰晶类别的实际分布表现出相对较小的 χ。

在确定冰或水的类别时有些模糊不清。在液水、三维冰晶和二维平板状冰晶类别的边界线附近有两类未知类型(未知 1 和 2)，此处云粒子类型的清晰判别似乎是困难的。未知 1 作为可能包含三维冰晶和二维平板状冰晶的类型被引入。根据定义，它存在于 $10\%>\delta>3\%$。这种类型由蓝线 1，绿曲线 3 和品红线 4 确定。未知 2 被认为含有液态水或三维冰晶。它的面积由红色曲线 2，绿色曲线 3 和紫色线 5 决定。未知 1 和 2 型的实际数量很少。

值得注意的是，CALIOP 在 2007 年 11 月 27 日之前的指向接近星下点方向(约 0.3°)，之后，激光雷达指向偏离了星下点 3°。三维冰和水的出现频率在两个时期没有显著差异，但在侧视点时期，二维平板状冰晶的出现频率急剧下降。无疑有必要在这个时期引入另一个标准来对二维平板状冰晶进行分类。

对于只有 CloudSat 探测到的云部分，还需要区分云粒子类型。引入经验确定的 Z_e 和温度二维图来推断云粒子类型似乎是有效的。通过对比 CALIOP 对 CloudSat 和 CALIOP 探测到云的粒子分类的结果，完成了对 CloudSat 二维图的开发[16]。

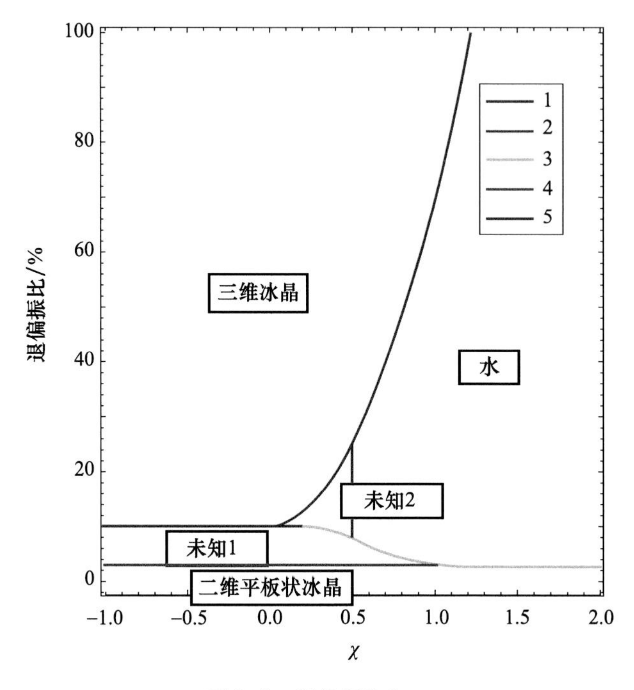

图 8.2 退偏振比与 χ

8.3.3 云微物理的反演

通过云掩码算法和云粒子分类算法分别对云粒子相态进行云检测和识别，即可进行云微物理反演。本小节将介绍利用 CloudSat 和 CALIPSO 反演冰云微物理的方法。为了获得微物理信息，我们首先需要建立 94 GHz 和 0.532 μm 或 1.064 μm 的可观测数据与云微物理的关系。利用米氏理论可以得到均匀球形粒子的单次散射特性。米氏理论的输入参数是使用波长、半径和此波长的复折射指数。当粒子是非球形时，米氏理论可能是不适用的，需要可以处理非球形粒子的散射理论。对于任意形状的粒子，用于表征散射计算的参数增加，出现了随之而来的困难。由于目前已知的解析解仅限于均匀球体、无限圆柱体、椭球体和单体聚合物，而没有任意形状的解析解，因此必须在散射计算中引入一定程度的近似。这也是正确的，我们必须近似粒子的形状，因为不可能处理无限数量的形状变化。因此，选择了两种适用于云雷达和激光雷达波长的散射理论。

还应注意，单次散射近似并不总是足以解释来自那些主动传感器的信号。当光学厚度较大时，会发生多次散射。在这种情况下，雷达或激光雷达方程应视为随时间变化的辐射传输方程。蒙特卡罗方法或一些半解析方法都可以应用。前者只要使用足够数量的光子，一般都是精确的，它还可以处理三维辐射传输方程。它的计算成本非常高。相反，后一种方法在适用性上有一定的局限性，但总体上比前一种方法快得多。

8.3.3.1 云雷达的离散偶极子近似(DDA)

当粒子很小或与波长相当时，离散偶极子近似(DDA)[31]、T 矩阵方法[22]和时域有限差分方法[48]被称为数值方法，这里介绍其中的 DDA。DDA 最初被开发用于星际尘埃粒子的散射。

后来文献[5]的工作扩展了该方法。粒子被许多子体积单元所分割,这些子体积单元的大小被认为小于波长。然后用点电偶极子代替亚体积元。这个粒子可以用 N 个偶极子的集合来近似。偶极间距设为 d,目标物的化学成分用偶极子的极化率 α 表示。α 是由 Clausius-Mossotti 关系或晶格色散关系(LDR)确定的,方法的选择与估计值的精度有关[7],入射电场在第 J 个偶极子 P_j 处产生的偶极矩为:

$$\boldsymbol{P}_j = \alpha_j\left(\boldsymbol{E}_{\mathrm{inc},j} - \sum_{k\neq j}^{N}\boldsymbol{A}_{jk}\boldsymbol{P}_k\right) \tag{8.6}$$

其中,等号右边第二项表示除第 j 个偶极子外 $N-1$ 个偶极子在第 j 个偶极子处产生的电场。此项可由式(8.7)进一步得到。由于充分考虑了各偶极子处电场的延迟效应,该表达式是精确的。

$$\boldsymbol{A}_{jk}\boldsymbol{P}_k = \frac{\exp(ikr_{jk})}{r_{jk}^3}\left\{k^2\boldsymbol{r}_{jk}\times(\boldsymbol{r}_{jk}\times\boldsymbol{P}_k) + \frac{1-ikr_{jk}}{r_{jk}^2}\times\left[r_{jk}^2\boldsymbol{P}_k - 3\boldsymbol{r}_{jk}(\boldsymbol{r}_{jk}\cdot\boldsymbol{P}_k)\right]\right\} \tag{8.7}$$

其中 k 为波数,r_{jk} 为第 j 个偶极子到第 k 个偶极子之间的距离,由 $r_j - r_k$ 估计。式(8.6)包含 $3N$ 个未知参数。由于入射电场和偶极子极化率已知,可以构造 $3N$ 对线性方程组,并有可能求解这些方程组。一旦得到 N 个偶极子的散射场之和,就可以直接确定整个粒子的散射和吸收特性,包括消光和后向散射截面。这就是 DDA 的理论背景。

DDA 解的误差来自:(1)用有限数量的偶极子来近似目标边界和公式。(2)忽略比电偶极子更高阶的误差[24]。对于可以得到解析解的球形粒子和两个接触的球体[6]以及采用收敛解的六边柱和板形粒子[23,24],对 DDA 的精度进行了评估。使用 m、λ 和 d 推导出有效性标准(式(8.8)):

$$2\pi\frac{d}{\lambda}|m| < 0.5 \tag{8.8}$$

由式(8.8)可知,当粒子比波长大时,N 应该足够大才能得到准确的结果。由式(8.8)还可以看出,DDA 适用于介电常数 m 比较小的计算,对于虚部 m 很大的吸波材质,精度会受到影响。还注意到计算应该对每个方向进行。也就是说,随机指向粒子的 DDA 计算成本变得非常高。在满足标准且考虑到粒子随机指向的情况下,可以得到 Z_e 的 DDA 解,误差在百分之几以内[23]。只要满足式(8.8),DDA 的适用性不仅限于均匀粒子,还可以适用于非均匀粒子,如含烟尘颗粒的云粒子。

通常用式(8.9)定义的尺度参数 X 来判定散射理论的适用范围,r_{eq} 是融化后的等效半径。

$$X \equiv \frac{2\pi r_{eq}}{\lambda} \tag{8.9}$$

DDA 可处理的 X 超过 100,并且由于可用计算机内存和速度的增加,X 也在增加[50]。粒径为 2000 μm 的云粒子,对于 94 GHz X 约为 4,这远远低于 DDA 的限制。这里 Z_e 可以用 λ、K 值和雷达后向散射系数 σ_{bk} 表示。

$$Z_e = \frac{\lambda^4}{\pi^5}\frac{1}{|K|^2}\sigma_{bk} \tag{8.10}$$

Z_e 的单位为[$\mathrm{mm}^6/\mathrm{m}^3$],$K = \left|\frac{m^2-1}{m^2+2}\right|$,$m$ 对应 94 GHz 水滴在 10℃处的复折射指数。$|K|^2$ 为 0.75[43]。σ_{bk} 由粒径分布函数 $\mathrm{d}n(r_{eq})/\mathrm{d}r_{eq}$ 估算,且后向散射截面为 $C_{bk}(r_{eq})$[23]。

$$\sigma_{bk} = \int\frac{\mathrm{d}n(r_{eq})}{\mathrm{d}r_{eq}}C_{bk}(r_{eq})\mathrm{d}r_{eq} \tag{8.11}$$

$$C_{bk} \equiv 4\pi \frac{dC_{sca}(\theta)}{d\Omega}\bigg|_{\theta=180°} \tag{8.12}$$

式中 $dC_{sca}(\theta)/d\Omega$ 表示差分散射截面，θ 和 Ω 分别表示散射角和方位角。冰云和水云采用修正 Gamma 分布函数和对数正态分布。

粒径分布函数通常采用改进的伽马分布函数或对数正态分布函数进行参数化。有效半径 r_{eff} 定义为：

$$r_{eff} = \frac{\int r_{eq}^{3} \frac{dn(r_{eq})}{dr_{eq}} dr_{eq}}{\int r_{eq}^{2} \frac{dn(r_{eq})}{dr_{eq}} dr_{eq}} \tag{8.13}$$

通常使用对数形式的 Z_e 来代替 Z_e；$dBZ_e = \log_{10}(Z_e)$。冰云的 Z_e 一般取决于 r_{eff}、冰水含量(IWC)和粒子形状。我们现在考虑水平平面方向的六边形冰柱(二维柱)，水平平面方向的六边形冰片(二维片状)，随机指向的子弹花状和球形冰晶来估计 Z_e(图 8.3)[36]。这里 IWC 设定为 1 g m^{-3}。当粒子粒径较小时，不同形状之间的 Z_e 值差异不大。二维片状冰晶产生的 Z_e 最大并且与球形和子弹花形冰晶产生 Z_e 相似。

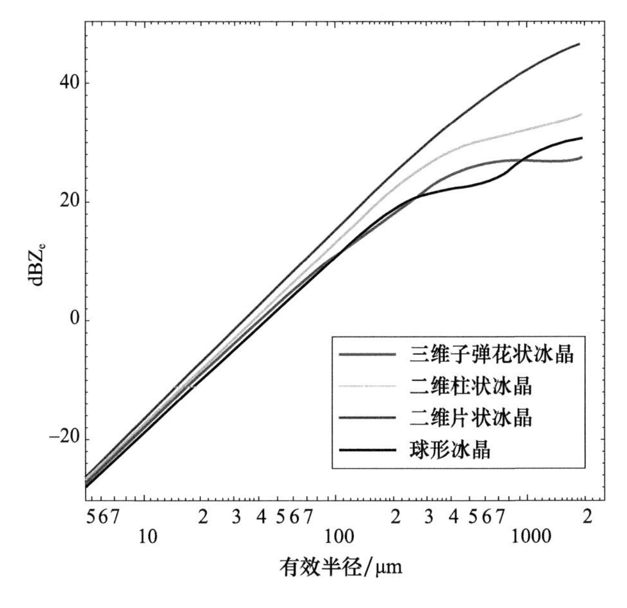

图 8.3 Z_e 对 r_{eff} 的依赖性，并对四种不同的冰粒子模型的 Z_e 进行了比较(IWC 假定为 1 g/m^3)

随着粒径增大，Z_e 总体上增大，不同形状之间的 Z_e 差异增大。这些差异在 $r_{eff}>100$ μm 时变得显著，例如，在 $r_{eff}=200$ μm 和 $r_{eff}=1$ mm 处，分别对应 6 dB 和 16 dB 的差异。综上所述，在 $r_{eff}<100$ μm 时，粒径是确定 Z_e 的关键参数；$r_{eff}>100$ μm 时，非球度起主要作用。

8.3.3.2 用于激光雷达分析的物理光学

应用 DDA 对冰云激光雷达信号进行分析是一个难点，因为波长在 0.532 μm 情况下，X 值在 $r_{eff}=5$ μm 时为 68，在 $r_{eff}=1$ mm 时为 11810。由于在 $X \gg 100$ 时无法通过 DDA 计算，因此需要基于几何光学的散射理论来模拟冰粒子的激光雷达信号。当粒子的粒径远大于波长时,推导出了几何光学。在几何光学中,散射波之间的相互作用被忽略,假设粒子内部点的入射场产生的散射电场表现为独立的电场,入射波被反射和折射,这样的过程在粒子内部重复多次,反射率和透过率根据菲涅耳定律确定,这些量决定了波的一部分留在粒子内,另一部分反射到粒子外。

为了对上述过程进行追踪,基于蒙特卡罗方法的射线追踪方法得到了广泛的应用[20]。用基尔霍夫衍射理论模拟了衍射效应,它取决于粒子的粒径、感兴趣/合理的波长和散射角,且与 m 无关。合成的散射波是由反射和折射产生的电场和由衍射产生的电场之和[3,20]。

这种方法存在一个关键问题。用射线追踪法估计式(8.12)中后向散射截面时,C_{bk} 由下式估算:

$$C_{bk} \cong 4\pi \frac{\Delta C_{sca}(\theta)}{\Delta\Omega}\bigg|_{\Delta\Omega\to 0,\theta\cong 180^\circ} \tag{8.14}$$

其中,计算反方向进入有限立体角 $\Delta\Omega$ 的光子数,得到 C_{bk}。根据定义,当 $\Delta\Omega$ 减少到一半时,只有 $\Delta C_{sca}(\theta)$ 变成一半,才能确定 C_{bk}。但是,将光束追踪法应用于如二维片状的六边形片状冰晶时,C_{bk} 估计值会随着 $\Delta\Omega$ 的减少而增加。因此,确定 C_{bk} 是不可能的[15]。

为了解决这一问题,引入了物理光学。Iwasaki 和 Okamoto[15] 展示了薄矩形的 C_{bk},以及激光束入射方向平行于矩形表面的法线。在这种情况下,可以用基尔霍夫衍射理论得到该形状的解析解[1]。

$$C_{bk} = 4\pi \left|\frac{R_{fre}S}{\lambda}\right|^2 \tag{8.15}$$

式中,R_{fre} 为菲涅尔公式计算的矩形反射率,S 为矩形面积。Borovoi 等[2] 进一步扩展了该方法以处理更真实的冰粒子形状。在他们的方法中,一个粒子只由水平表面组成,并且假设粒子内部的电磁场用几何近似表示,一旦估计了表面上的电场,就可以使用一般多面体粒子的衍射解析表达式来计算从粒子向外辐射的散射波[11]。与简单的光束追踪方法不同,物理光学方法可以为 C_{bk} 提供收敛解。当 $X<100$ 时物理光学法的误差可能会变大。

Intrieri 等[14] 将 35 GHz 雷达和 10 μm 波长的 CO_2 激光雷达相结合,对冰粒子粒径进行了首次分析。他们没有考虑到激光雷达信号的衰减订正。Okamoto 等人[25] 提出了一种使用 94 GHz 雷达和 0.532 μm 激光雷达的正演算法,在[26] 中对地基数据进行了全面的描述、误差分析和分析。94 GHz 和激光雷达信号的衰减用一致性方式来订正。在第一个版本中,没有将退偏振比作为输入参数。然后 Okamoto 等[29] 在他们的算法中引入了退偏振比来分析 CloudSat 和 CALIPSO 信号。反演算法简单描述如下。

本研究考虑两种类型的冰粒子的混合物作为冰模型,即三维冰晶和二维片状冰晶。三维冰晶由 50% 的二维柱状和 50% 的子弹花状冰晶组成。允许改变二维片状冰晶的质量混合比占总 IWC 的比重。首先,我们使用 DDA 创建了三维冰晶和二维片状冰晶在 94 GHz 处的 Z_e 和消光系数的查算表,以及使用物理光学方法创建了在 0.532 μm 处的 β、消光系数和退偏振比的查算表,用于某一固定 IWC 的各种 r_{eff}。

由于当 IWC 不变时,雷达和激光雷达信号对 r_{eff} 依赖性是非常不同的,因此通过结合雷达和激光雷达信号来推导云微物理是可能的。利用 Z_e、β 和退偏振比对关注的网格进行了 r_{eff}、IWC 和二维片状冰晶质量混合比的反演。同时反演 94 GHz 和 0.532 μm 处的消光。当由于激光雷达的强衰减或粒子粒径太小而无法被 CloudSat 探测到而导致 CloudSat 和 CALIPSO 的激光雷达信息仅有其中之一可用时，必须引入一些额外的处理。例如，当激光雷达信号完全衰减，仍能观测到云雷达信号时，利用上层信号估计激光雷达信号，采用 Levenberg Marquardt (列文伯格－马奈尔特)方法对整个云层进行激光雷达-雷达反演[37]。

8.4 CloudSat 和 CALIPSO 的分析：个例研究

在本节将展示一个云掩码、云粒子类型和冰云微物理的个例研究。这里使用的数据是于2007 年 7 月 3 日在北纬 31°至 38°之间采集的。来自 CloudSat 的 Z_e 原始数据如图 8.4a 所示，且前一节描述的云掩码后的结果见图 8.4b，约 8～12 km 处有高云，2 km 以下亦有低云。同样，CALIPSO 探测到的云如图 8.5a 所示，云掩码结果的 β 值如图 8.5b 所示。由气溶胶、微粒和噪声引起的 CALIPSO 信号通过云掩码去除。CloudSat 与 CALIPSO 激光雷达探测到的云面积基本一致，只是在 13 km 左右的地方，CALIPSO 激光雷达探测到的一些薄云没有被 CloudSat 探测到。一些由 CALIPSO 激光雷达在 1 km 左右探测到的低层云，也没有被 CloudSat 探测到。

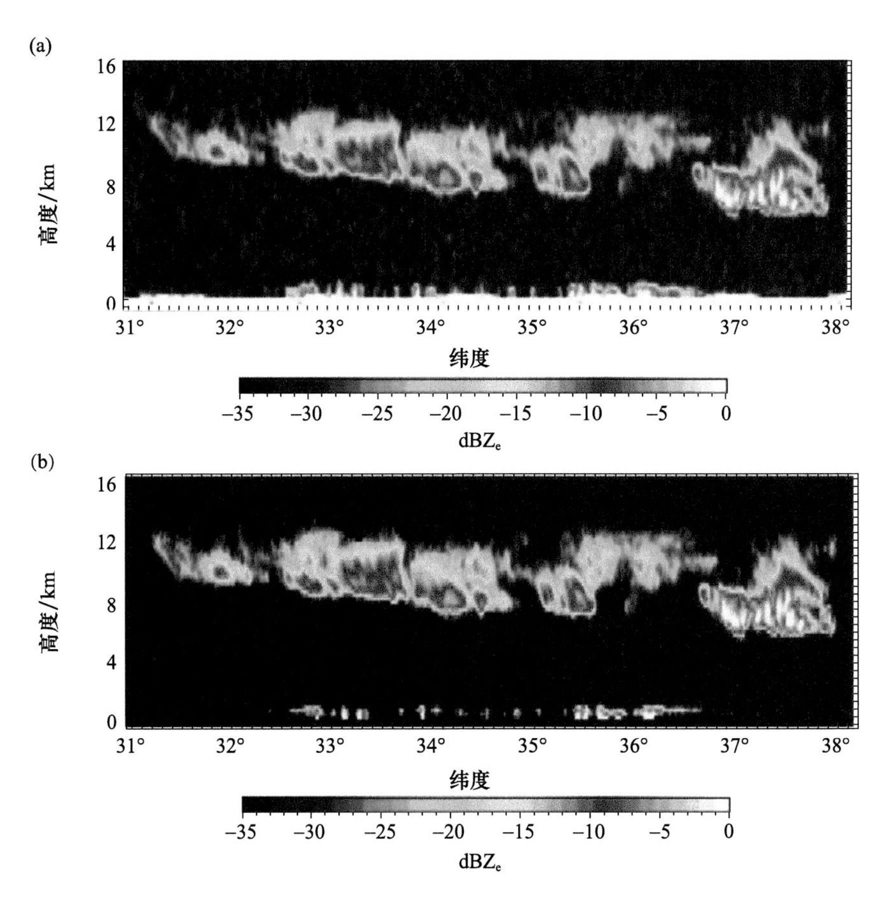

图 8.4 (a)CloudSat 观测到的和(b)CloudSat 应用云掩码后观测到的雷达反射率因子 Z_e 的高度-纬度剖面

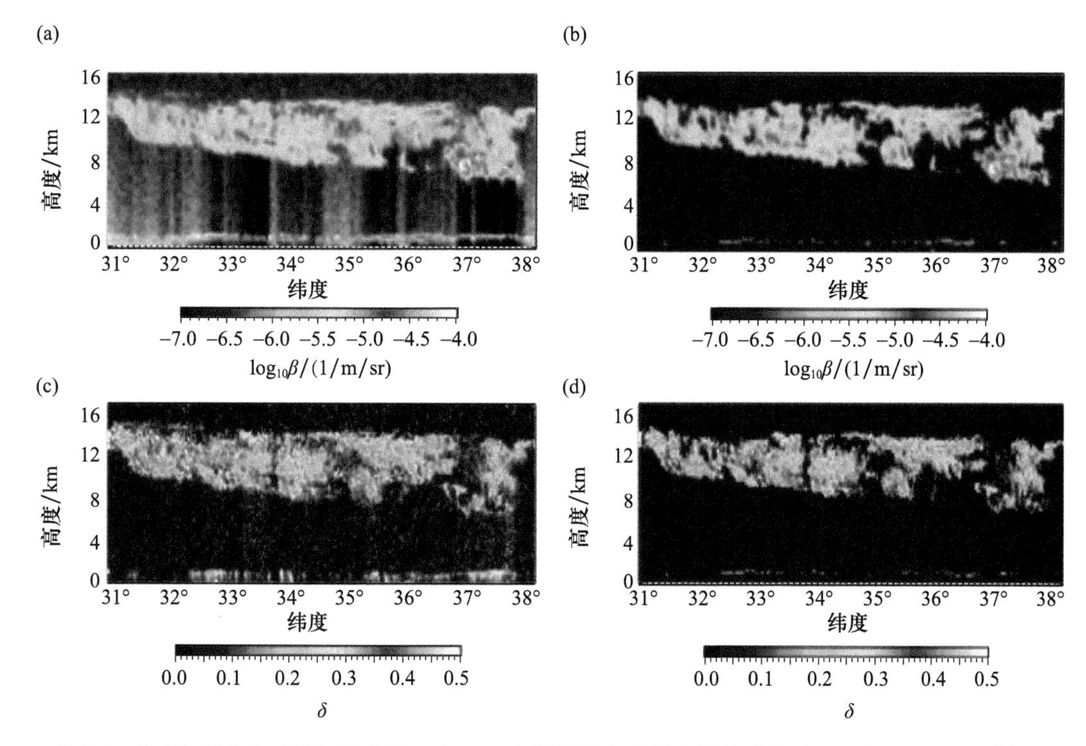

图 8.5　(a)与图 8.4a 相同，但是通过 CALIPSO 探测的衰减后向散射系数 β_{att}；(b)CALIPSO 云掩码；(c)CALIPSO 探测的退偏振比；(d)与(c)相同，但在 CALIPSO 掩码之后

云中粒子类型是针对上述云体推导出来的(图 8.5)。冰云在 8～12 km 处被识别出来。云粒子类型主要是三维冰晶(绿色)，以及少部分二维片状冰晶(蓝色)和未知类型。在 1 km 处发现了暖水云(温度>0℃)(图 8.6)。

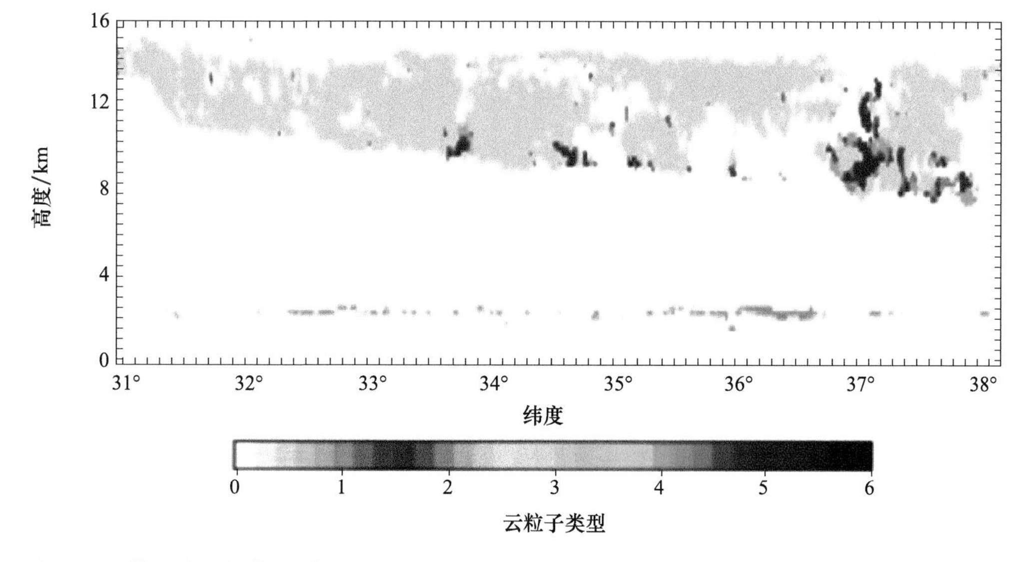

图 8.6　云粒子类型的高度-纬度图(三维冰晶、二维片状冰晶、暖水云分别用绿色、蓝色和红色表示)

选择 CloudSat 和 CALIPSO 探测的云部分作为 C3 掩码数据。然后根据云粒子分类方案的识别结果，将识别出的云分别去掉液态和未知类型。最后进行了冰相微物理的反演。图 8.7a 显示了

r_{eff}的反演结果。随着海拔高度的降低，由于粒子增长和水汽可用率的影响，r_{eff}值有增加的趋势。云顶部分由 11 km 及以上 50～70 μm 的粒子，以及 8～9 km 处约 150 μm 的粒子组成。IWC 的分布与 r_{eff}略有不同，地基雷达-激光雷达分析显示，IWC 最大值的位置比 r_{eff}略高[26]。

Cesana 等[4]通过 CALIPSO 对三个全球云冰和云水分数产品进行了比较，发现三者之间存在较大差异。与 NASA 标准产品和 GOCCP 产品相比，KU 型的冰云和水云分数最小。三种产品的差异是由于对噪声、水平和垂直分辨率、完全衰减像素、云-气溶胶区分处理方法不同。这些差异源于对星载激光雷达信号多重散射的理解不足。同时指出，CALIPSO 目前还没有关于水云微物理的全球产品。

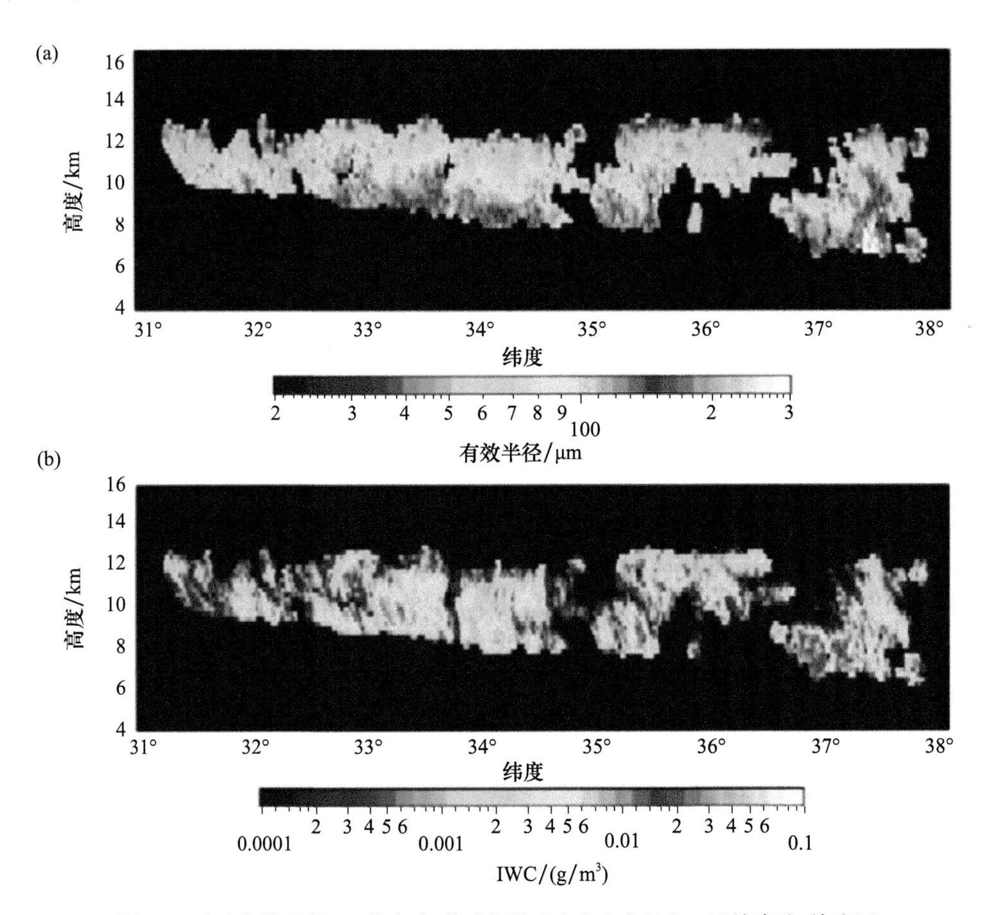

图 8.7 (a)有效半径 r_{eff}的高度-纬度图和(b)冰水含量(IWC)的高度-纬度图

8.5 总结和讨论

本章介绍的算法最初是利用地基云雷达和激光雷达开发和测试。如前几节所述，地基仪器和星载仪器之间存在着本质的区别。例如，在海拔 1 km 处地基激光雷达观测的是半径约为 1 m 的区域与 CALIPSO 激光雷达观测半径(90 m)相比要小得多。这些差异导致我们对在太空中观测到的多重散射理解不足。为了克服传统地基激光雷达的局限性，Okamoto 等[30]开发了一种多视场多散射偏振激光雷达（MFMSPL）系统。该系统被设计用来探测来自光学较厚云层的激光雷达信号，传统的激光雷达无法穿透，它也被设计用来研究类似的激光雷达信号，这些信号受到星载激光雷达（如 CALIPSO）探测的多重散射的严重影响。对水云进行了实际观测，观测到的退偏振比与 CALIPSO 的相似。目前，MFMSPL 被用于评估 CALIPSO 激光雷达数据的算法。

2019 年，JAXA-ESA 合作任务 EarthCARE 将启动。以下四个传感器将被搭载：94 GHz 多普勒云雷达（CPR）、0.355 μm 高光谱分辨率激光雷达（ATLID）、7 通道多光谱成像仪（MSI）和宽带辐射计（BBR）[13]。与 CloudSat 和 CALIPSO 相比，EarthCARE 任务有几个优势。EarthCARE-CPR 提供来自太空的第一个多普勒信息，它的最小灵敏度大约比 CloudSat 雷达（-30 dBZ）低 7 dB（-37 dBZ），ATLID 提供了 CALIPSO 激光雷达未直接观测到的消光信息。因此，EarthCARE-CPR 和 ATLID 可以比 CloudSat 和 CALIPSO 探测到更多的高薄卷云和水云。此外，利用陆基/舰载多普勒云雷达和激光雷达[38,39]反演大气垂直运动和云粒子下落末速度的方法是相似的。这些新性能将大大加快我们对云以及云微物理和大气垂直运动之间关系的认识，从而通过使用新数据集评估云参数化，提高预测气候变化的能力。

致谢　这项工作由 JSPS KAKENHI 资助编号 JP17H06139 和 JP15K17762 支持。CloudSat 和 CloudSat 配置的 ECMWF 数据是从 CloudSat 数据处理中心（http://www.cloudsat.cira.colostate.edu）下载的，CAILIPSO 数据是从大气科学数据中心（https://eosweb.larc.nasa.gov）下载的。

参考文献*

[1] Born, M., and E. Wolf. 1975. *Principles of Optics*. Oxford: Pergamon.

[2] Borovoi, A., A. Konoshonkin, N. Kustova, and H. Okamoto. 2012. Backscattering Mueller matrix for quasi-horizontally oriented ice plates of cirrus clouds: application to CALIPSO signals. *Optics Express* 20: 28222-28233. https://doi.org/10.1364/OE.20.028222.

[3] Cai, Q., and K. Liou. 1982. Polarized light scattering by hexagonal ice crystals: theory. *Applied Optics* 21 (19): 3569-3580.

[4] Cesana, G., et al. 2016. Using in situ airborne measurements to evaluate three cloud phase products derived from CALIPSO. *Journal of Geophysical Research-Atmospheres*. 121. https://10.1002/2015JD024334.

[5] Draine, B. T. 1988. The discrete-dipole approximation and its application to interstellar graphite grains. *Astrophysics Journal* 333: 848-872.

[6] Draine, B. T., and P. J. Flatau. 1994. Discrete-dipole approximation for scattering calculations. *Journal of the Optical Society of America* A11: 1491-1499.

[7] Draine, B. T., and J. Goodman. 1993. Beyond Clausius-Mossotti: wave propagation on a polarizable point lattice and the discrete dipole approximation. *Astrophysics Journal* 405: 685-697.

* 参考文献沿用原版书中内容，未改动

[8] Dufresne,J-L. ,and S. Bony. 2006. An assessment of the primary sources of spread of global warming estimates from coupled atmosphere-ocean models. *Journal of Climate*. https://10. 1175/2008JCLI2239. 1.

[9] Hagihara,Y. ,H. Okamoto,and R. Yoshida. 2010. Development of a combined CloudSat-CALIPSO cloud mask to show global cloud distribution. *Journal of Geophysical Research* 115:D00H33. https://doi:10. 1029/2009JD012344.

[10] Hagihara,Y. ,H. Okamoto,and Z. J. Luo. 2014. Joint analysis of cloud top heights from CloudSat and CALIPSO:new insights into cloud top microphysics. *Journal of Geophysical Research-Atmospheres*119:4087-4106. https://doi:10. 1002/2013JD020919.

[11] Heffels,C. ,D. Heitzmann,E. D. Hirlemann,and B. Scarlett. 1995. Forward light scattering for arbitrary sharp-edged convex crystals in Fraunhofer and anomalous diffraction approximations. *Applied Optics*34(28):6552-6560.

[12] Hirakata,M. ,H. Okamoto,Y. Hagihara,T. Hayasaka,and R. Oki. 2014. Comparison of global and seasonal characteristics of cloud phase and horizontal ice plates derived from CALIPSO with MODIS and ECMWF. *Journal of Atmospheric and Oceanic Technology* 31: 2114-2130. https://10. 1175/JTECH-D-13-00245. 1.

[13] Illingworth,A. J. ,H. W. Barker,A. Beljaars,M. Ceccaldi,H. Chepfer,N. Clerbaux,J. Cole,J. Delanoë,C. Domenech,D. P. Donovan,S. Fukuda,M. Hirakata,R. J. Hogan,A. Huenerbein,P. Kollias,T. Kubota,T. Nakajima,T. Y. Nakajima,T. Nishizawa,Y. Ohno,H. Okamoto,R. Oki,K. Sato,M. Satoh,M. W. Shephard,A. Velázquez-Blázquez,U. Wandinger,T. Wehr,and G. van Zadelhoff. 2015. The EarthCARE satellite:the next step forward in global measurements of clouds,aerosols,precipitation,and radiation. *Bulletin of the American Meteorological Society* 96:1311-1332. https://doi. org/10. 1175/BAMS-D-12-00227. 1.

[14] Intrieri,J. M. ,G. L. Stephens,W. L. Eberhard,and T. Uttal. 1993. A method for determining cirrus cloud particle sizes using lidar and radar backscatter technique. *Journal of Applied Meteorology*32:1074-1082.

[15] Iwasaki,S. ,and H. Okamoto. 2001. Analysis of the enhancement of backscattering by nonspherical particles with flat surfaces. *Applied Optics* 40(33):6121-6129.

[16] Kikuchi,M. ,H. Okamoto,K. Sato,K. Suzuki,G. Cesana,Y. Hagihara,N. Takahasi,T. Hayasaka,and R. Oki. 2017. Development of algorithm for discriminating hydrometeor particle types with a synergistic Use ofCloudSat and CALIPSO. *Journal of Geophysical Research :Atmospheres*122. https://dx. doi. org/10. 1002/2017JD027113.

[17] Lhermitte,R. 1987. A 94-GHz Doppler radar for cloud observations. *Journal of Atmospheric and Oceanic Technology*4:36-48.

[18] Li,J. -L. F. ,D. Waliser,C. Woods,J. Teixeira,J. Bacmeister,J. Chern,B. -W. Shen,A. Tompkins,W. -K. Tao,and M. Köhler. 2008. Comparisons of satellites liquid water estimates with ECMWF and GMAO analyses,20th century IPCC AR4 climate simulations,and GCM simulations. *Geophysical Research Letters* 35:L19710. https://doi:10. 1029/2008GL035427.

[19] Mace,G. G. ,Q. Zhang,M. Vaughan,R. Marchand,G. Stephens,C. Trepte,and D. Winker. 2009. A description of hydrometeor layer occurrence statistics derived from the first year of mergedCloudSat and CALIPSO data. *Journal of Geophysical Research* 114:D00A26. https://doi:10. 1029/2007JD009755.

[20] Macke, A. ,J. Mueller,and E. Raschke. 1996. Single scattering properties of atmospheric ice crystals. *Journal of the Atmospheric Sciences* 53:2815-2825.

[21] Marchand,R. ,G. G. Mace,T. Ackerman,and G. Stephens. 2008. Hydrometeor detection usingCloudsat-an

Earth-orbiting 94-GHz cloud radar. *Journal of Atmospheric and Oceanic Technology*25: 519-533. https://doi:10.1175/2007JTECHA1006.1.

[22] Mishchenko, M. I. 1993. Light scattering by size-shape distributions of randomly oriented axially symmetric particles of a size comparable to a wavelength. *Applied Optics* 32:4652-4666.

[23] Okamoto, H. 2002. Information content of the 95-GHz cloud radar signals: theoretical assessment ofnon-sphericity and error evaluation of the discrete dipole approximation. *Journal of Geophysical Research*107 (D22):4628. https://doi:10.1029/2001JD001386.

[24] Okamoto, H., A. Macke, M. Quante, and E. Raschke. 1995. Modeling of backscattering by non-spherical ice particles for the interpretation of cloud radar signals at 94GHz. An error analysis. *Contributions to Atmospheric Physics* 68(4):319-334.

[25] Okamoto, H., S. Iwasaki, M. Yasui, H. Horie, H. Kuroiwa, and H. Kumagai. 2000. 95-GHz cloud radar and lidar systems: preliminary results of cloud microphysics. *Proceedings of SPIE* 4152:355-363.

[26] Okamoto, H., S. Iwasaki, M. Yasui, H. Horie, H. Kuroiwa, and H. Kumagai. 2003. An algorithm for retrieval of cloud microphysics using 95-GHz cloud radar and lidar. *Journal of Geophysical Research*108 (D7):4226. https://doi:10.1029/2001JD001225.

[27] Okamoto, H., T. Nishizawa, T. Takemura, H. Kumagai, H. Kuroiwa, N. Sugimoto, I. Matsui, A. Shimizu, S. EMori, A. Kamei, and T. Nakajima. 2007. Vertical cloud structure observed from shipborne radar and lidar: midlatitude case study during the MR01/K02 cruise of the research vesselMirai. *Journal of Geophysical Research* 112:D08216. https://doi:10.1029/2006JD007628.

[28] Okamoto, H., T. Nishizawa, T. Takemura, K. Sato, H. Kumagai, Y. Ohno, N. Sugimoto, A. Shimizu, I. Matsui, and T. Nakajima. 2008. Vertical cloud properties in the tropical western Pacific Ocean: validation of the CCSR/NIES/FRCGC GCM by shipborne radar and lidar. *Journal of Geophysical Research*113: D24213. https://doi:10.1029/2008JD009812.

[29] Okamoto, H., K. Sato, and Y. Hagihara. 2010. Global analysis of ice microphysics from CloudSat and CALIPSO: incorporation of specular reflection in lidar signals. *Journal of Geophysical Research*115: D22209. https://doi:10.1029/2009JD013383.

[30] Okamoto H., K. Sato, T. Makino, T. Nishizawa, N. Sugimoto, Y. Jin, and A. Shimizu. 2016. Depolarization ratio of clouds measured by multiple-field of view multiple scattering polarization lidar. *EPJ Web of Conferences* 119:11007. https://doi.org/10.1051/epjconf/201611911007.

[31] Purcell, E. M., and C. R. Pennypacker. 1973. Scattering and absorption of light bynonspherical dielectric grains. *Astrophysics Journal* 186:705-714.

[32] Rossow, W. B., and R. A. Schiffer. 1991. ISCCP cloud data products. *Bulletin of the American Meteorological Society*72(1):2-20. https://doi.org/10.1175/1520-0477(1991)072<0002:ICDP>2.0.CO;2.

[33] Rossow, W. B., and Y. Zhang. 2010. Evaluation of a statistical model of cloud vertical structure using combinedCloudSat and CALIPSO cloud layer profiles. *Journal of Climate* 23:6641-6653. https://doi:10.1175/2010JCLI3734.1.

[34] Sassen, K. 1991. The polarization lidar technique for cloud research: a review and current assessment. *Bulletin of the American Meteorological Society* 72(12):1848-1866.

[35] Sassen, K., and S. Benson. 2001. A midlatitude cirrus cloud climatology from the Facility for Atmospheric Remote Sensing. Part II: Microphysical properties derived from lidar depolarization. *Journal of the Atmospheric Sciences* 58:2103-2112.

[36] Sato,K. ,and H. Okamoto. 2006. Characterization of Ze and LDR ofnonspherical and inhomogeneous ice particles for 95-GHz cloud radar:its implication to microphysical retrievals. *Journal of Geophysical Research*111:D22213. https://doi:10. 1029/2005JD006959.

[37] Sato,K. ,and H. Okamoto. 2011. Refinement of global ice microphysics using spaceborne active sensors. *Journal of Geophysical Research* 116:D20202. https://doi:10. 1029/2011JD015885.

[38] Sato,K. ,H. Okamoto,M. K. Yamamoto,S. Fukao,H. Kumagai,Y. Ohno,H. Horie,and M. Abo. 2009. 95-GHz Doppler radar and lidar synergy for simultaneous ice microphysics and in-cloud vertical air motion retrieval. *Journal of Geophysical Research* 114:D03203. https://doi:10. 1029/2008JD010222.

[39] Sato,K. ,H. Okamoto,T. Takemura,H. Kumagai,and N. Sugimoto. 2010. Characterization of ice cloud properties obtained by shipborne radar/lidar over the tropical western Pacific Ocean for evaluation of an atmospheric general circulation model. *Journal of Geophysical Research* 115:D15203. https://doi:10. 1029/2009JD012944.

[40] Scotland,R. M. ,K. Sassen,and R. J. Stone. 1971. Observations by lidar of linear depolarization ratios by hydrometeors. *Journal of Applied Meteorology* 10:1011-1017.

[41] Stephens,G. L. ,et al. 2002. The CloudSat mission and the A-TRAIN:a new dimension to space-based observations of clouds and precipitation. *Bulletin of the American Meteorological Society*83:1771-1790. https://doi:10. 1175/BAMS-83-12-1771.

[42] Stephens,G. L. ,et al. 2008. CloudSat mission:performance and early science after the first year of operation. *Journal of Geophysical Research* 113:D00A18. https://doi:10. 1029/2008JD009982.

[43] Tanelli,S. ,et al. 2008. CloudSat's cloud profiling radar after two years in orbit:performance,calibration and processing. *IEEE Transactions on Geoscience and Remote Sensing* 46(11):3560-3573. https://10. 1109/TGRS. 2008. 2002030.

[44] Vaughan,M. A. ,D. M. Winker,and K. A. Powell. 2005. CALIOP Algorithm Theoretical Basis Document: Part 2; Feature detection and layer properties algorithms. Release 1. 01,NASA Langley Research Center Doc. PC-SCI-202 Part 2,87 pp. [Available online athttp://calipsovalidation. hamptonu. edu/PC-SCI-202 Part2rev1x01. pdf].

[45] Waliser,D. ,et al. 2009. Cloud ice:a climate model challenge with signs and expectations of progress. *Journal of Geophysical Research* 114:D00A21. https://doi:10. 1029/2008JD010015.

[46] Winker,D. M. ,W. H. Hunt,and M. J. McGill. 2007. Initial performance assessment of CALIOP. *Geophysical Research Letters*34:L19803. https://doi:10. 1029/2007GL030135.

[47] Winker,D. M. ,et al. 2009. Overview of the CALIPSO mission and CALIOP data processing algorithms. *Journal of Atmospheric and Oceanic Technology* 26:2310-2323. https://10. 1175/2009JTECHA1281. 1.

[48] Yang,P. ,and K. Liou. 1995. Finite-difference time domain method for light scattering by small ice crystals in three-dimensional space. *Journal of the Optical Society of America A* 13:2072-2085.

[49] Yoshida,R. ,H. Okamoto,Y. Hagihara,and H. Ishimoto. 2010. Global analysis of cloud phase and ice crystal orientation from Cloud-Aerosol Lidar and Infrared Pathfinder Satellite Observation(CALIPSO)data using attenuated backscattering and depolarization ratio. *Journal of Geophysical Research*. 115: D00H32. https://doi:10. 1029/2009JD012334.

[50] Yurkin,M. A. ,and A. G. Hoekstra. 2007. The discrete dipole approximation:an overview and recent developments. *Journal of Quantitative Spectroscopy and Radiative Transfer* 106:558-589. https://doi: 10. 1016/j. jqsrt. 2007. 01. 034.

第9章　利用微波观测资料的海洋/陆地及晴朗/多云条件下的大气水汽廓线

Filipe Aires

9.1　引言

水汽是地球水循环的一个重要组成部分：它约占大气湿度的99%并且限制了降水总量。作为大气能量的主要来源，水汽强烈影响天气(短时间尺度)和气候(长时间尺度)。它是主要的温室气体，吸收的热量比二氧化碳更多。水汽对云、环流和气候敏感性的作用非常重要，特别是在热带地区[11]，水汽循环及其相关潜热约占热带到两极热输送的50%。此外，水汽是数值天气预报模式的一个重要变量，因此水汽的监测对天气预报质量有很大影响。

大气中水汽的估计可以使用原位地观测约束的数值模式来完成。但是，结果表明，由于卫星数据的全球覆盖范围和很高的观测频率(特别是在结合使用几种仪器时)，它的使用对数值天气预报质量有非常重要的影响。在气象站密度较低的地区，如南半球，情况尤其如此。此外，仅基于卫星观测构建长时间序列是气候学界分析长期趋势和验证全球气候模式的重要工具[53,56]。

红外卫星观测已广泛用于反演大气水汽，如红外大气探测干涉仪(IASI)[4,13,46]或大气红外探测器(AIRS)[49]。然而，这些探测不能穿透云层，因此仅限于晴空场景。在183.311 GHz水汽线附近的卫星被动微波观测可以提供准确的水汽廓线监测，其良好的时间和空间采样率可供业务使用。当云层存在时，它们是红外探测测量的一个很好的补充。此外，红外和微波两种类型设备的协同观测可更好地用于反演大气水汽廓线[1,9,43,49]或近红外反射[34]。

在海洋上空，现在被动微波测量通常被同化进数值模式系统中，它们提供了有趣的大气廓线分析能力。然而，在陆地上，它们并没有被充分利用。首先，地表的微波发射率通常比海洋发射率高得多，其结果是地表对测量信号的贡献要大得多；其次，地表辐射比海洋表面辐射具有更强的空间异质性；此外，由于地表辐射依赖于大量的变量(如土壤湿度、植被和雪的特性)，因此更难以建模。然而，目前正在努力利用、反演和同化陆地上的微波辐射[3,30]，这主要得益于直接从卫星观测计算出来的地表发射率数据库的发展[6,37,38,40,54]。

为了利用微波仪器反演大气水汽，已经开发了各种反演算法。SSM/I(特殊传感器微波仪器)的观测主要在海洋上被使用[10,27,50]。在文献[17]中,建立了针对海洋和陆地表面的算法,Grody 等[24]以及 Jethva 和 Srinivasan[28]使用 AMSU(高级微波探测)在海洋上空的观测来反演水汽;在文献[23,52]中,TRMM(热带降雨测量任务)微波成像仪(TMI)的观测数据被用于热带地区,主要是海洋表面。

常用的反演技术是一维变分法,这是一个迭代过程,使用辐射传输方程及其雅可比矩阵来最小化质量判据(通常是模拟与观测的均方根误差),文献[12]中描述了这种方法。在这项研究中,使用神经网络(NN)方法来代替的几个原因是:它在计算方面是快速的;易于实现和测试;它的灵活性允许测试几种仪器配置;它能够利用初猜信息(如一维变分)[3];它已被证明对许多来自红外或微波观测或两者融合的大气和地表产品有效。这里开发的处理链用于海洋和陆地上空晴朗或多云/无降水的情况。这种神经网络反演是在利用 RTTOV 代码[45]和 ECMWF 分析对大量辐射传输模拟进行训练。它利用地表比辐射率的先验信息来帮助解决地表比辐射率的反演问题[6]。在反演之前,它还依赖于一种创新的方法来校准观测[5]。这是必要的,因为辐射传输模拟和卫星观测之间有系统的差异(两者在反演过程中应是一致的)。一个单独的神经网络用于四种不同的配置:陆地/海洋和晴朗/多云条件。需要降水和云的判别。ECMWF 的温度廓线也被用作先验信息。

本章提出了一种用于卫星微波观测反演大气湿度的处理链。该算法适用于海洋和陆地上的观测数据,适用于晴朗、多云/无降水的情况,适用于包含水汽微波探空仪的三个平台,包括热带云卫星任务(Megha-Tropiques mission),并给出了微波反演结果的理论评价。然后,该算法将在两个平台上进行测试:AQUA 任务(AMSR-E 和 HSB 仪器)和 MetOp(气象业务)平台(带有不同的配套仪器(AMSU-A 和 MHS)),在这两个平台上测试近实时方案。

第 9.2 节介绍了本研究中考虑的三个卫星平台的微波仪器,并根据其主要特点对其探测能力进行了初步评估;第 9.3 节介绍了这项工作中使用的各种数据集;第 9.4 节描述了反演方案;第 9.5 节讨论了理论结果,并与 AQUA 和 MetOp 平台的结果进行了比较;第 9.6 节对水汽(WV)估计值进行了评价;第 9.7 节讨论了卫星观测空间内的数据验证问题;最后,在第 9.8 节给出了结论。

9.2 微波仪器

这里介绍了三个卫星探测任务的水汽探测仪,以及在同一平台上可用的其他微波辐射计。通过加权函数和雅可比矩阵的计算以及对仪器噪声的考虑,对仪器的探测能力进行了初步分析。表 1 总结了这些仪器的主要特性。

9.2.1 AQUA

AQUA 平台的被动微波测量由两个辐射计组成:巴西湿度探测仪(HSB)(只在 2002 年 5 月至 2003 年 2 月期间使用)和先进微波扫描辐射计-地球观测系统(AMSR-E)。HSB 交叉轨道探测仪是一个与 AMSU-B 几乎相同的复制品,有四个测湿通道(而不是 AMSU-B 的五个),其中 3 个通道位于 183.31 GHz(±1.0、±3.0 和±7.0)强水汽吸收线附近,第 4 个通道是位于 150 GHz 的窗口通道[32]。在一次扫描中，HSB 在±49.5°之间采集了 90 个 1.1°的扫描，天底点处的扫描区域直径为 13.5 km。AMSR-E 是一个双偏振辐射计，工作频率分别为 6.9、10.7、18.7、23.8、36.5 和 89 GHz，用于反演包括水汽、云液态水、降水、海面温度、近地面风速和土壤湿度的地面和大气变量。这种仪器有一个圆锥扫描天线，在 1445 km 的波段上以 55°的恒定入射角提供多通道观测。AMSR-E 的空间分辨率在 6.9 GHz 时约为 60 km，在 89 GHz 时约为 5 km[31]。

表 9.1　本研究涉及的三个卫星任务的仪器特性：Megha-Tropiques、AQUA 和 MetOp

平台	热带卫星云			AQUA			MetOp		
仪器	**Saphir**			**HSB**			**MHS**		
扫描方式	跨轨			跨轨			跨轨		
空间分辨率	星下点 10 km			星下点 13.5 km			星下点 16 km		
频率/GHz(左) 带宽/MHz(中) 噪声/K(右)							89.0	2800	0.22
				150.0	4000	1.0	157.0	2800	0.34
	183±0.2	2×200	2.03						
	183±1.1	2×350	1.53	183.311±1	2×500	1.0	183.311±1	2×500	0.51
	183±2.8	2×500	1.37	183.311±3	2×1000	1.0	183.311±3	2×1000	0.40
	183±4.2	2×700	1.25						
	183±6.8	2×1200	1.06	183.311±7	2×2000	1.2	190.311	2200	0.46
	183±11	2×2000	0.99						
仪器	**Madras**			**AMSR-E**			**AMSU-A**		
扫描方式	锥形			锥形			锥形		
空间分辨率	40×67 km(低频) to 6×10 km(高频)			60～5 km			星下点 48 km		
频率/GHz(左) 噪声/K(右)	18.7H		0.46	6.9V		0.3	23.8		0.30
	18.7V		0.53	6.9H		0.3	31.4		0.30
	23.8V		0.48	10.8V		0.6	50.3		0.40
	36.5H		0.44	10.8H		0.6	52.8		0.25
	35.5V		0.49	18.7V		0.6	53.596±0.115		0.25
	89.0H		0.63	18.7H		0.6	54.4		0.25
	89.0V		0.58	23.8V		0.6	54.94		0.25
	157.0H		1.75	23.8H		0.6	55.5		0.25
	157.0V		1.65	36.5V		0.6	57.290344($-f_0$)		0.25
				36.5H		0.6	$f_0\pm0.217$		0.40
				89.0V		1.1	$f_0\pm0.3222\pm0.048$		0.40
				89.0H		1.1	$f_0\pm0.3222\pm0.022$		0.60
							$f_0\pm0.3222\pm0.010$		0.80
							$f_0\pm0.3222\pm0.0045$		1.20
							89.0		0.50

注释：$f_0\pm y\pm z$；f_0 为中心频率(单位为 GHz)。如果 y 出现(以 GHz 为单位)，中心频率不被感知，但在中心频率两侧的两个波段被感知；y 是 f_0 到两个通带中心的距离。如果 z 出现，它是两个通带的宽度(以 GHz 为单位)。

9.2.2 MetOp

MetOp 卫星于 2006 年 10 月 19 日发射，是欧洲第一颗专门用于气象业务的极轨卫星。MetOp 卫星携带两个被动微波探测仪，微波湿度探测仪(MHS)和用于温度探测的 AMSU-A(1 和 2)。

与 DMSP(国防气象卫星计划)和 NOAA(国家海洋和大气局)轨道飞行器上的旧仪器类似，MHS 提供了在 183.31 GHz 水汽吸收线、±1、±3 和±7 GHz 和 190 GHz 以及在 89 和 157 GHz 的两个窗口通道上的测量结果，这些窗口通道可使微波穿透大气层深入到地球表面，每个条带由 90 个连续的独立像素组成，每 2.67 s 扫描一次，MHS 像素在天底点的直径约为 16 km。

AMSU-A 专门用于大气温度廓线的反演，在 50～60 GHz 的 O_2 波段有 12 个探测通道，以及其他 3 个探测通道，分别在 23.8、31.4 和 89 GHz 处。这是一种交叉轨迹扫描辐射计，从天底点扫描±48.3°，每条扫描线共有 30 个 3.3°的地球视场，提供在天底点 48 km 的理论空间分辨率。扫描范围约为 2000 km，仪器在 8 s 内实现一次扫描。MHS 和 AMSU 扫描是同步的。每个 AMSU-A 像素被 3×3 个 MHS 像素覆盖，这有助于它们的协同使用[8]。

9.2.3 热带云卫星

Saphir 是一个被动微波探测仪，在 183.31GHz 吸收线周围有 6 个通道，位于±0.2、±1.1、±2.8、±4.2、±6.8 和±11 GHz(见表 1)。与目前运行的水汽微波仪器相比，它拥有更多的探测通道，理论上可以更好地描述水汽垂直廓线。请注意，它不包括 90 和 150 GHz 附近的窗口通道，与高级微波探测单元 B(AMSU-B)类似，但这些频率在下文描述的 Madras 伴星仪器上可用。Saphir 有一个交叉轨道观测构造，从天底点到±42.96°的每条扫描线有 130 个像素。这意味着在天底点为 10 km 的像素尺度随着扫描角度的增加而增加。可以注意到，Saphir 的仪器噪声比 MHS 更大，但预计 Saphir 中额外的通道将能够更好地描述水汽的垂直分布。

Madras 是一个被动微波成像仪，在垂直和水平偏振探测九个频段的辐射：18.7，23.8，36.5，89 和 157 GHz，除了仅有垂直探测的 23.8 GHz，它具有锥形探测构造：探测的入射角恒定在 53.5°(即约 45°的星载角)，这使它更容易利用偏振信息。像素的大小是恒定的，但扫描几何形状的差异使数据与来自 Saphir 的交叉轨道测量的融合变得复杂。其扫描范围为±65°且刈幅宽度约 1700 km。

9.2.4 探测能力的初步评估

某一仪器通道的加权函数定义为：

$$W(v)=\int\frac{\partial t}{\partial \ln(P)} \tag{9.1}$$

其中 v 为频率，t 为到大气顶部(TOA)的传输信号，P 为气压。积分是在垂直方向上进行的，从大气底部到顶部。分析这个物理量是监控仪器探测特性的标准方法。图 9.1 展示了不同仪器在不同频率下的水汽权重函数，例如晴空条件下的 Saphir(左)、HSB(中)和 MHS(右)(这些数据在有云的情况下会发生明显变化)。这些结果是用标准热带大气计算的。在给定的气压/高度下，某一通道的水汽权重函数，表示在该压力/高度下，当水汽受到扰动时，位于 TOA 的探测亮温的变化。它们在这里以相对湿度的变化表示，且以 1 km 进行归一化（K/km 相对湿度的单位变化）。通道离线中心越远，它探测的大气层结越低。Saphir 在 183.311 GHz 附近的通道比其他仪器多，而且这些通道的带宽明显比其他仪器窄。距离线中心±11 GHz 的通道距离地表较近，但来自地表本身的贡献仍然有限。因此，预计这种仪器将无法很好地描述低层大气。通道的带宽有两个含义：一方面，带宽越窄，权重函数越小（在一定程度上），廓线的垂直分辨率越高；另一方面，仪器噪声与通道带宽的平方根成反比，直接影响反演精度。对于较干燥的大气，权重函数在较低的高度达到峰值（大气中的不透明度较低）。注意，权重函数有很大的重叠，这意味着来自不同通道的信息将不是独立的。这可以通过计算本研究中考虑的三种探测器 HSB、MHS 和 Saphir 的每个通道观测的主成分数量来说明。当进行这种主成分分析时，可以看到三个（四个）主成分占 99%（99.5%）以上的卫星观测数据集的总变率。这两个数字意味着在 6 个 Saphir 测量值中有 4 段独立信息（HSB 和 MHS 仪器的结果是相同的）。

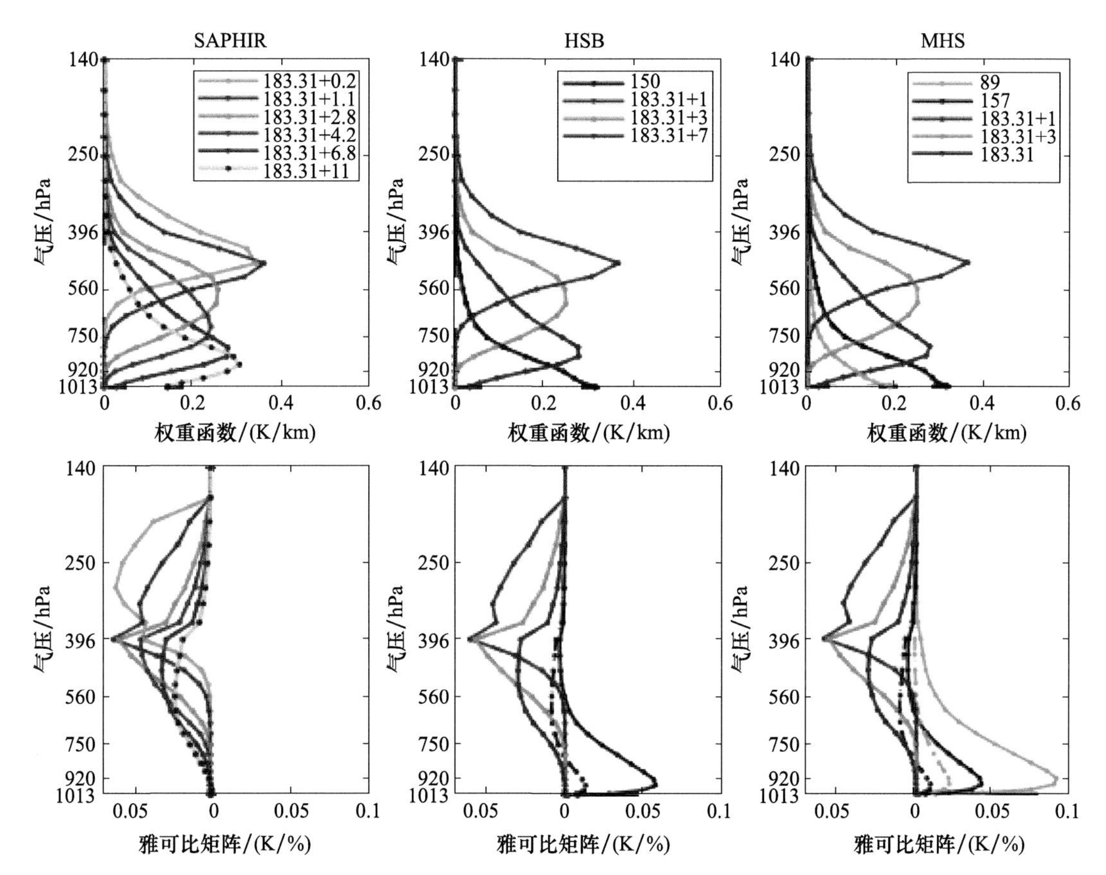

图 9.1 标准热带晴空条件下 Saphir（左列）、HSB（中列）和 MHS（右列）的权重函数（上半部分）和雅可比矩阵（下半部分）。给出了海洋条件（连续线）和陆地表面（虚线）的雅可比矩阵。雅可比矩阵用 K/% 表示，因为水汽用相对湿度（%）表示

雅可比矩阵给出了分析仪器探测能力的另一种方法，即在每个通道上测量的亮温（T_B）相对于相关的地球物理变量（在本例中为相对湿度 R_H）的一阶导数：$\frac{\partial T_B}{\partial R_H}$。实际上，由于用相对湿度表示的水汽无法得到解析雅可比矩阵，我们使用“输入摄动”方法对它们进行估计，其中

R_H被摄动 10%，独立地对每个大气层进行估算，并计算在 T_B 中引起的摄动。雅可比矩阵是观测信息含量的直接度量，用于大气反演。雅可比矩阵与加权函数相关，但对进行反演的大气层结积分。这意味着雅可比矩阵依赖于所选大气的离散化，而不是权重函数。另一个不同之处是，雅可比矩阵可以计算每个大气变量，而权重函数则计算所有变量在同一时间的积分贡献（这当然代表更少的信息）。图 9.1（下部）表示使用 RTTOV 辐射传输模型[45]计算的这种雅可比矩阵。本研究将使用该代码构建卫星观测综合数据集。与辐射计的仪器噪声相比，这些雅可比矩阵可以依据廓线反演来解释：当第一个相对湿度 R_H 的猜测足够接近真实廓线时，每个通道 T_B 的扰动由向量 $\Delta T_B = \frac{\partial T_B}{\partial R_H} \cdot \Delta R_H$ 给出。这将与辐射计的仪器噪声进行比较。这些计算是利用标准热带条件在陆地上空（虚线内发射率为 0.9）和海洋上空（连续线内发射率为 0.6）进行的。如果看不到连续的线，就意味着它与海洋的情况相同。加权函数对地表类型（即发射率）不敏感，反之通道的雅可比矩阵对地表敏感：可以看出，对于 HSB 的 150 GHz 和 MHS 的 89 和 157 GHz，陆地表面的雅可比矩阵比海洋表面的雅可比矩阵要低。因为 Saphir 没有一个对地表真正敏感的通道，所以没有影响（但是在 Madras 的地表敏感通道上可以观察到影响，这里未做展现）。对雅可比矩阵的检查还显示，Saphir 仪器可能会提供更多关于大气上部的信息，因为它的通道位于更靠近线中心的地方。然而，该通道上较大的辐射不确定度可能限制其潜力，仅从 Saphir 只能获得有关低层大气的有限信息。因此，必须将 Saphir 的观测结果与 Madras 的测量结果结合起来，以估计大气低层的水汽含量。这将通过第 5 节的理论反演统计得到证实。

9.3 数据库和辐射传输代码

9.3.1 ECMWF 业务分析

为了实现卫星观测的 RT 模拟建立了一个多种地表和大气条件的数据库。使用由欧洲中期天气预报中心（ECMWF）综合预报系统的 6 h 业务全球分析[51]提供的常规 1.125°格点（以下简称 ANA 数据库）上的大气廓线和地表特性。晴空卫星观测和无线电探空仪数据都被同化在 ANA 数据库中。为了进行准确的 RT 模拟，需要保存以下信息：在 1000 至 1 hPa 的 21 个气压级别上的温度、水汽和臭氧廓线、云廓线（云量、液水和冰水）和地表特性（10 m 水平风、2 m 气压和温度、地表温度、对流和大范围降水以及总云量）。

下文中，仅考虑非降水情况，以限制由于水滴上行辐射的散射而在模拟中可能出现的偏差：在降水场中使用 0.1 mm 的阈值来检测降水。去除了 1000 m 以上的情况，以尽量减少地形的影响。同时，舍弃了距离海岸 100 km 以内的像素，以减少不同视场下不同频率通道之间的不一致性。

为了使 ECMWF 数据与实际卫星观测数据相吻合，采用双线性空间插值方案对分析场进行插值。对于每个卫星观测值，只考虑 ECMWF 格点中具有相同地表类型（海洋或陆地）的相邻区域来计算双线性插值。只有在时差小于一个半小时的情况下，一致性才被接受。

9.3.2 全球陆地表面微波发射率

对于大气廓线，地表敏感的微波观测目前主要用于海洋上空。在陆地上空，地表发射率很难估计：它通常很高，限制了与大气贡献的对比，在空间上变化很大，而且建模复杂。

地表面微波发射率的参数化方法最近得到了发展[39]。对于一年中的每个地点和时间，它提供了从 19 到 100 GHz 的陆地表面微波发射率的真实初猜估计，包括所有扫描条件、入射角和偏振。它基于从 SSM/I[37,38] 计算的 19，37 和 85 GHz 发射率的气候月平均地图。它最初是为 19 到 85 GHz 之间的频率设计的，但测试证明它在 5 GHz 和 190 GHz 之间表现良好[6]。发射率估计值的标称空间分辨率为 0.25°×0.25°。对结果进行了全面评估，均方根(RMS)误差通常在 0.02 以内，只有积雪覆盖区域例外，在那里发射率特征的高时空变异性难以捕捉。基于此参数化的工具已为 EUMETSAT 数值模式卫星应用设施(SAF)开发[6]。该工具已接入 RTTOV 辐射传输代码。

9.3.3 卫星观测数据库

对反演方法的评估将基于 AMSR-E/HSB 和 AMSU-A/MHS 的现有观测结果。此外，为了实现准确的反演，必须对观测结果进行校准，在这里采用了基于学习数据库的创新校准程序。建立了 AMSR-E/HSB 和 AMSU-A/MHS 的组合数据集，可用于校准和评价，用于校准的学习数据库占总数据库的 5%以下。

AMSR-E 与 HSB 覆盖范围的配置过程在文献[5]中描述，它假设两个传感器的覆盖范围是相同的，对 AMSR-E 的观测是从 2A 级数据中提取的，该数据提供了标称分辨率观测到较粗分辨率视场的重采样。在本研究中，选择的分辨率为 21 km。1B 级数据库提供校准和地理定位的 HSB 观测，以及场景信息。该配置方法对每个 AMSR-E 椭圆覆盖范围内的所有 HSB 场景进行平均，接受最大时间差为 70 s，并假设每个 HSB 场景的信息集中在其中心。最终的 AQUA 卫星数据库由两个月的观测数据(2002 年 9 月和 2003 年 1 月)组成，其中包含 16 个配置的亮温，以及来自 MHS 数据库的场景信息(陆地分数)。

由于 AMSU-A 和 MHS 的扫描机制相似(见 9.2.2 节)，两者之间的配置过程要简单得多。最终的 MetOp 卫星数据库由 2 个月的观测数据(2007 年 7 月和 2008 年 1 月)组成，其中包含 20 个同步观测的亮温数据。

9.3.4 RTTOV 辐射传输

本研究提出的反演方案是基于辐射传输模型，所选择的模型是反演方案的一部分，并在迭代反演算法中直接使用。此外，它还被用来创建统计算法中的学习数据库(用于反演和校准)。RTTOV-9.3 辐射传输模型可以模拟 Megha-Tropiques、AQUA 和 MetOp 平台上的微波仪器。该模型最初由 ECMWF[21] 开发，现在得到了 EUMETSAT 数值模式-SAF 的支持[35,45]，它能够对给定大气状态向量的卫星红外和微波辐射计的辐射进行快速模拟。

为了比较实际卫星观测和辐射传输模拟，RTTOV 使用所有分析信息(见 9.3.1 节)和地表发射率数据库(见 9.3.2 节)运行，而在海洋上空，发射率是由 FASTEM-3[14] 地表发射率模型计算。为了保持地表发射率的空间变异性，对 1.125°网格点的 ANA 数据库内插到每个像素点为 0.25°的发射率数据库。基于这些输入值,RTTOV 被用来对每个微波仪器的 T_B 进行模拟。为了进行比较和建立校准学习数据库,本研究计算了与 ANA 数据库和地表发射率数据库一致的每个真实观测值的 RT。

9.4 反演方案

本研究提出的反演方案由识别被处理场景类型的分类程序和两个处理步骤组成:第一步利用统计神经网络(NN)模型对卫星观测数据进行校正;第二步基于另一个神经网络,该神经网络使用校准数据库执行实际反演(图 9.2)。下一节将描述这种方法的选择。

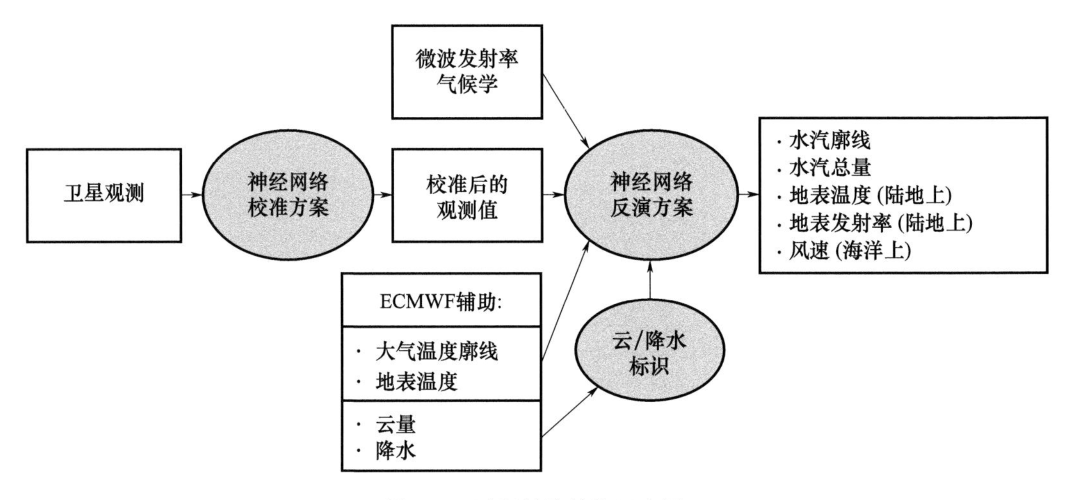

图 9.2 过程链的结构示意图

9.4.1 一般策略

为了训练 NN 用于遥感目的,可以使用两种策略。首先,训练可以使用由真实卫星观测和配置的 WV 廓线组成的学习数据库,例如,来自 ECMWF 的业务分析。这种类型的方案被称为"经验"反演,因为没有使用 RT 模型来解决反演问题。其次,训练也可以使用由 ECMWF 业务分析的相同 WV 廓线及其模拟卫星观测组成的学习数据库,由 RT 模型提供,而不是实际观测。这种类型的反演被称为"物理"反演,因为使用了 RT 模型。第一种方法只涉及真实观测的一种转换:它在一个独特的程序中混合了校准和反演。第二种方法明确涉及两个转换:数据校准和实际反演。由于 RT 模拟(用于训练反演方案)与真实观测数据之间存在系统差异,因此有必要对真实卫星观测数据进行校正。这里首选第二个,因为第一个假设 WV 配置文件(来自 ECMWF)是真实的,但事实并非如此。在这里,我们更喜欢依赖 RT 模型[5]。在这种方法中,没有引入符合性或分辨率误差,而经验方法则需要将 ECMWF 廓线和卫星观测数据重合。

由于各种原因,人们选择了用于水汽反演的神经网络。其中之一是,当大气中水汽含量增加时,权重函数趋于升高。这意味着卫星观测数据与某一海拔高度的水汽含量之间的依赖关系随实际大气状况而变化。神经网络是一种非线性方法,这意味着它的输出/输入关系是依赖于状态的，训练将教会神经网络适应其反演的水汽含量。

9.4.2 先验分类

反演链基于四种不同的配置：(1)海上无云场景：CF/O；(2)陆地无云场景：CF/L；(3)海上多云场景：CL/O；(4)陆地多云场景：CL/L。这里只考虑非降水情况。这意味着该链需要两个分类过程：一个确定条件是多云还是晴空，一个消除降水场景。

在反演方案中研究了三种方法来识别晴空和多云场景：(1)使用来自 ECMWF 业务分析的云量作为先验标志；(2)使用来自地球同步卫星上 VIS/IR 观测的独立分类器；(3)基于微波观测开发的专用云分类器。第三种选择已经在文献[7]中进行了研究：微波云分类器已被设计出来，以确定一个场景是晴空，还是有低、中或高云层覆盖。使用 MSG-Seviri(Meteosat 第三代旋转增强可见和红外成像仪)云分类对该统计方法进行了训练[18,19]。在目前的研究中，选择了 ECMWF 业务分析的云量，尽管我们知道它的局限性。未来将考虑使用微波专用云分类器。每种大气条件都被划分为晴朗(云量低于 2%)或多云(云量超过 2%)，不考虑部分覆盖。

对于降水标志也考虑了各种选择：(1)使用 ECMWF 业务分析的降水估计(即短期预测)；(2)使用文献中的微波降水标志，如 Alishouse 等[10]、Ferraro 等[22]、Weng 等[55]或 Hong 和 Heygster[25]；(3)使用同一卫星平台上其他仪器反演的降水。初步试验表明，ECMWF 的先验降水分析的质量足以达到这一目的。接下来，如果 ECMWF 的降水估算值(对大尺度和对流降水)高于 0.1 mm，则该场景将被删除。这是一个保守的估计，因为 ECMWF 的再分析过度估计了降水，特别是出现了低值。

9.4.3 神经网络校准方案

这里将只简要地说明，因为这个校准方法已经在文献[5]中描述过。针对 AQUA(AMSR-E/HSB)和 MetOp(AMSU-A/MHS)数据开发了该程序。

9.4.3.1 校准模型

神经网络(NN)在遵循某些特定的统计约束情况下，已经被广泛用于执行从一个空间到另一个空间的非线性转换。在此校准过程中，多层感知器(Multi-Layer Perceptron，MLP)模型[26,44]被用作非线性映射模型。它由三层神经元组成：第一层编码输入(即观测到的亮温)，第三层编码输出(即校准的亮温)，第二层称为隐层，用于增加神经网络模型的复杂性。这个隐层中的神经元数量越多，神经网络就越复杂。训练神经网络以再现样本数据库所描述的行为，即由真实观测 T_{Bs} 和相关校准 T_{Bs} 组成的学习数据集。真正的 AMSRE/HSB(分别是 AMSU-A/MHS)观测结果和第 9.3 节中提出的同步 T_B RTTOV 模拟结果构成了这里的学习数据集。神经网络被设计用来“投射”或校准对 RTTOV 模拟空间的真实观测。

四种神经网络对应前面描述的四种情况：NN：CF/O；CF/L；CL/O；CL/ L。这四个学习数据库中的每一个都包含大约 10 万种情况，这在很大程度上足以训练四个独立的神经网络。每一个校准案例都将与不同的反演模型相关联(见 9.4.4 节)。校准程序的目标是减少 RT 误差(例如偏差、可变性范围、结构、饱和度)，但应该记住，学习数据库中仍然存在其他误差来源，校准程序不能通过在真实和模拟卫星 T_{Bs} 之间建立一个完美的桥梁来抑制所有误差。

9.4.3.2 真实与模拟卫星观测的比较

对于四种配置(CF/O、CF/L、CL/O 和 CL/L)以及 AQUA 和 MetOp 平台,模拟 T_{Bs} 与初始真实观测值和校准观测值之间的 RMS 误差差异如图 9.3 所示。所有模拟和校准 T_{Bs} 之间的差异大于这些 T_B 差异的三个标准差的情况都被过滤掉(对应少于 0.5%的情况)。这些情况中的大多数都对应于降水场景,被降水过滤器遗漏了。校准程序从来不会降低统计数据,而且常常将模拟和观测卫星测量之间的差异缩小几度。对于陆地场景,无云和多云情况以及所有仪器的每个通道都是如此。对于海洋场景,对于无云和多云的影响也非常有趣,除了较低的频率(≤10 GHz)在晴空场景的影响接近于零。校准过程不会降低统计数据。请注意,模拟和初始观测之间的很大一部分差异可能是由于 RT 模拟,而不是仪器标校问题,模拟误差可能来自输入参数的误差(例如,ECMWF 的分析、地表发射率)以及 RT 代码的缺陷。ECMWF 分析中的误差可能是由云的错位造成的。这可以在海洋上得到最好的观测结果,在 89 GHz 的多云情况下,强发射信号并不因为使用校准而消失(图 9.3)。

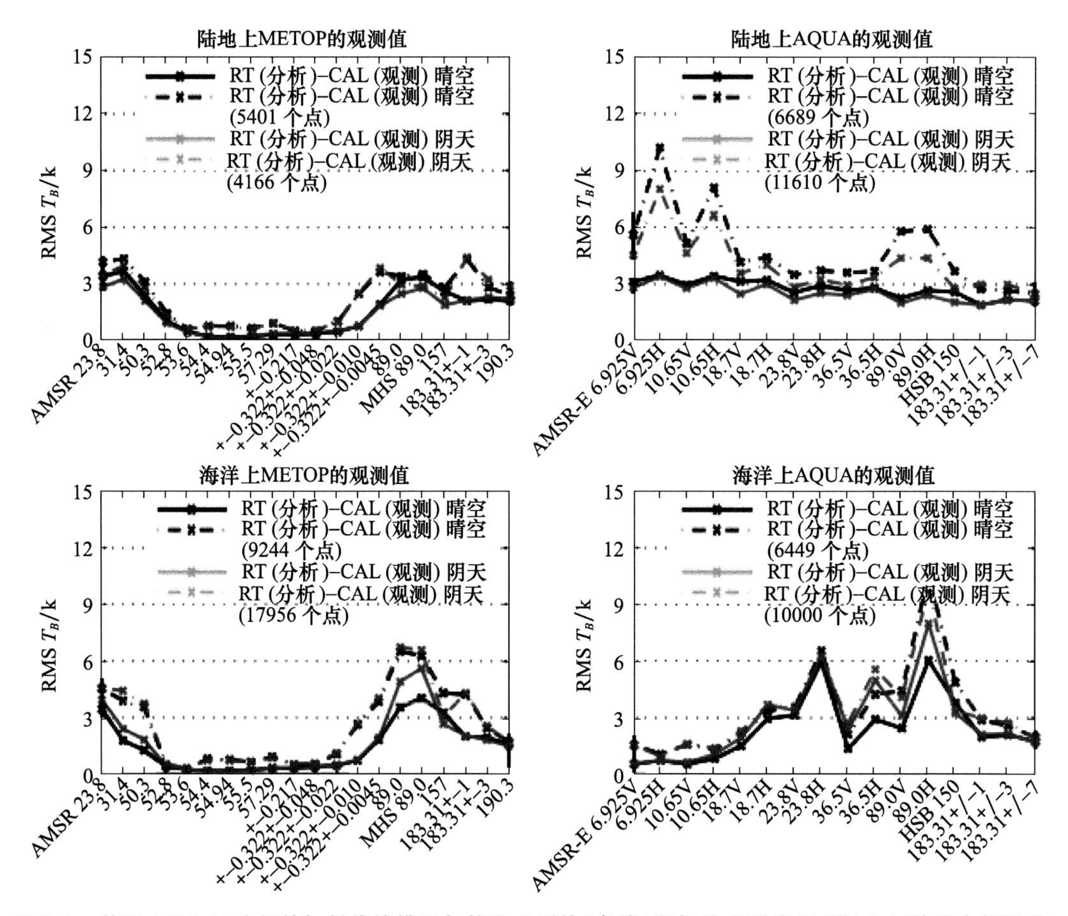

图 9.3 使用 ECMWF 分析的辐射传输模拟与校准观测值(虚线)和初始观测值(实线)之间的均方根误差。考虑不同的情况:陆地(上)和海洋(下),MetOp(左)和 AQUA(右),晴空(黑线)和阴天(灰线)

例如,陆地上低 AMSR-E 频率(在 6 和 10 GHz,尤其是 H 偏振)的初始误差很可能是由于发射率的外推误差(陆地表面发射率来自 19~85 GHz 的卫星观测)。在海洋上空,AMSR-

E 在 H 偏振校准前后的误差可能与辐射传输模型模拟在 H 偏振时特别敏感的海面发射率时的局限性有关。

通过分析其他统计结果(未展示)以检查质量校准:(1)使用校准数据时误差分布较窄;(2)校准真实观测值时相关性仅略有增加,这意味着校准转换对大多数样本在统计上接近线性;(3)误差散点图表明校准过程的质量对所有 T_B 范围都是令人满意的。校准的角度依赖性已经检查过,并且非常有限(未展示):无论扫描角度如何,都采用相同的校准程序。

9.4.4　神经网络反演方案

神经网络(NN)技术在发展遥感应用的高计算效率算法方面非常成功。例如,应用 NN 算法基于 AMSU-A 和-B 观测[2]同时反演海面上 1000～1 hPa 的温度和湿度大气廓线。在这一节中,该神经网络反演算法与神经网络标定方案结合使用。

9.4.4.1　神经网络模型

该反演方案是由一个 MLP 神经网络组成的,与 9.4.3 节中描述的类似。该神经网络模型由三层组成:第一层编码反演方案的输入(即卫星观测和先验信息),第三层代表输出(即反演的产品),隐层用于控制模型的复杂度。如 9.4.3 节所述,对于四种配置(CF/O、CF/L、CL/O 和 CL/L)以及 Megha-Tropiques、AQUA 和 MetOp 平台,需要考虑不同的反演方式。NN 模型的输入为:

(1)Megha-Tropiques、AQUA 和 MetOp 微波仪器在 RT 模拟场中的校准 T_B(第 3.3 节);

(2)ECMWF 业务分析给出的大气温度廓线的先验信息(见 9.3.1 节);

(3)微波发射率气候学的先验地表发射率(见 9.3.2 节)和来自 ECMWF 分析的地表温度。

为了独立于其他相关信息,更好地评估卫星观测的信息内容,在反演中没有直接使用水汽(WV)廓线的先验信息。尽管如此,ECMWF 的业务分析被用来提供一些关于水汽的先验信息,根据水汽总量(TCWV)对情况进行分类:TCWV≤20 kg/m^2;20 kg/m^2＜TCWV≤35 kg/m^2;或 35 kg/m^2＞TCWV。与学习数据库相比,这限制了反演动态范围的缩小,这通常是在统计映射中观察到的。每个类别和前面考虑的各种反演配置的每个 NN 都使用一个专用的 NN。此外,对不同扫描角度(5°、15°、25°、35°、45°和 55°)的反演进行了训练:在特定扫描角度下真实数据的实际反演将对两个扫描角度最接近的 NN 的结果进行线性插值。NN 的总数由四种配置组成:(晴天/阴天和陆地/海洋)×3 个平台×3 个湿度范围×6 个入射角,总共 216 个 NN。

反演方案的中间输出是从地表到大气顶部的 43 个水汽含量的大气廓线和 TCWV,以及陆地的表面温度和发射率、海洋的表面风速。注意,由于 Megha-Tropiques 和 MetOp 不提供对该参数敏感的低频观测数据,因此无法在海洋上反演得到表面温度。我们决定使用相对湿度(在 0%～100%),因为从数值天气预报中心可以获得温度廓线,并且这个变量可以方便地比较干湿大气、低层和上层大气。Megha-Tropiques 数据的 NN 体系结构为:第一层为 15＋43＋15＋1＝74(或 15＋43＝58)个神经元,隐层为 50 个神经元,且 43＋1＋15＋1＝60(或 43＋1＋1＝45)用于地面(或海洋)反演。出于实际原因,WV 反演是在 RTTOV 定义的 43 个大气层结上进行的(特别是为了便于下一段所述的后验检验)。然而,Madras/Saphir、AMSR-E/HSB或AMSU-A/MHS仪器不能提供如此高的垂直分辨率(见9.2.4节),反演方案的最终反演定义在六个较厚的高度层上,由地表、920 hPa、750 hPa、560 hPa、400 hPa、250 hPa和大气顶部划分。从43层到6层的映射是将每个较薄层的水汽含量整合到较厚层。对这些较厚层的数量和界限进行优化,以最小化反演不确定度:考虑43个原始层的所有可能组合为6个较厚层,对每一层的反演结果进行估计,最后选择不确定度最低的6个厚垂直层。在反演统计数据中,所有结果都将在这六个层中显示。

在反演之后,执行后验检验。它利用反演结果和ECMWF的先验信息作为RTTOV模型的输入。将这些模拟的T_{Bs}与观测值进行比较,当差异大于一个固定的阈值(对应于先前执行的统计值的三个标准差)时,结果将被丢弃。这是业务反演方案中的一个标准过程。

9.4.4.2 学习和测试数据集

使用一年的ECMWF业务分析(见9.3.1节)来构建学习数据集。在所有这些地表和大气情况下使用RT模式在计算上要求太高,这也会是低效的,因为在这个数据集中会有冗余,改为使用统一的抽样程序减少了全年的数据。随着数据的精简,只保留了1%的可用点。对于RT模拟,使用第4.2节描述的云标志,所有降水场景都被删除。对Madras/Saphir、AMSR-E/HSB、AMSU-A/MHS仪器进行了RT模拟。

生成的数据集分为学习数据集(代表80%的情况)和一般化数据集(对应剩下的20%)。下一节给出的结果将使用一般化数据集,使用反向传播算法进行训练[44]。

9.5 理论评估

在本节中,使用两种方法评估Megha-Tropiques、AQUA和MetOp平台上微波仪器的信息内容:首先,直接衡量在模拟观测方面反演方案的性能;第二,采用经典的信息内容分析[41,42]。第一种方法提供的不确定性估计通常过于乐观(原因将在下一节给出),信息内容分析也是对现实世界的简化,但它通常被用作比较仪器配置的工具。进行这些理论评价的优点是不需要实际的观测,并且可以进行比较。在接下来的章节中,将使用AQUA和MetOp仪器的真实观测结果对相同的反演方案进行评估。

9.5.1 利用模拟观测的理论反演不确定性

完整的神经网络反演方案已应用于测试数据集(而不是学习数据集),针对三个平台(Megha-Tropiques、AQUA和MetOp),包括四种配置(无云/多云情况、陆地/海洋之上)。图9.4表示反演六层大气廓线时相对湿度的均方根误差(单位为%)。对于任何配置和任何平台(无展示)都没有偏差。正如预期的那样,陆地反演比海洋反演更加困难,特别是较低高度的大气层结,误差增加了两倍(对于高于500 hPa的层,反演误差保持相同)。对于所有个例,在晴朗和多云的条件下,结果是相似的。MetOp反演比AQUA和Megha-Tropiques更精确。这可以解释为与HSB或Saphir相比,MHS通道的仪器噪声更低(表9.1)。它们都包括一个水

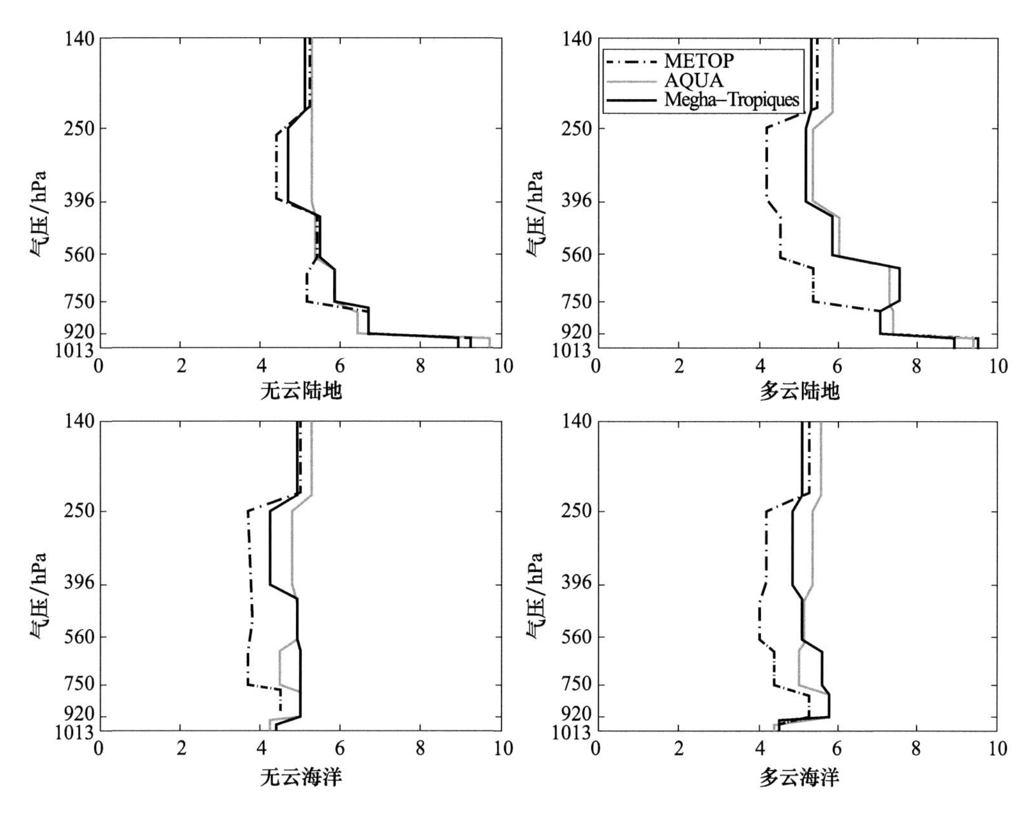

图 9.4　基于模拟资料使用神经网络反演水汽廓线的理论 RMS 误差。单位为相对湿度百分比。提供了 MetOp(灰色点虚线)、AQUA(灰色连续线)和 Megha-Tropiques(黑色)数据的统计,以及四种配置:无云/多云和陆地/海洋的情况

汽探测仪和一个带窗口通道的成像仪。Saphir 仪器的改进与增加通道有关,但它的仪器噪声比 HSB 高,这与图 9.5b 的结果一致。

表 9.2 给出了海洋和陆地上 TCWV(即 TCWV 的直接反演,而不是廓线积分)反演的理论 RMS 误差,这些统计数据是对三个平台的估计。在可能的情况下,将它们与用作反演方案输入的第一猜测的不确定性进行比较,并在神经网络的开发中指定。在海洋上空,晴空和多云情况下的 TCWV 均为 1.6～1.7 kg/m^2,三个平台的精度相近。对于陆地表面,AQUA 反演略好于具有相似特征的 Megha-Tropiques 和 MetOp 反演。

表 9.2　海洋和陆地表面 Megha-Tropiques、AQUA 和 MetOp 平台的 TCWV 理论反演误差

			反演误差		
			MT	AQUA	MetOp
海洋	TCWV/(kg/m^2)	Clr	1.7	1.6	1.6
		Cld	1.7	1.7	1.7
陆地	TCWV/(kg/m^2)	Clr	2.1	1.8	2.0
		Cld	3.3	2.9	3.3

在所有情况下,反演误差都比初猜误差好,这意味着反演确实带来了额外的有用信息。

理论反演误差低估了真实反演的不确定性,因为没有考虑现实世界的情况下可能出现的

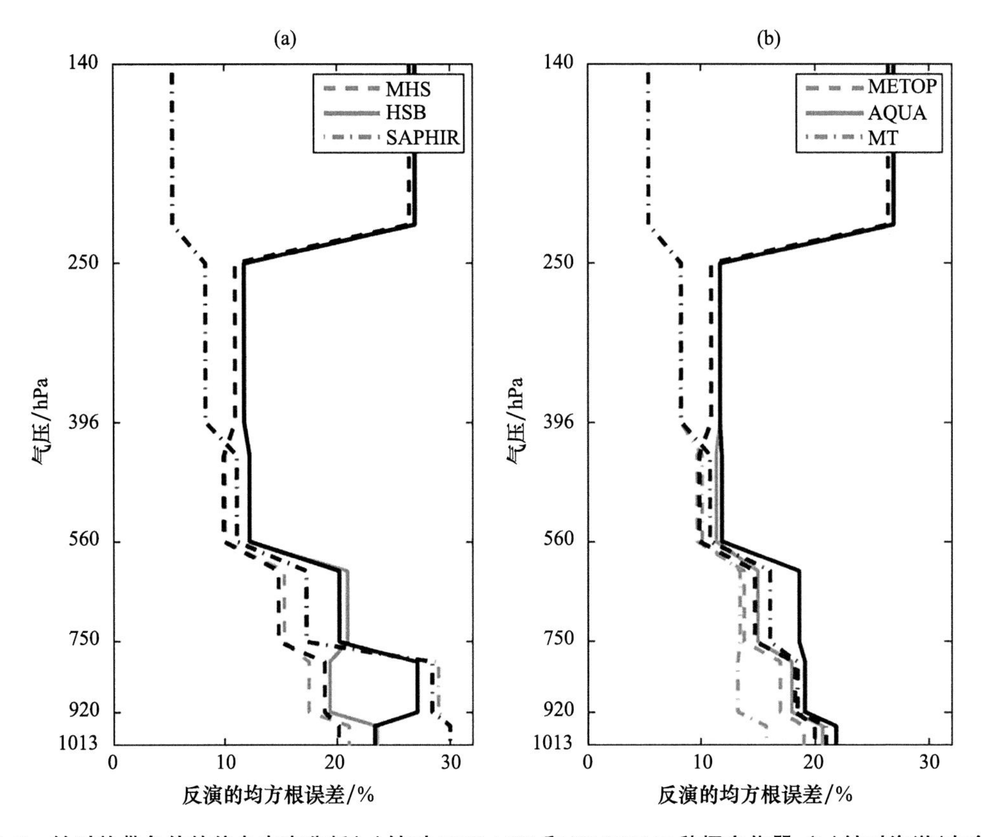

图 9.5　针对热带条件的信息内容分析(a)针对 MHS、HSB 和 SAPHIR 三种探空仪器;(b)针对海洋(灰色)和大陆(黑色)热带情况的 MetOp、AQUA 和 Megha-Tropiques 三个平台的仪器组合

所有不确定性来源:

(1)介绍的两种被考虑的仪器之间或与 ECMWF 业务分析的先验信息之间没有重合误差;

(2)没有使用 RT 误差,因为这是一个很难测量的量;

(3)先验信息(如 ECMWF 分析的温度廓线)的高斯性质是一种简化;

(4)统计是在多样化的大数据集上执行的,但这些大气状况来自 ECMWF 的业务分析,这意味着廓线比真实的廓线更平滑,或者如果分析不正确,它们可能会有偏差;

(5)此外,在本研究中,先验被认为对所有情况有相同的统计数据,这是一种简化。例如,在干燥或潮湿的情况下,水汽的先验不确定性应该是不同的;

(6)总的来说,用于进行这些统计的数据集与用于训练 NN 的数据集是完全一致的,当然,当使用真实世界的观测时,这就不那么准确了。

这些理论不确定性评估是完美反演条件的代表,但任何额外的不确定性来源将使它们偏离真实估计。特别是,校准误差的影响预计将非常显著。关于这些理论不确定性估计的重要一点是,比较仪器配置的反演能力是可能的。

9.5.2　信息内容分析

在经典的信息内容分析[41,42]中,RT 是围绕第一猜测解 f_0 线性化的:

$$(T_{B\varepsilon}-T_{B0})=A\cdot(f-f_0)+\varepsilon, \tag{9.2}$$

其中 f 为待反演的地球物理变量(在本例中为大气 WV 廓线),f_0 为第一猜测,$T_{B\varepsilon}$ 为微波仪器观测到的亮温,T_{B0} 为第一猜测解 f_0 对应的亮温,A 为 RT 模型的雅可比矩阵(第 2.4 节),以及 ε 为仪器噪声(其他不确定性来源也可包含在此项中)。如果假设问题中考虑的变量遵循高斯分布,那么贝叶斯反演可以使用迭代过程:

$$f=f_0+(A^t\cdot S_\varepsilon-1\cdot A+S_f^{-1})^{-1}\cdot A^t\cdot S_\varepsilon^{-1}\cdot(T_{B\varepsilon}-T_{B0}) \tag{9.3}$$

其中 S_ε 为仪器噪声的协方差,S_f 为第一猜测误差的协方差矩阵。迭代从第一猜测解 f_0 开始,用式(9.3)依次更新,直到达到收敛。这是数值天气预报中心使用的一种传统反演方法。

对于所做的任何反演,有一个不确定性估计总是非常重要的。在式(9.3)的反演方法中,反演的不确定性由反演误差协方差矩阵给出:

$$Q=(A^t\cdot S_\varepsilon^{-1}\cdot A+S_f^{-1})^{-1} \tag{9.4}$$

根据仪器噪声信息、考虑通道的 RT 雅可比矩阵和第一猜测提供的先验信息,可以使用这个表达式来衡量反演的质量。

等式(9.4)右侧需要的一个重要信息是 S_f,即第一猜测误差的协方差矩阵。这表示要反演的变量的先验信息,在反演之前。在本试验中,水汽选择了 30%的先验不确定性,与大气层结之间没有相关性。考虑到热带地区水汽的变化范围,这意味着实际上没有考虑先验信息,因此估计的不确定性将只描述来自卫星观测的反演误差。

另一个必要信息是辐射传输的线性化 A,它等于第 2.4 节给出的雅可比矩阵。图 9.1 给出了标准热带情况下水汽探测通道的灵敏度。注意,这里只考虑水汽中的雅可比矩阵,假设其他变量已知。

式(9.4)中的 S_ε 代表表 1 所示的仪器噪声。在这个协方差矩阵中也可以考虑并引入 RT 误差,但它们很难描述,因此在下面忽略了它们。也可以引入校准误差,但为了比较各种仪器配置和评估内在反演方案的能力,已决定不考虑它们。由于我们在这里感兴趣的是比较仪器的探测能力,因此引入这些额外的参数不会改变结论。

图 9.5a(左)为三个探空仪 MHS、HSB 和 Saphir 在海洋(灰线)和陆地(黑线)表面上的反演理论 RMS 误差,R_H 单位为%。陆地/海洋的差异只存在于较低层(气压 600 hPa)。从雅可比矩阵的初步分析和各自的仪器不确定性中可以预期,MHS 提供了大多数大气层结中最精确的大气廓线反演。Saphir 对高于 400 hPa 的气压非常有效,因为它的 183.31 GHz±0.2 通道(见图 9.1):不确定度可以达到 5%,这可能是过于乐观和过度依赖简化假设。相反,Saphir 没有地表敏感通道,其对低层大气的不确定度接近 30%,即无信息(本试验中先验信息的指定不确定度为 30%)。虽然 Saphir 比其他探空仪有更多的通道,但缺乏地表敏感通道和接收机噪声限制了 Saphir 仪器单独使用的能力。然而,一旦结合它们的配套仪器的观测结果(图 9.5b),就能发现三个任务为水汽廓线提供了相似的信息内容,Megha-Tropiques 在更高层(气压>400 hPa)和海洋上空的低层具有探测优势。Madras 仪器为低层大气提供的信息显著改善了大约 600 hPa 以下水汽的反演结果,并证明了在我们的反演链中使用 Saphir/Madras 组合的正确性。

9.6 水汽反演结果评估

在本节中,将通过反演链对水汽进行评价,并将其与 ECMWF 分析和无线电探空仪测量结果进行比较。

首先，应该清楚的是，ECMWF的分析已经是对大气状态的一个很好的估计，它考虑了来自卫星数据(包括AQUA和MetOp观测)和无线电探空仪的大量信息。这些分析可以用作参考，但卫星反演的目标是改进分析，即当将反演到的廓线作为辐射传输(RT)模型的输入时，应该提供比使用分析的模拟更接近卫星观测的模拟。任何背离分析的偏差都被视为反演中的错误或改进之处，考虑到这些因素，有必要在卫星观测空间内验证反演结果，这将在第7节中进一步阐述。

验证卫星水汽测量是一项非常困难的任务。这里进行了与探空仪的相互比较，但只进行了很短的一段时间。这限制了处理探空仪的多重问题(干偏差、代表性、高海拔的不确定性或不同探空仪类型)的可能性。在这里没有使用任何GNSS(GPS和俄罗斯GLONASS)测量，这些测量可以在连续的基础上提供相当精确的总水汽测量值，现在已经成为事实上的参考[20,36,43,47]。我们没有使用ARM(大气辐射测量)计划的地基激光雷达和微波辐射计测量值，这样的相互比较本身就代表了一项研究，我们决定在这里专注于评估反演质量的创新诊断，特别是对 T_B 场。

虽然算法在全球范围内适用，但它们是在热带地区(30°N－30°S)上空经过特别测试的，在那里，水汽对云、环流和气候敏感性的作用是非常重要的。结果将主要使用AQUA反演来呈现。下面评述的统计数字是根据第3.4节所述的2个月卫星观测的系统抽样选择处理的。

9.6.1 柱内水汽总量

柱内水汽总量(TCWV)的反演可以用两种不同的方法进行。首先，可以训练专门的神经网络直接反演TCWV；其次，可以从神经网络反演的水汽廓线中垂直积分得到TCWV。第二种方案在我们的测试中似乎更令人满意(没有呈现)。这个结果可能令人惊讶，是由于直接反演可能被认为更简单。一种可能的解释是，廓线反演在反演过程中使用了更多的神经元，因此该反演方案能够从观测中获得更多的信息。

图9.6显示了2002年9月7日上午，从世界时00时到11时40分，由AQUA观测资料和相应的ECMWF分析得到的热带地区的TCWV图，图中AMSR-E的轨迹与HSB的观测结果匹配在一起。这张图(以及下面的图)显示了下降轨道和上升轨道的反演结果：第一部分定位在地图的西部，在当地夜间被观测到，地图的东部对应于当地白天的上升轨道。缺失的像素对应未处理的数据(降水场景或高仰角)。

反演的TCWV的空间结构与分析非常相似(图9.6)。考虑到反演方案中没有使用关于水汽的先验信息，这是非常令人鼓舞的，即使反演使用了对温度的初步猜测。然而，可以注意到一些差异，例如在印度南部的印度洋，或巴西西部的印度洋。海洋和陆地表面(例如，非洲西部或中美洲)之间有很好的连续性，这证实了陆地和海洋反演方案是稳健的，彼此一致：这是一个关键点，因为到目前为止卫星水汽信息的反演仅限于陆地。由于考虑了地表对微波发射率的限制，在大陆上空的反演是可能的。反演场表现出稍多的空间噪声，但这是可以预期的，因为分析本质上是平滑的。

图9.7(左)散点图给出了分析得到的TCWVs与直接反演结果进行的比较，数据覆盖了海洋和陆地。结果表明是有限偏差(0.39 kg/m^2)，误差的标准差为3.6 kg/m^2，反演略微低估了非常高的TCWV，特别是在陆地上空(回归的斜率略低于1)。这是统计反演方案的常规行为：它们倾向于抑制反演的地球物理变量的变化。请注意，这种变化的减少是有限的，部分原因是三个神经网络反演模型在三个TCWV范围上进行了训练，这是由先验ECMWF分析给出的。

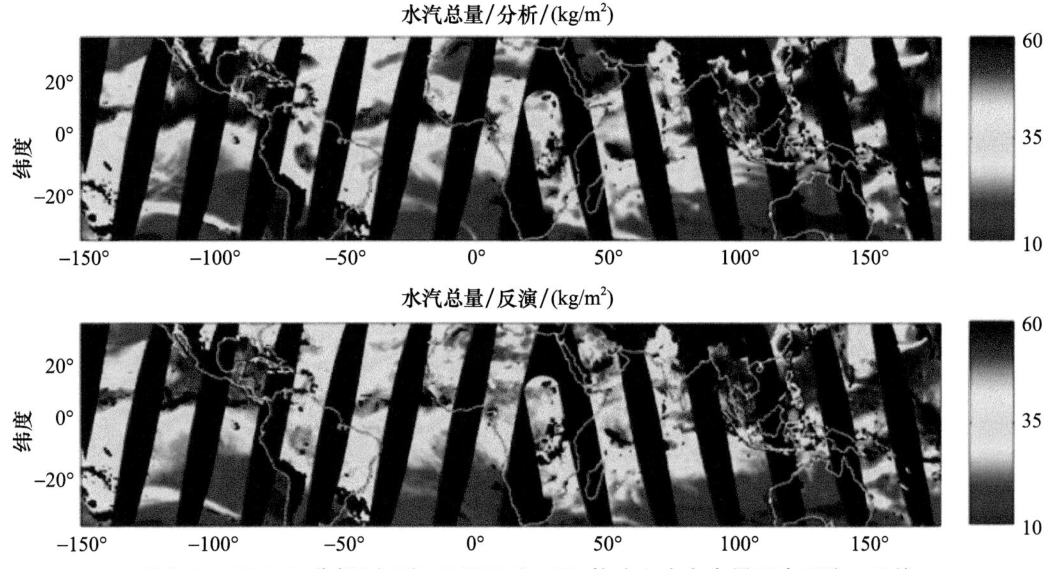

图 9.6　ECMWF 分析(上)和 AMSRE 和 HSB 轨迹上水汽含量同步反演(下)的水汽柱总量示例(2002 年 9 月 7 日)

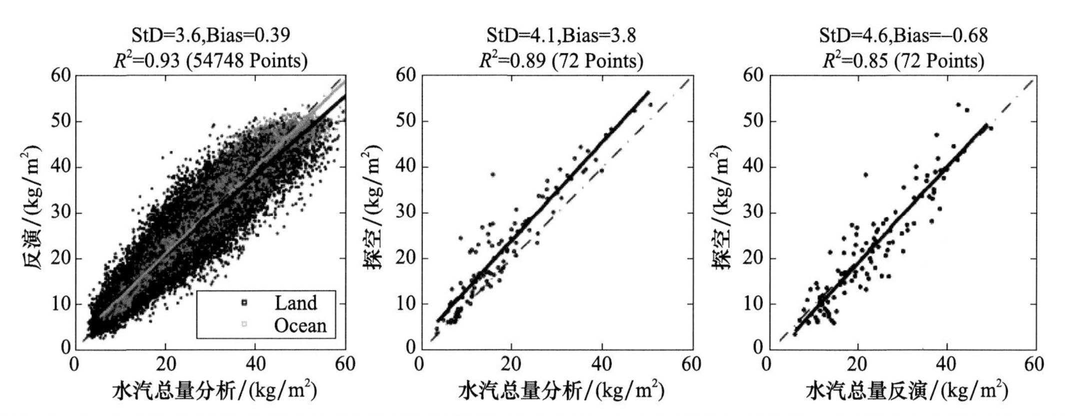

图 9.7　柱内水汽总量散点图:对于陆地(黑色)和海洋(灰色),AQUA 反演与 ECMWF 分析(左),探空仪与分析(中),探空仪与反演(右)。图上提供了标准差(StD)、偏差和相关系数 R^2,并为不同地表绘制了简单的线性回归拟合。与无线电探空仪不同,分析和反演是在海洋和陆地上进行的。虚线表示二者一致。对于左边的图表(与分析的对比),总体 StD=3.6 kg/m² 可以拆分为:海洋个例为 3.1(RMS=3.12)、陆地个例为 4.1(RMS=4.12)

为了进一步检查反演质量,从 ERA40/ERA-Interim 再分析同化过程中使用的 ECMWF 业务探测数据档案中提取了 2002 年 9 月和 2003 年 1 月的无线电探空仪数据[51],对温度和湿度测量值进行了质量控制,以丢弃不完整的廓线(温度的阈值为 30 hPa,湿度为 350 hPa),并使用气候学垂直外推应用到大气顶部。无线电探空仪与最近的 AMSR-E 观测值之间的最大距离必须是±0.3°,最大时差为±90 mn。丢弃了带有散射特征的 AMSR-E 场景,发现匹配的 72 根(廓线)符合上述一致性标准,对于夜间和白天的观测,它们在无云和多云的情况下几乎平均分布。无线电探空仪的测量结果只来自大陆地区或小岛,因此图 9.7 中右边的两个子图可能偏向于陆地情况或沿海地区。对于每个子图,误差的统计结果包括标准偏差、偏差和相关系数 R^2 一并进行了提供。

这些结果表明，AQUA 反演和分析在与探空仪进行比较时的结果一样好：两者的标准差 StD 相关系数 R^2 都在相同的范围内（StD 为 4.1 和 4.6 kg/m^2，R^2 为 85%～89%）。反演的偏差小于分析的偏差，考虑到分析中同化了探空仪观测结果，这是令人惊讶的。与探空仪差值的 StD 与从分析中反演的偏离程度相当，加强了与分析比较的意义。在夜间和无云情况的反演中观察到 2.1 kg/m^2 的负偏差，而其他情况的偏差低于 0.5 kg/m^2。均方根误差（RMS）与观测时间或云量没有很强的联系。目前，只有在非降水条件下，才能对海洋区域的 TCWV 进行准确估计。为了进行比较，Deblonde 和 Wagneur[16] 获得了 SSM/I TCWV 估计值与小岛屿无线电探空仪观测值之间的偏差为－0.2 kg/m^2、标准差为 3.7 kg/m^2。反演结果与 ECMWF 在海洋上的分析结果吻合较好。在海洋上，我们的结果给出了 0.3 kg/m^2 的偏差和 2.8 kg/m^2 的均方根误差。注意，2002－2003 年，RS80 探测仪由于众所周知的湿度偏差原因频繁施放，在数值模式中心已经测试了几种校正方法来纠正这种偏差，但这是一项艰巨的任务，我们的研究结果可能会随着其他年份使用其他探空仪而改变。

在之前的类似条件下，SSM/I 反演与数值模式分析之间的对比结果显示出较大的差异，偏差为 1.08 kg/m^2、均方根误差为 3.25 kg/m^2[15]。Aires 等[3] 利用 SSM/I 观测数据开发了类似的反演方案，在陆地上晴空个例为 3.8 kg/m^2、多云案例为 4.9 kg/m^2。在文献[52]中，为 TMI 观测建立了半统计反演方案。与探空仪相比，均方根误差为 3.5 kg/m^2（当统计数据中剔除 5%最糟糕的案例时为 2.5 kg/m^2），但这一统计是针对海洋案例进行的，且仅针对热带地区。

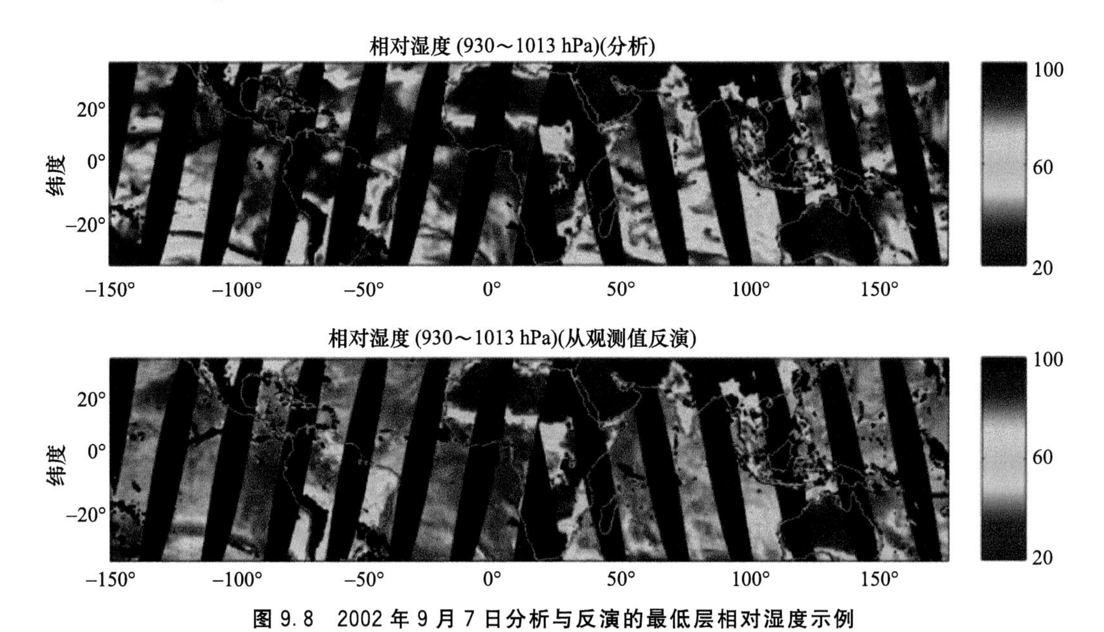

图 9.8　2002 年 9 月 7 日分析与反演的最低层相对湿度示例

（ECMWF 分析（上图），AMSR-E 和 HSB 轨迹上同步的 AQUA 反演（下图））

9.6.2　大气水汽廓线

利用 AMSU-B、MHS、HSB 或 Saphir 等仪器反演大气水汽廓线 WV 是一个真正的挑战。尽管它们的权重函数（图 9.1）在垂直柱中采样相对较好，但它们重叠明显，这意味着自由度或独立信息块的数量相当有限（约 3～4）。当这些卫星观测数据在数值模式方案中被同化时，来自大气环流模式的先验信息有助于约束反演的垂直结构。在没有先验垂直信息的情况下，在直接反演中估计垂直结构是非常困难的。

与图 9.6 类似，图 9.8 给出了从分析和反演得到的 920～1013 hPa 的相对湿度。请注意，由于地表的影响，这一层尤其难以反演，特别是在陆地上。对比图显示，反演得到了西太平洋急流的主要结构，从海洋到陆地的过渡是平稳的，即使是较低的这一层。即使在陆地之上 WV 层的中尺度结构也可以被正确估计，例如在苏丹、非洲或巴西东部。这特别令人鼓舞，因为在大陆上空使用微波观测很困难，特别是对流层低层的地表敏感通道[6]。同时这是意料之中的，因为在算法中对地表发射率进行了很好的约束。然而，甚至在海洋上，可以发现一些明显的差异，例如在南太平洋和印度洋，分析中出现的 WV 结构没有被反演重现。

图 9.9 给出了反演结果与 ECMWF 分析差异的 RMS 廓线。这些“偏差”并不是错误，因为不能判断哪一个更接近实际，这种表述是该评估的挑战之一。在海洋上的反演结果似乎比在陆地上的反演结果与分析结果更接近。这可能与这样一个事实有关，即分析同化了更多来自海洋微波传感器的观测结果(无论是晴天还是多云情况)，提供了更好的一致性。偏差似乎总体上接近理论不确定性估计值(即相对湿度百分比的 RMS 误差在 5%～10%，见图 9.4)。对于海洋案例，理论估计值高于偏差值。此外，偏差有一个平坦的垂直误差结构，理论误差在对流层低层较大、在高层较小。

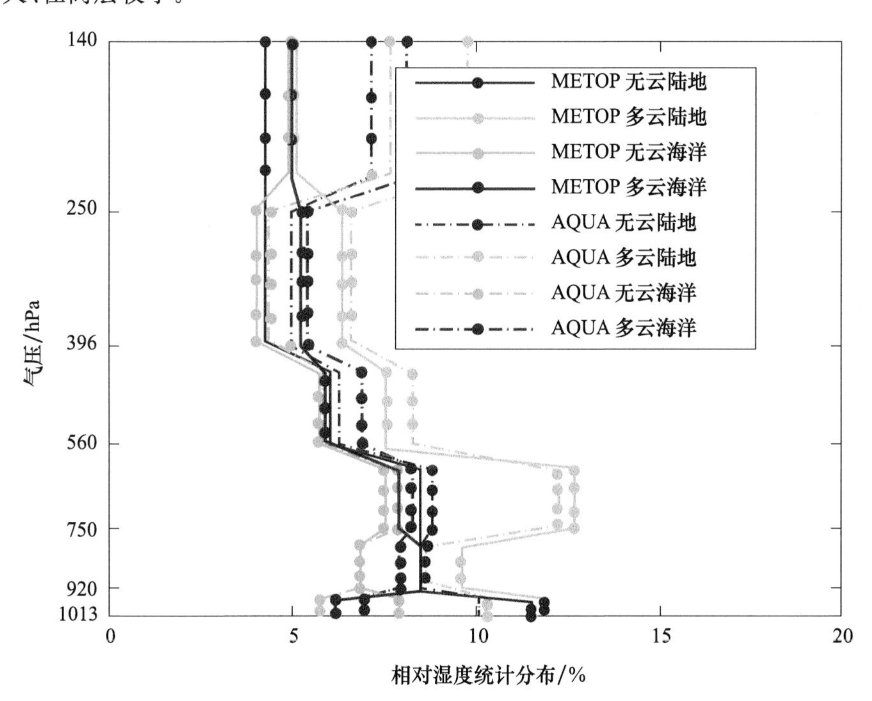

图 9.9 ECMWF 分析与从 AQUA(点状)和 MetOp(连续线)观测反演结果的相对湿度 RMS 偏差：具体偏差：无云/陆地(红色)、多云/陆地(绿色)、无云/海洋(青色)和多云/海洋(蓝色)

对于陆地上和海洋上的估计值，陆地上较低层的误差更高，但理论误差(约 12%)大于偏差(小于 10%)。在陆地上，微波观测不能轻易地被低层大气同化分析，因为难以处理地表贡献。它会导致对 RH 廓线的描述不太准确。这种情况在从真实观测到分析的信息不太可靠的情况下是最糟糕的。在这种配置中，分析不会完全符合实际情况，因此不应该作为反演的目标。

还应明确,这些统计数据取决于时空一致性处理时使用的阈值。为了保持足够的数据样本,在这些统计中允许正负 1.5 h 的时间差。如果用 10 min 来代替,RMS 廓线可能会减少一到两个百分点。RMS 廓线只到 250 hPa,因为卫星提供的超过该值的信息较少(见图 9.1 中的权重函数)。正如预期的那样,在海洋上较低层的反演结果比在陆地上更好。总的来说,MetOp 和 AQUA 反演的统计结果相似。通常,由于 AMSR-E 成像仪提供了更多的窗口通道,AQUA 平台在接近地面时性能更好。相反的情况通常发生在较高层,可能是由于 MetOp 上搭载的 MHS 比 AQUA 上的 HSB 仪器噪声更低。总体统计数据与分析的偏差小于 10%,令人满意。对于低于 49°的角度,反演统计数据是稳定的(统计数据是 49°以内的所有角度的积分,更高的角度被排除)。

另一种诊断方法是比较水汽廓线结果的变化。图 9.10 显示了来自 ECMWF 分析和使用 AQUA 与 MetOp 反演 WV 廓线的统计可变特征。通过对两套微波仪器的神经网络反演,分析得到了 90%以上的变化特性,垂直结构的可变性被很好地被再现,与 WV 廓线分析结果相比,反演结果的变化略低。另一个原因是反演得到的廓线水平分辨率比 ECMWF 分析更高,而 ECMWF 分析的数据更平滑,分辨率为 1.125°。训练数据集的粗分辨率如何影响 WV 变率需要在未来进行研究(参见 Kahn 等[29],了解水汽变化的尺度范围缩放)。

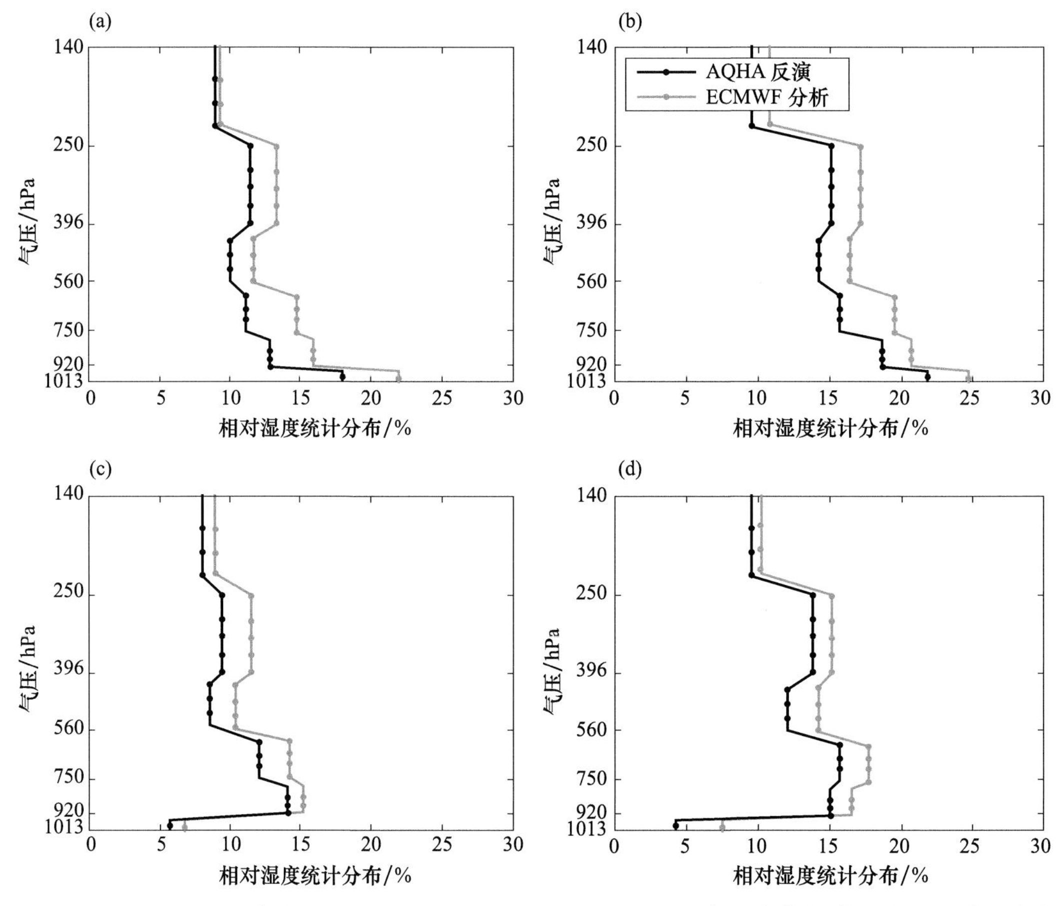

图 9.10 ECMWF 分析(灰线)和从 AQUA(黑线)观测反演的相对湿度廓线的统计“离散度”:(a)晴天陆地(6689 点);(b)多云陆地(11610 点);(c)晴天海洋(6449 点);(d)多云海洋(30000 点)。包括四种不同情况(晴天/多云和陆地/海洋。这些统计数据来自两个月的样本(2002 年 9 月和 2003 年 1 月)

在图 9.11 中，也将反演的水汽廓线与无线电探空仪进行了比较，得到了与图 9.7 中相同的原位测量结果。将相应的 ECMWF 分析添加到比较中。对于 560 hPa 以下的较低层，虽然反演结果和实地测量之间的差异有限，特别是接近地表处，但分析结果总体上比反演结果更接近探空仪测量值。由于在分析中已经同化了无线电探空仪测量值，它们的预期结果非常接近。然而，这说明了比较这些不同类型测量的困难：无线电探空仪提供的是距离发射地点 100 km 远的当地信息，但它们的平均值与 20 km×20 km 的微波传感器像素相比，空间代表性较差，这是由于对流层低层[48]的陆地上 WV 廓线受强烈的空间梯度影响。反演结果与探空仪的对比结果表明，除了低层之外，相对湿度的百分比差异在 15%以内[28]。之前的一项研究使用了 RaCCI/LBA 外场试验的无线电探空仪和 HSB 反演的水汽，在晴空条件下显示出类似的结果，而在多云条件下则有更明显的差异[33]。

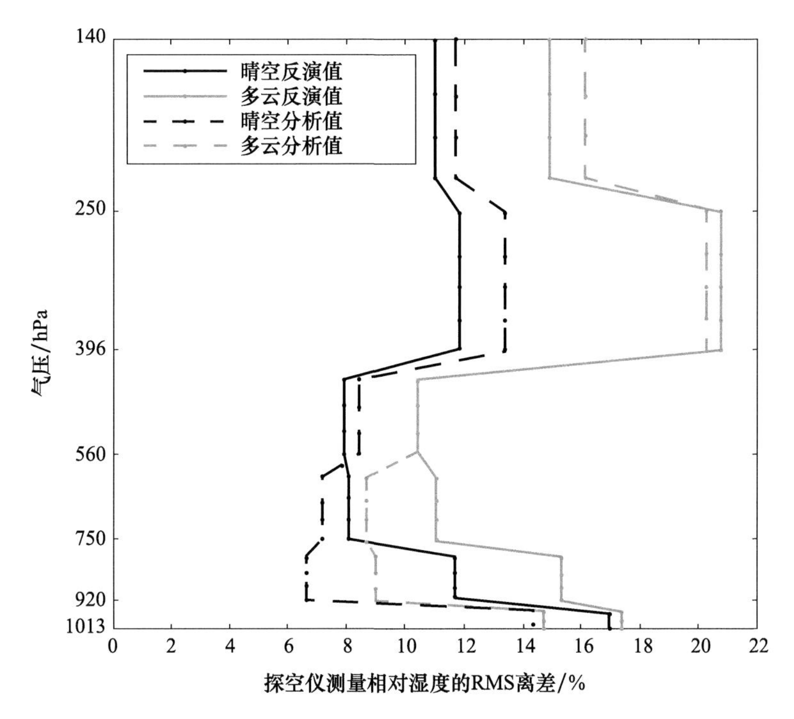

图 9.11　探空仪测量相对湿度与 ECMWF 分析(虚线)和同步 AQUA 反演相对湿度(连续线)的 RMS 离差。统计使用陆地晴空(灰色)和多云(黑色)数据

9.7　亮温：空间验证

在基于物理过程的反演方案中，反演过程的质量标准是基于当使用反演的大气和地表参数是实际观测值与 RT 模拟值之间的差值。反演的目标是将观测结果与模拟结果之间的差异最小化。因此，当使用大气状态的第一猜测或先验信息、反演参数时，可以通过比较观测结果与 RT 模拟之间的差异，来衡量反演结果较第一猜测或先验信息的改进。如果反演的变量使这些统计数据变差，则反演没有改善先验信息，或者存在数学偏差，无法执行反演。最终反演

结果的精度在很大程度上取决于辐射传输计算的精度。反演的目的是在反映真实世界的辐射传输模拟的基础上进行准确的反演。注意，观测结果和模拟结果之间的差异不仅与反演的质量有关，卫星观测包括仪器噪声与系统偏差和漂移(取决于传感器)，RT 也会产生误差。在反演过程之前的校准步骤的目标是限制这两种误差来源，但校准从来都不是完美的。此外，观测和分析之间的空间和时间差异也很关键。

下面将介绍几种诊断方法，以便对卫星反演的 T_B 场进行评估。

9.7.1 亮温对比

第 4.3 节中描述的反演方案反演得到的地球物理变量被用作 RT 模型(即 RTTOV)的输入值。这些数据包括 WV 廓线、地表温度和发射率，以及海洋表面风速。为了比较反演提供的潜在相对改进，还使用 ECMWF 的初始分析进行了模拟。

图 9.12 给出同时考虑了陆地(上图)和海洋(下图)的 MetOp 仪器(AMSU-A 和 MHS，左)和 AQUA 仪器(AMSR-E 和 HSB，右)的统计结果，为晴空(黑色)和多云(灰色)条件提供

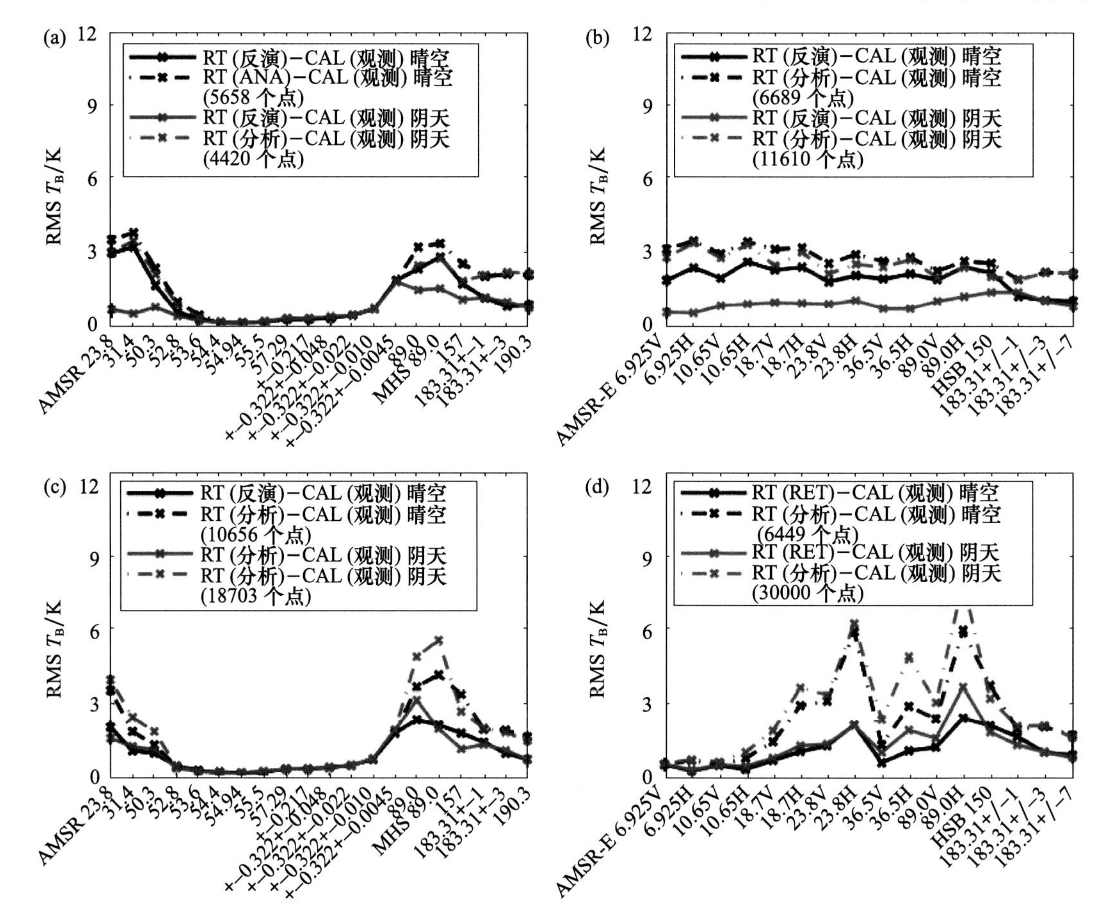

图 9.12 2 个月内的模拟和观测后的均方根误差：(a)陆上 METOP；(b)陆上 AQUA；(c)海洋 METOP；(d)海洋 AQUA。左图为 MetOp 上的 AMSU-A 和 MHS，右图为 AQUA 上的 AMSR-E 和 HSB，上图为陆地下图为海洋，黑色代表晴朗和灰色代表阴天。MetOp 为 2007 年 7 月和 2008 年 1 月，MetOp 为 2002 年 9 月和 2003 年 1 月。其中，连续线表示使用反演产品进行的模拟，虚线表示使用 ECMWF 分析和 TELSEM 表面发射率进行的模拟

了RMS统计数据，统计数据代表了使用 RT 模型模拟的反演(连续线)结果和分析(虚线)结果，模拟 T_B 之间的差异是由反演方案获取的参数唯一产生的。与使用先验信息(虚线)的模拟相比，使用反演产品(连续线)的模拟总是具有较低的 RMS:这清楚地表明，与分析结果相比，反演的变量代表了 T_B 场的改进。当将反演结果替代 ECMWF 分析作为 RT 代码的输入时，模拟的 T_B 与校准的观测结果更加接近，这正是反演方案的目标。特别地，通过反演用于 AQUA 和 MetOp 仪器四种配置的 WV 廓线，改进了 183 GHz 通道的模拟。WV 探测通道的统计数据在所有情况下(陆地/海洋、晴朗/多云、AQUA/MetOp)都是相似的，这意味着无论条件如何，反演都非常可靠。由于陆地上的表面温度和发射率更好，海洋上的表面风更优，窗口通道得到了改善。在大陆地表，多云条件下对地表敏感通道获得了更好的统计数据:这是由于多云条件下地表温度的变化较小。请注意，部分改进可能与反演和卫星数据之间更好的空间/时间重合有关(通过构造，反演在与卫星观测相同的时间和位置进行)。

图 9.13 给出了 AQUA 上 HSB 在 23.8 GHz 下垂直偏振的标定观测(中间图)，以及来自 ECMWF 分析(上图)和反演(下图)的模拟结果。该通道对 TCWV 非常敏感，使用反演提高了与观测数据的对比效果。例如，利用模拟的反演结果，可以较好地再现太平洋中 150°E 和

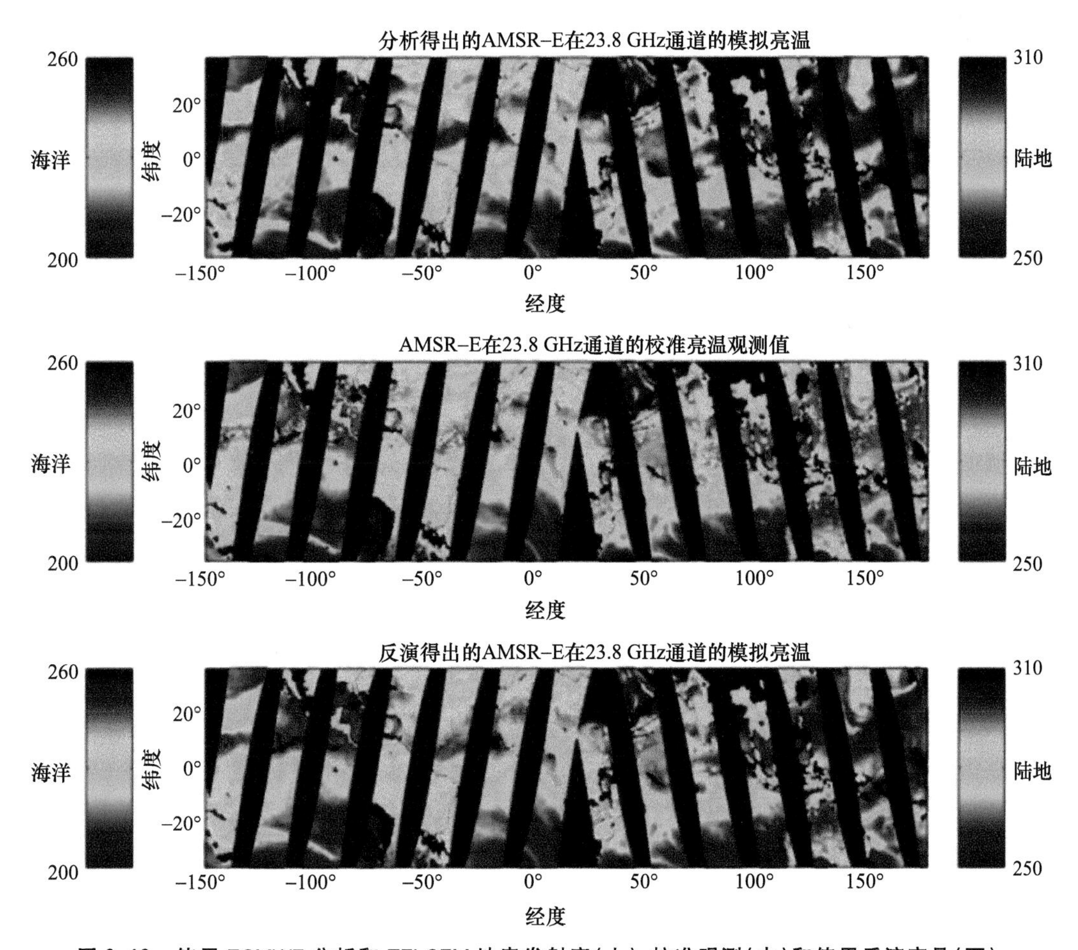

图 9.13 使用 ECMWF 分析和 TELSEM 地表发射率(上)、校准观测(中)和使用反演产品(下)。在 AMSR-E 和 HSB 轨道上的模拟(2002 年 9 月 7 日)在 23.8 GHz 垂直极化中的 TB 示例

20°N 周围或 140°W 和 20°S 附近的观测结构。在海洋上空，从分析得到的模拟似乎在非常潮湿的大气中有很强的正偏差(从观测所得可达 5 K)，而在纬度为 10°的反演中则没有这样的结果。此图可与图 9.6 相关联，其中反演值远低于这些地区的分析预测结果。由于 RTTOV 模型被数值模式中心广泛用于预报，辐射传输模型正偏差的可能性不足以解释这一现象。此外，通过我们的校准，可以减少观测和模拟 T_B 之间的偏差。

图 9.14 给出了 AQUA 上 HSB 在 183.81±3 GHz 通道的校准观测值、校准观测值与利用反演结果模拟得到的观测值之间的差异。该通道与对流层低层的水汽高度相关。当使用分析结果时，模拟表现出与观测的显著差异，在海洋和陆地上通常有 3K 的差异，不能仅仅用时间不匹配解释这些巨大的差异，因为在一些地区，如南大西洋或非洲，时间差异小于 30 min。当将反演结果用于模拟时，无论是海洋还是陆地，与观测结果的一致性都有显著提高，没有出现海洋和陆地之间的明显不连续。

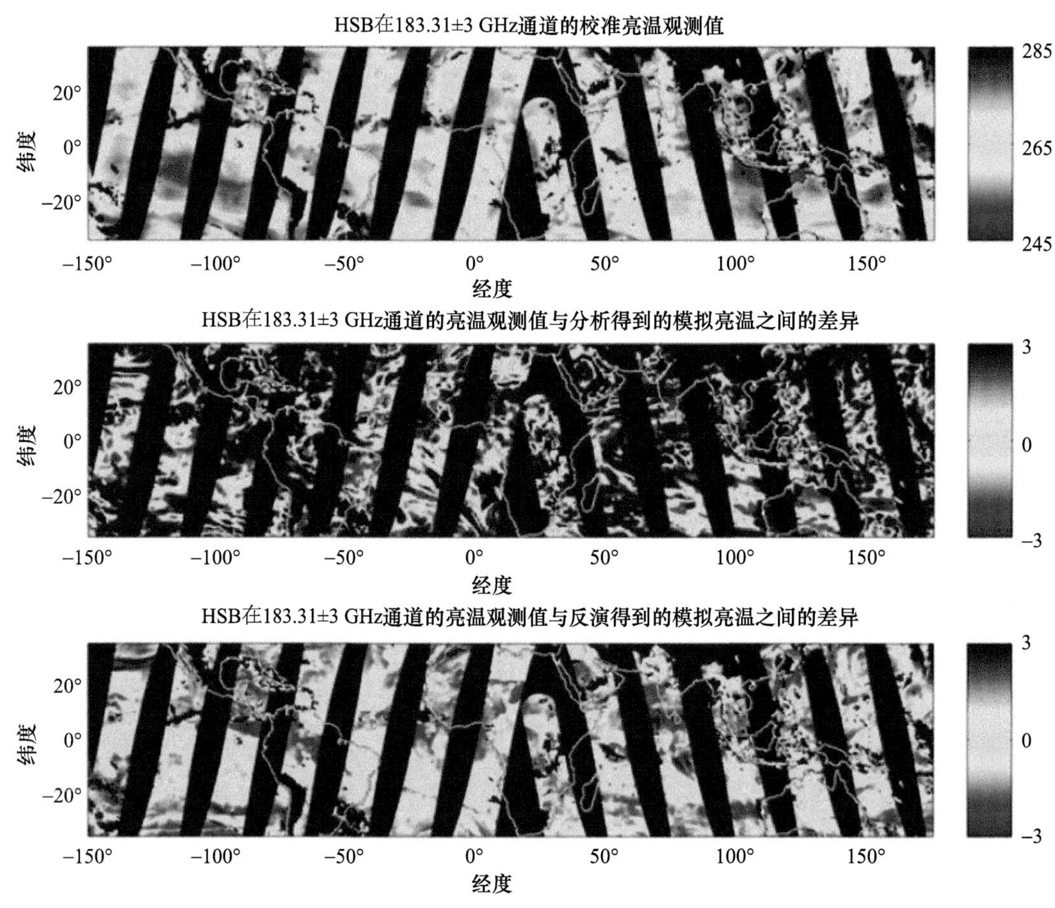

图 9.14 183.81±3 GHz 通道 T_B 的校准观测值示例(2002 年 9 月 7 日)(上图)，观测与使用 ECMWF 分析和 TELSEM 地表发射率的模拟的偏差(中图)，观测与使用 AMSR-E 和 HSB 轨道上反演产品的模拟的偏差(下图)

9.7.2 每个反演变量的贡献

为了评估每个反演变量对 T_B 场改进的影响，当只使用部分反演变量作为 RT 代码的输入时，进行了类似的实验。与图 9.12 类似，图 9.15 显示了 RMS 统计数据，但针对 AQUA 观测(AMSR-E 和 HSB),对于不同的 RT 输入:(1)ECMWF 初始分析(蓝色);(2)所有反演变量(黑色);(3)仅反演所得大气湿度廓线(红色);(4)仅反演所得地表参数(绿色)。对于 183 GHz 通道和两个平台,正如预期的那样,WV 的反演是减少与观测值差异的主要因素,地表对信号的影响非常小或为零。即使对于窗口通道,在海洋上空情况下,水面反演的影响也是有限的。首先,分析结果中海洋表面参数的质量较好,与实测数据的偏差较小;其次,由于海洋的发射率较低,海洋表面对辐射的贡献比陆地少。在陆地上,对于所有达到 100 GHz 的通道,TS 和发射率的反演都得到了显著的改善(晴空案例中剩余均方根更高,这可能是由于在晴空条件下 TS 的较大变异性,如前所述)。

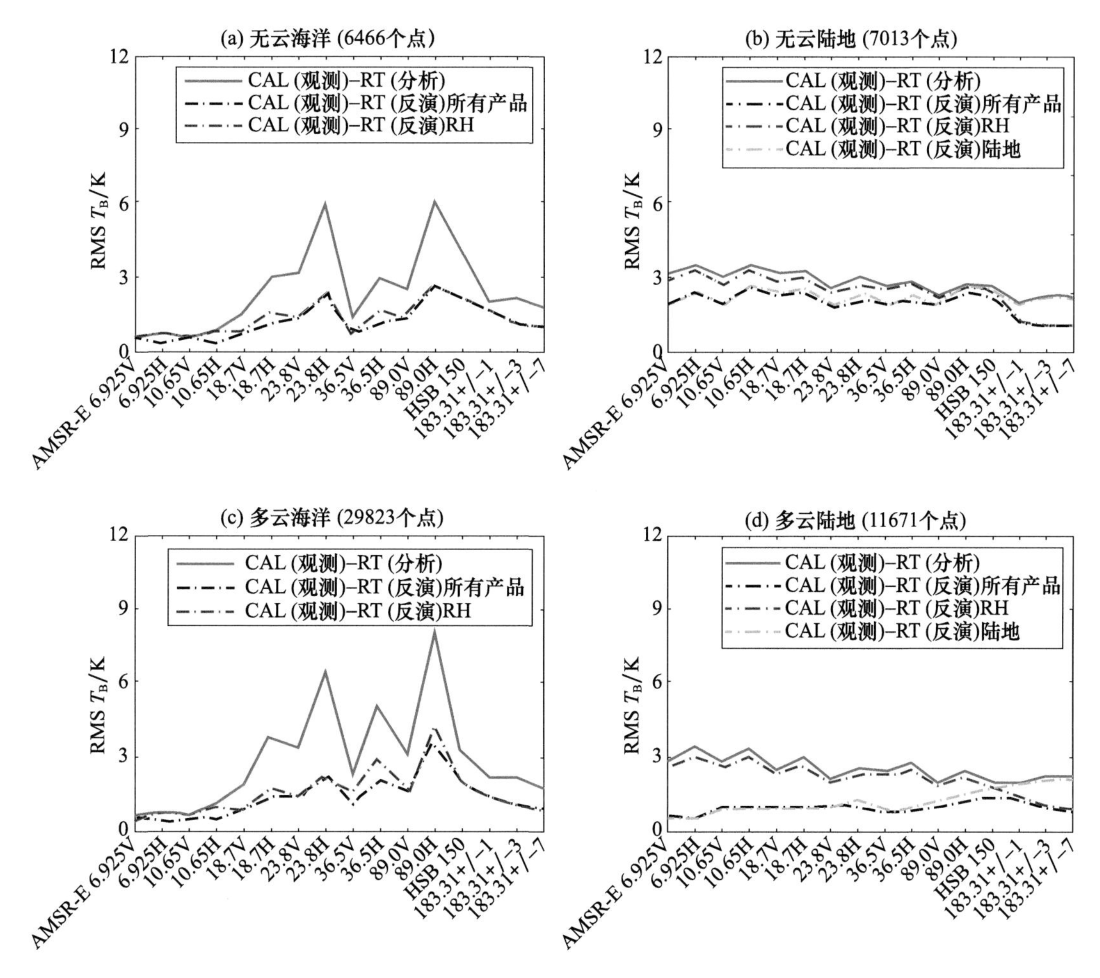

图 9.15 在 2 个月中(2002 年 9 月和 2003 年 1 月),AQUA 上 AMSR-E 和 HSB 的模拟和观测 TB 之间的 RMS 误差:无云/海洋(a)、无云/陆地(b)、多云/海洋(c)和多云/陆地(d)。蓝色实线表示使用 ECMWF 分析和 TELSEM 地表发射率的模拟结果,红色虚线表示相同的模拟结果,只是引入了相对湿度廓线的反演,绿色线为只引入了地表发射率和温度的反演结果(仅陆地),黑色虚线表示使用所有反演产品的模拟结果

9.7.3 亮温场改进的评价

可以设计各种诊断来自动监测反演结果的质量:这些标准可以帮助过滤掉收敛得不够理

想的反演结果，以及评估反演结果的可信度。

对反演相对于分析所提供改进的评估方法如下：

$$|RT(ana)-CAL(obs)|-|RT(ret)-CAL(obs)|$$

其中 RT(ana)是用于分析的 RT 模拟，RT(ret)用于反演，CAL(obs)用于校准的观测结果。当这个参量为正时，与分析结果相比，反演结果减少了与观测值的差异。

该参量在 HSB 的 150 GHz 和 183.31±3 GHz 处的直方图中(图 9.16)偏向左侧，表明在超过 50%的情况下，与校准的观测相比，反演改善了 RT 模拟。这对于四种场景(陆地或海洋、晴空或多云)是正确的，改进大小能达到 10 K。对于 183.31±3 GHz，四种场景的分布是相同的，这意味着我们的反演改进类型对于多云/晴空案例以及陆地/海洋表面均相似，对于相对重要的通道(但对更少的页面 150 GHz 通道)。相反，150 GHz 通道的分布是不同的：在陆地案例中它们更对称、稍微向左偏，在海洋案例中更向左偏、尾部更长。这意味着，对海洋案例的改善可能会更明显(可以达到 10 K)，而陆地案例的重要性降低(即达到 5 K)。请注意，在我们的惯例中，不够好的案例/样本是得到了改进的。然而，较好的案例不一定是分析结果的退化：传统上，在业务中心，这种诊断是用来测试反演是否改善了分析，如果它使分析退化，反演结果就被直接丢弃(只有一小部分卫星数据实际用于同化系统)。

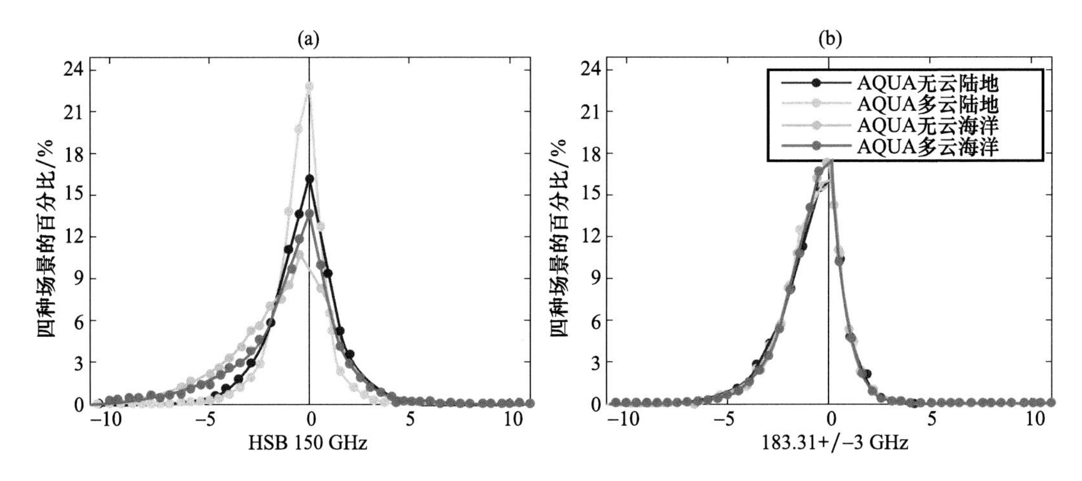

图 9.16 在 150 GHz(a)和 183.81±3 GHz(b)条件下，利用 ECMWF 分析和 AQUA 反演产品，模拟得到的 T_B 误差与观测值之间差异的归一化分布。统计数据为两个月(2002 年 9 月和 2003 年 1 月)，其中：无云/陆地(红色)、多云/陆地(绿色)、无云/海洋(黑色)、多云/海洋(蓝色)。负值为使用反演产品的模拟最接近实际观测的情况

图 9.16 提供了关于反演结果的潜在改进程度的信息。图 9.17 展现了 AQUA 上搭载的所有 AMSR-E/HSB 通道在使用反演产品而不是分析产品时情况得到改善的百分比。无论何种场景情况(陆地/海洋和晴空/多云)或通道，50%以上的案例都得到了改善。在多云条件下，用于陆地的窗口通道的改善率最高。注意，对于陆地和海洋样本，改进情况的百分比是相同的，这证实了反演在大多数情况下，在陆地和海洋上提供了相似的性能。183±3 和±7 GHz 通道中较高的改善率说明了我们的算法能够反演对流层下部的大气湿度，不论是否有云层覆盖，这是红外传感器无法获得的关键信息。

提供的统计数据表明，在大多数情况下，反演结果改进了分析产品，在 T_B 场（几个开氏度）具有相当显著的影响。这证明，在反演过程中有可能同化或使用越来越多的微波观测资料，以便改进天气预报，特别是在陆地上。

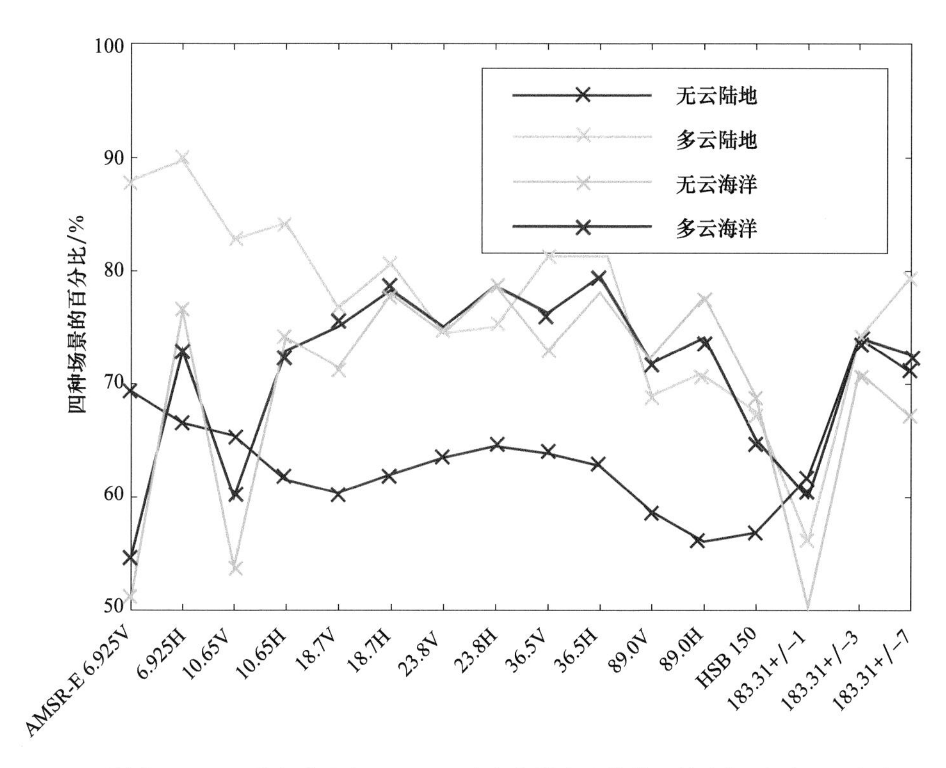

图 9.17 AQUA 反演与 ECMWF 分析产品和 TELSEM 地表发射率改善情况的比例，考虑了它们与观测 T_B 场的偏离。图中数据为两个月(2002 年 9 月和 2003 年 1 月)的统计结果其中：无云/陆地(红色)、多云/陆地(绿色)、无云/海洋(青色)、多云/海洋(蓝色)

9.8 结论

我们已经开发了一种用于在晴空和多云条件下（降水案例未被处理）反演海洋和陆地上大气水汽廓线的算法。反演方法是基于神经网络开展的反演，该算法已被训练用于从 AQUA、MctOp 和 Mcgha-Tropiques 上的微波仪器中进行水汽廓线反演。所述反演链包括一种校准方案以及一种获得地表发射率第一估计值的工具，反演结果在海洋（晴空和多云情况下的 STD=1.7 kg/m²）和陆地（晴空和多云情况下的 STD 分别为 2.2 和 3 kg/m²），甚至对低层大气的描述都表现出令人满意的理论性能，这是非常令人鼓舞的，考虑到在同化过程中传统方法反演低层大气时使用地表敏感通道的困难，采用模拟数据和传统的信息含量分析方法对不确定度进行了评估。

在 AQUA 和 MetOp 平台上对算法性能进行了测试，WV 廓线和柱内总含量与 ECMWF 分析以及无线电探空仪测量结果进行了比较。在陆地和海洋上，在晴天和多云情况下，柱内总水汽含量的估计误差是可比较的，与无线电探空仪相比，包括其自身的不确定性在内的 STD 误差约为 4.5 kg/m^2。在没有任何先验水汽信息的情况下，在 6 个大气层结上反演了水汽廓线，在陆地上较低层相对湿度的最大 RMS 误差为 20%；反演的水汽场没有表现出任何与海洋/陆地或多云/晴空转变有关的明显不连续。微波反演是红外估计的一个非常有前景的补充，特别是当探测和窗口通道一起使用时，微波观测在有云存在时的探测是非常意义的。

使用模拟亮温对反演产品进行评估，反演过程结束后，将反演所得参数作为 RT 代码的输入，并将模拟结果与观测结果进行比较。这些差异提供了对反演性能的直接测试，并可用于过滤掉不成功的反演结果。统计评估表明，与大多数情况下的分析结果相比，反演结果得到了改进，并且 T_B 场的改进是显著的（几个开氏度）。T_B 场的这种改善是地球物理变量（如湿度廓线）改善的结果。然而，使用原位测量的验证对于湿度反演的直接验证也是必要的。

这项工作的一个重要视角是结合使用微波和红外观测（如来自 IASI）的水汽反演方案。在这两种类型的观测之间存在着良好的协同作用[1,8]，神经网络方法是利用它的一个很好的候选方案。虽然这项研究排除了降水情况，但将反演扩展到所有天气条件对未来的水循环研究很重要。

致谢 我们要感谢法国空间机构 CNES 在 Megha-Tropiques 任务的准备阶段对这方面工作的一些支持。我们还要感谢 Megha-Tropiques 任务的所有科学和技术团队。这项工作已经与几个合作者 Catherine Prigent，Frédéric Bernardo，Francis Marquisseau 和 Hélène Brogniez 等人合作了多年。

参考文献*

[1] Aires，F. 2011. Measure and exploitation of multi-sensor and multi-wavelength synergy for remote sensing：Part I-Theoretical considerations. *Journal of Geophysical Research* 116：D02301. https://doi. org/10. 1029/2010JD014701.

[2] Aires，F. ，and C. Prigent. 2007. Sampling techniques in high-dimensional spaces for satellite remote sensing databases generation. *Journal of Geophysical Research* 112：D20301. https://doi. org/10. 1029/2007JD008391.

[3] Aires，F. ，C. Prigent，W. B. Rossow，and M. Rothstein. 2001. A new neural network approach including first-guess for retrieval of atmospheric water vapor，cloud liquid water path，surface temperature and emissivities over land from satellite microwave observations. *Journal of Geophysical Research* 106(D14)：14887-14907.

[4] Aires，F. ，A. Chedin，N. Scott，andW. B. Rossow. 2002. A regularized neural network approach for retrieval of atmospheric and surface temperatures with the IASI instrument. *Journal of Applied Meteorology* 41(2)：144-159.

[5] Aires，F. ，F. Bernardo，H. Brogniez，and C. Prigent. 2010. An innovative calibration method for the inversion of satellite observations. *Journal of Applied Meteorology and Climatology* 49(12)：2458-2473.

* 参考文献沿用原版书中内容，未改动

[6] Aires,F. ,C. Prigent,F. Bernardo,C. Jiménez,R. Saunders,and P. Brunel. 2011. A tool to estimate land surface emissivities at microwaves(TELSEM) frequencies for use in numerical weather prediction. *Quarterly Journal of the Royal Meteorological Society* 137:690-699.

[7] Aires,F. ,F. Marquisseau,C. Prigent,and G. Sèze. 2011. A land and ocean microwave cloud classification algorithm derived from AMSU-A and-B, trained using MSG-SEVIRI infrared and visible observations. *Monthly Weather Review* 139(8):2347-2366. https://doi. org/10. 1175/MWR-D-10-05012. 1.

[8] Aires,F. ,M. Paul,C. Prigent,B. Rommen,and M. Bouvet. 2011. Measure and exploitation of multi-sensor and multi-wavelength synergy for remote sensing: Part II-An application for the retrieval of atmospheric temperature and water vapour from METOP. *Journal of Geophysical Research* 116:D02303. https://doi. org/10. 1029/2010JD014702.

[9] Aires,F. ,O. Aznay,C. Prigent,M. Paul,and F. Bernardo. 2012. Synergistic multi wavelength remote sensing versus a posteriori combination of retrieved products: application for the retrieval of atmospheric profiles using MetOp-A. *Journal of Geophysical Research* 117(D18). https://doi. org/10. 1029/2011JD017188.

[10] Alishouse,J. ,S. Snyder,V. Jennifer,and R. Ferraro. 1990. Determination of oceanic total precipitable water from the SSM/I. *IEEE Transactions on Geoscience and Remote Sensing* 28:811-816.

[11] Bony,S. ,B. Stevens,D. M. W. Frierson,C. Jakob,M. Kageyama,R. Pincus,T. G. Shepherd,S. C. Sherwood,A. P. Siebesma,A. H. Sobel,M. Watanabe,and M. J. Webb. 2015. Clouds,circulation and climate sensitivity. *Nature Geoscience* 8:261-268.

[12] Boukabara,S. -A. ,et al. 2011. MiRS: An all-weather 1DVAR satellite data assimilation and retrieval system. *IEEE Transactions on Geoscience and Remote Sensing* 49(9):3249-3272.

[13] Collard,A. D. 2007. Selection of IASI channels for use in numerical weather prediction. *Quarterly Journal of the Royal Meteorological Society* 133(629):1977-1991.

[14] Deblonde,G. ,and S. J. English. 2001. Evaluation of the FASTEM-2 fast microwave oceanic surface emissivity model. In *Technical Proceedings of ITSC-XI*,Budapest,20-26 Sept 2000,67-78.

[15] Deblonde,G. ,and N. Wagneur. 1997. Evaluation of global numerical weather prediction analyses and forecasts using DMSP special sensor microwave imager retrievals. 1. Satellite retrieval algorithm inter-comparison study. *Journal of Geophysical Research* 102:1833-1850.

[16] Deblonde G. ,W. Yu,L. Garland,and A. P. Dastoor. 1997. Evaluation of global numerical weather prediction analyses and forecasts using DMSP special sensor microwave imager retrievals,2,analyses/forecasts inter-comparison with SSM/I retrievals. *Journal of Geophysical Research* 102:1851-1866.

[17] Deeter,M. N. 2007. A new satellite retrieval method for precipitable water vapor over land and ocean. *Geophysical Research Letters* 34(L02815). https://doi. org/10. 1029/2006GL028019.

[18] Derrien,M. ,and H. Le Gléau. 2005. MSG/SEVIRI cloud mask and type from SAFNWC. *International Journal of Remote Sensing* 26:4707-4732.

[19] Derrien,M. ,and H. Le Gléau. 2010. Improvement of cloud detection near sunrise and sunset by temporal-differencing and region-growing techniques with real-time SEVIRI. *International Journal of Remote Sensing* 31(7):1765-1780.

[20] Diedrich,H. ,R. Preusker,R. Lindstrot,and J. Fischer. 2015. Retrieval of daytime total columnar water vapour from MODIS measurements over land surfaces. *Atmospheric Measurement Techniques* 8:823-836. https://doi. org/10. 5194/amt-8-823-2015.

[21] Eyre,J. R. 1991. A fast radiative transfer model for satellite sounding systems. *ECMWF Research Depart-*

ment Technical Memorandum, 176(Available from the librarian at ECMWF).

[22] Ferraro, R., F. Weng, N. Grody, and A. Basist. 1996. An eight-year(1987-1994) time series of rainfall, clouds, water vapor, snow cover, and sea ice derived from SSMI/I measurements. *Bulletin of the American Meteorological Society* 77:891-905.

[23] Gentemann, C. L., F. J. Wentz, C. A. Mears, and D. K. Smith. 2004. In situ validation of tropical rainfall measuring mission microwave sea surface temperatures. *Journal of Geophysical Research* 109: C04021. https://doi.org/10.1029/2003JC002092.

[24] Grody, N., J. Zhao, R. Ferraro, F. H. Weng, and R. Boers. 2001. Determination of precipitable water and cloud liquid water over oceans from the NOAA 15 advanced microwave sounding unit. *Journal of Geophysical Research* 106:2943-2953. https://doi.org/10.1029/2000JD900616.

[25] Hong, F., and G. Heygster. 2005. Detection of tropical deep convective clouds from AMSU-B water vapor channels measurements. *Journal of Geophysical Research* 110(D05205). https://doi.org/10.1029/2004JD004949.

[26] Hornik, K., M. Stinchcombe, and H. White. 1989. Multilayer feedforward networks are universal approximators. *Neural Networks* 2:359-366.

[27] Jackson, D. L., and G. L. Stephens. 1995. A study of SSM/I-derived columnar water vapor over the global oceans. *Journal of Climate* 8:2025-2038. https://doi.org/10.1175/1520-0442(1995)008<2025:ASOS-DC>2.0.CO;2.

[28] Jethva, H., and J. Srinivasan. 2004. Role of variation in vertical profiles of relative humidity from AMSU-B data. *Geophysical Research Letters* 31(23). https://doi.org/10.1029/2004GL021098.

[29] Kahn, B. H., J. Teixeira, E. J. Fetzer, A. Gettelman, S. M. Hristova-Veleva, X. Huang, A. K. Kochanski, M. Köhler, S. K. Krueger, R. Wood, and M. Zhao. 2011. Temperature and water vapor variance scaling in global models: comparisons to satellite and aircraft data. *Journal of the Atmospheric Sciences* 68(9):2156-2168. https://doi.org/10.1175/2011JAS3737.1.

[30] Karbou, F., F. Aires, C. Prigent, and L. Eymard. 2005. Potential of AMSU-A and-B measurements for atmospheric temperature and humidity profiling over land. *Journal of Geophysical Research* 110(D97109). https://doi.org/10.1029/2004JD005318.

[31] Kawanishi, T., T. Sezai, Y. Ito, K. Imaoka, T. Takeshima, Y. Ishido, A. Shibata, M. Miura, H. Inahata, and R. W. Spencer. 2003. The advanced microwave scanning radiometer for the Earth Observing System(AMSR-E), NASDA's contribution to the EOS for global energy and water cycle studies. *IEEE Transactions on Geoscience and Remote Sensing* 41:184-194.

[32] Lambrigtsen, B. H., and R. V. Calheiros. 2002. The humidity sounder for Brazil-an international partnership. *IEEE Transactions on Geoscience and Remote Sensing* 41:352-361.

[33] Lima, W. F. A., and L. A. T. Machado. 2006. Analise do sensor HSBna estimativa do conteúdo integrado de vapor d'àgua durante o experimento RaCCI/LBA. *Revista Brasileira de Meteorologia* 21(2):1-8.

[34] Lindstrot, R., M. Stengel, M. Schröder, J. Fischer, R. Preusker, N. Schneider, T. Steenbergen, and B. R. Bojkov. 2014. A global climatology of total columnar water vapour from SSM/I and MERIS. *Earth System Science Data* 6:221-233. https://doi.org/10.5194/essd-6-221-2014.

[35] Matricardi, M., F. Chevallier, and S. Tjemkes. 2001. An improved general fast radiative transfer model for the assimilation of radiance observations. *ECMWF Research Department Technical Memorandum*, 345. http://www.ecmwf.int/publications.

[36] Mears, C. A., J. Wang, D. Smith, and F. J. Wentz. 2015. Inter-comparison of total precipitable water measurements made by satellite-borne microwave radiometers and ground-based GPS. *Journal of Geophysical Research* 120:2492-2504. https://doi.org/10.1002/2014JD022694.

[37] Prigent C., W. B. Rossow, and E. Matthews. 1997. Microwave land surface emissivities estimated from SSM/I observations. *Journal of Geophysical Research* 102:21867-21890.

[38] Prigent, C., F. Aires, and W. B. Rossow. 2006. Land surface microwave emissivities over the globe for a decade. *Bulletin of the American Meteorological Society* 1572-1584. https://doi.org/10.1175/BAMS-87-11-1573.

[39] Prigent, C., E. Jaumouille, F. Chevallier, and F. Aires. 2008. A parameterization of the microwave land surface emissivity between 19 and 100 GHz, anchored to satellite-derived estimates. *IEEE Transactions on Geoscience and Remote Sensing* 46:344-352.

[40] Prigent, C., F. Aires, D. Wang, S. Fox, and S. Harlow. 2017. Sea-surface emissivity parameterization from microwaves to millimeter waves. *Quarterly Journal of the Royal Meteorological Society* 143(702):596-605. https://doi.org/10.1002/qj.2953.

[41] Rodgers, C. D. 1976. Retrieval of atmospheric temperature and composition from remote measurements of thermal radiation. *Reviews of Geophysics* 14(4):609-624.

[42] Rodgers, C. D. 1990. Characterization and error analysis of profiles retrieved from remote sounding measurements. *Journal of Geophysical Research* 95(D5):5587-5595.

[43] Roman, J., R. Knuteson, T. August, T. Hultberg, S. Ackerman, and H. Revercomb. 2016. A global assessment of NASA AIRS v6 and EUMETSAT IASI v6 precipitable water vapor using ground-based GPS Suomi Net stations. *Journal of Geophysical Research* 121: 8925-8948. https://doi.org/10.1002/2016JD024806.

[44] Rumelhart, D. E., G. E. Hinton, and R. J. Williams. 1986. Learning internal representations by error propagation. In*Parallel Distributed Processing*: *Explorations in the Microstructure of Cognition*, ed. D. E. Rumelhart, J. L. McClelland, and the PDP Research Group, Vol. I, Foundations, 318-362. Cambridge: MIT Press.

[45] Saunders, R. W., M. Matricardi, and P. Brunel. 1999. An improved fast radiative transfer model for assimilation of satellite radiance observations. *Quarterly Journal of the Royal Meteorological Society* 125: 1407-1425.

[46] Schlüssel, P., and M. Goldberg. 2002. Retrivela of atmospheric temperature and water vapour from IASI measurements in partly cloudy situations. *Advances in Space Research* 29(11):1703-1706.

[47] Schröder, L., A. Richter, D. V. Fedorov, L. Eberlein, E. V. Brovkov, S. V. Popov, C. Knöfel, M. Horwath, R. Dietrich, A. Y. Matveev, M. Scheinert, and V. V. Lukin. 2017. Validation of satellite altimetry by kinematic GNSS in central East Antarctica. *The Cryosphere* 11:1111-1130. https://doi.org/10.5194/tc-11-1111-2017.

[48] Steinke, S., S. Eikenberg, U. Löhnert, G. Dick, D. Klocke, P. Di Girolamo, and S. Crewell. 2015. Assessment of small-scale integrated watervapour variability during HOPE. *Atmospheric Chemistry and Physics*15:2675-2692. https://doi.org/10.5194/acp-15-2675-2015.

[49] Susskind, J., C. D. Barnet, and J. M. Blaisdell. 2003. Retrieval of atmospheric and surface parameters from AIRS/AMSU/HSB data in the presence of clouds. *IEEE Transactions on Geoscience and Remote Sensing* 41(2):390-409.

[50] Tjemkes, S. A., G. L. Stephens, and D. L. Jackson. 1991. Spaceborne observation of columnar water vapor: SSM/I observations and algorithm. *Journal of Geophysical Research* 96(10): 10941-10954. https://doi.org/10.1029/91JD00272.

[51] Uppala, S. M., P. W. Kallberg, A. J. Simmons, U. Andrae, V. da Costa Bechtold, M. Fiorino, J. K. Gibson, J. Haseler, A. Hernandez, G. A. Kelly, X. Li, K. Onogi, S. Saarinen, N. Sokka, R. P. Allan, E., Andersson, K. Arpe, M. A. Balmaseda, A. C. M. Beljaars, L. van de Berg, J. Bidlot, N. Bormann, S. Caires, F. Chevallier, A. Dethof, M. Dragosavac, M. Fisher, M. Fuentes, S. Hagemann, E. Holm, B. J. Hoskins, L. Isaksen, P. A. E. M. Janssen, R. Jenne, A. P. McNally, J. -F. Mahfouf, J. -J. Morcrette, N. A. Rayner, R. W. Saunders, P. Simon, A. Sterl, K. E. Trenberth, A. Untch, D. Vasiljevic, P. Viterbo, J. Woollen. 2005. The ERA-40 re-analysis. *Quarterly Journal of the Royal Meteorological Society* 131: 2961-3012.

[52] Wang, Y., Y. Fu, G. Liu, Q. Liu, and L. Sun. 2009. A new water vapor algorithm for TRMM microwave imager(TMI) measurements based on a log linear relationship. *Journal of Geophysical Research* 114 (D21304). https://doi.org/10.1029/2008JD011057.

[53] Wang, J., A. Dai, and C. Mears. 2016. Global water vapor trend from 1988 to 2011 and its diurnal asymmetry based on GPS, radiosonde, and microwave satellite measurements. *Journal of Climate*. https://doi.org/10.1175/JCLI-D-15-0485.1.

[54] Wang, D., C. Prigent, L. Killic, S. Fox, C. Jimenez, F. Aires, C. Grassoti, and F. Karbou. 2017. Surface emissivity at microwaves to millimeter waves over polar regions: parameterization and evaluation with aircraft experiments. *Journal of Atmospheric and Oceanic Technology* 34(5). https://doi.org/10.1175/JTECH-D-16-0188.1.

[55] Weng, F., L. Zhao, R. Ferraro, G. Poe, X. Li, and N. Grody. 2003. Advanced microwave sounding unit cloud and precipitation algorithms. *Radio Science* 38: 8086-8096.

[56] Wentz, F. J. 2017. A 17-Yr climate record of environmental parameters derived from the tropical rainfall measuring mission(TRMM) microwave imager. *Journal of Climate*. https://doi.org/10.1175/JCLI-D-15-0155.1.

第10章　基于遥感测量的云和降水模式研究进展

Takamichi Iguchi and Toshihisa Matsui

10.1 引言

天气和气候模式的终极目标是预测未来。良好的预测是满足各种社会需求所必需的，而不仅仅是对未来气候变化的预测。这些需求包括减轻自然灾害造成的损失和为农业活动提供咨询。短期和长期预报的一个重要问题是预测的可靠性，正确的答案最终可以通过未来的探测得到；然而，这时要修改预报中的错误就为时已晚了。因此，在发生之前，对未来作出尽可能精确的预报是至关重要的。通过对回顾时期的模拟，探测数据的积累为这些模式提供了一个极好的测试平台。引入所谓的后验或回溯测试来检查模式中的最佳配置，并分析它们的准确性。与探测结果的比较是验证模拟结果与确定误差和不确定性来源的关键。

地表降水是大气模式和探测的关注重点，因为它与人类在陆地表面的活动密切相关。与降水有关的液态水粒子在大气中生成和生长，随后被引力拉向地球表面。粒子形成前的碰并/聚合可以在云中被观察到，云粒子的生长特征对地面降水特征具有关键作用，降水特征包括降水位置、持续时间、累积降水量以及粒子的相态和类型等。云在大气模式中更好的表现，可以改进对地表降水、潜热/冷却和云辐射强迫等的模拟。

与探测地面降水的方法相比，探测云层的方法有限。地基原位探测仍然可以在高山[3]或建筑物[52]上有效地直接观测低层云。飞机探测甚至可以在高空云中提供实地观测[14]，尽管出于成本高和安全方面的考虑，其操作受到很大限制。气球探空仪测量具有捕获特征的合理能力[36,51]，尽管很难将探空仪放置在目标云层或降水中。除了这些实地观测，目前常用各种遥感探测，特别是地球观测卫星的发展是一个里程碑，其大大扩展了探测能力和覆盖范围，甚至可以覆盖远离陆地的海洋。

有效利用遥感探测数据对于通过后验模拟进一步发展和改进大气模式至关重要。如何恰当地处理遥感探测数据，对大气模拟结果进行评估是一个挑战。一般来说，合理的比较是基于这样一个准则：观测和模拟都提交了相同的物理量来相互比较。在与实地测量比较的情况下，观测到的物理量与物体的质量、体积、尺寸、速度、能量等直接相关。相比之下，遥感探测是通过测量物体发射或散射的辐射来间接观测物体。因此，总体上有两种方式来统一观测和模拟计算的物理量。第一种方法是将测量中检测到的原始信号转化为与物体的组成和结构直接相关的其他物理量，需要开发所谓的反演算法，通过求解反演问题来估计这些物理量。第二种方法是通过前向辐射传输模式计算大气模式输出的测量-可观测信号，然后将其与实测直接信号进行比较，有多种方式来处理关于测量的直接信号和模式模拟输出的匹配程度，以统一评估大气模式模拟的参量。

哪一种方案在观测和模拟之间产生更合适的对比可能完全取决于数值模式的模型、传感器和探测平台的类型、其他探测数据的可用性、目标现象等。由于探测数据和大气模式的组合是不受限制的，所以一种方案的优点和缺点是不确定的。使用这种比较的研究人员应该注意哪种方案更好，它的优点和缺点是什么[34]。

本章回顾了如何在大气模式中表示云和降水微物理，并介绍了基于云和降水遥感探测与模拟比较的最新研究实例。本章的结构如下：第 10.2 节简要回顾了云和降水建模和参数化；第 10.3 节将使用传统反演方法与基于信号新反演方法对观测和模拟结果进行了比较；总结和结论详见第四节。

10.2 大气模式中的云和降水参数化

10.2.1 大气中云和降水的模拟

云和降水微物理参数化的原型是由 Kessler 开发的[21,22]。大气中的水凝物使用两种不同类型进行模拟：云和降水。云粒子的重力沉降被忽略，而降水粒子有特定的下落速度，因此降水可以落向地面。云微物理过程，例如云粒子通过碰并自动转化为降水，使用两类粒子之间的质量转移和水汽来表示。Kessler[22]给出了云和降水粒子的质量密度连续性方程的微分形式，对这些方程进行数值求解，对质量密度的变化进行计算。

与现有的主流大气模式相比，文献[21,22]中对云和降水的建模和参数化设计较为原始，特别是在云和涡分辨率尺度上，但是，最新的模式沿用依然使用基本的概念和设计，大气水凝物粒子根据其密度、相态、生长过程等特征的不同可分为多种类型，然后，各类粒子的质量密度或混合比的连续性方程是数值求解的；引入质量加权平均下落末速度[48]，假定粒径分布(PSD)是由粒径的函数表示，一个水凝物类型中不同大小的粒子可以使用不同的下落速度描述，与云微物理过程相关的每个空间网格点上的水凝物类型之间的质量转化通常在每个时间步长内完成。这一假设使模式编程更容易，更适合通过并行计算确认加速数值。

大气水凝物的分类可以从粒子质量扩展到数量还有其他的高阶矩。另一方面，一种特殊的建模方法根据粒子的大小，将具有相同粒子特性但大小不同的粒子通过离散化进行分类。通过各种分类，类别的组合是无限的。随着类别数量的增加，需要求解更多的连续性方程，从而降低了计算速度。因此，开发人员需要决定应该从无穷无尽的组合中选择什么类别，以最有效地表示云和降水。迄今为止，在云和降水建模的历史上已经提出和发展了大量的分类模式。

10.2.2 云微物理参数化的典型模式

所谓的体积水云微物理参数化形成了最近大气模式的主流。液态的大气水凝物一般分为两类液相和冰/冰相，液相包括云滴和雨滴，冰相包括云冰、雪、霰和/或冰雹等类型。根据求解连续性方程的阶数体积微物理参数化进一步分为几种类型：只针对一个矩的方案称为一或单参体微物理，它只预测质量浓度；双参方案通常考虑水凝物粒子的质量和数量。

体积水微物理参数化中公式的一个例子可以在[45,46]中找到。水凝物类别的第 k 阶矩公式为：

$$M^k = \int_{x_1}^{x_2} x^k f(x) \mathrm{d}x \tag{10.1}$$

其中 x 为粒子质量，$f(x)$为数密度粒径分布函数（单位体积大气单位粒径间隔中的粒子数）；x_1 和 x_2 分别表示不同类别粒径分布的下限和上限。零阶矩为所分类水凝物的数浓度，一阶矩为质量密度。

体积云微物理方案通常对粒子谱分布（PSD）应用具有少量截距参数的特殊数学函数。因此，阶矩适合使用微分和积分的形式表示。该方法有效地解决了云微物理过程公式化中由于PSD变化可能出现的闭合问题。例如，文献[45]表明，粒子通过凝结和碰并增长的公式可能在阶矩的变化方程中包含一个高阶矩。

特殊数学函数的选择取决于方案的设计。通常，广义伽玛分布中的子集在公式化水凝物PSD时是首选的，数浓度粒径分布的典型伽马分布由三个参数决定[53]：

$$N(D) = N_0 D^{\mu} \exp(-\lambda D) \tag{10.2}$$

式中，D 为粒子直径；N_0、μ 和 λ 是分布的截距参数，如果 $\mu=0$，该分布与指数分布相同。式(10.1)中的任意阶矩都可以用这三个参数和伽玛函数表示，从而便于在解析计算中处理阶矩。从 Marshall 和 Palmer[30] 开始，伽玛分布就被广泛应用于观测到的水凝粒子 PSD 中。此外，伽玛分布拟合水凝物 PSD 的研究历史与使用天气雷达探测值的降雨反演技术的发展密切相关[1]。

迄今为止，许多类型的体积云微物理参数化已经开发出来，用于各种业务和研究目的的大气模式中。在 Khain 等[25] 中可以找到参数化及其参考的子集。天气研究和预报（WRF）模式[47]是一个主要的区域大气模式，在该模式中有大量可选择的微观物理参数化选项。

与使用特殊数学函数的体积云微物理参数化方案相比，谱-分档云微物理参数化方案提出通过直接离散水凝物 PSD 分布来避免闭合问题。大多数谱分档参数化采用传统的分类方法对冰相水凝物粒子进行分类，这与体积云微物理参数化中使用的方法类似。然而，一些谱分档参数化也将直接离散化应用于决定冰相粒子特征的参数，例如冰粒子的体积密度、纵横比等[12]。此外，双参谱-分档微物理[4]被开发出来，以克服在传统单参谱分档参数化[23]中离散的PSD的不连续问题。

与体积云微物理方案相比，已经开发了有限数量的谱-分档云微物理参数化方案，这个数量很小的原因是谱-分档微物理方案需要更大的计算资源，而它不能产生额外的可解析值，特别是在用于业务数值天气预报中。模拟 PSD 的信息由于变异性大、实用性低，往往未得到充分利用。虽然谱分档微物理通过求解大量的连续性方程直接计算了离散化水凝物 PSD 的变化，但它并不能保证比体积云微物理更好地模拟地表降水率。

10.3 大气模式中云微物理参数化产品与遥感探测的比较

10.3.1 与反演物理量的比较

通过使用反演技术的变量转换、时空插值和去噪等过程，从遥感探测中获得的直接辐射率或后向散射数据被处理成更高级别的数据产品。高级别产品通常由有用的地球物理参数组成。因此，探测数据很容易被用于物理现象的分析，特别是通过统计方法。

在进行大气模式的后验模拟结果与遥感探测结果比较时，通常探测数据使用这种更高级别的数据产品，模拟从预报变量中计算出与探测数据产品相同的地球物理参数。通常情况下，模拟数据或探测数据将其固有的空间分辨率被重新构建为另一个空间分辨率，因此可以计算出两者在同一空间坐标上的差值。该过程可以计算统计参数和分数，如总体偏差、相关系数、均方根误差和命中率，以便进行模拟结果与观测结果的系统偏差比较。通过改进这些统计参数和分数以减少误差是实现更好模拟结果的重要环节，通过比较这些统计参数和分数，可以定量地评价数值模式或同一模式中参数配置的性能优劣。抽样比较越多，预期得出的结论就越有力。该工作流程是一种长期建立的模式，以便用于改进大气模式中的云和降水参数化。

由于更高级别的数据产品是由研究部门认为非常重要的地球物理参数(如地表降水率)组成，因此只要产品中的数据是完全准确的，模拟结果与它们相匹配就成为最终目标。不幸的是，这种更高级别的产品中包含误差和不确定性，这些误差和不确定性不仅存在于遥感探测数据中，而且存在于从校准辐射率/后向散射数据到产生更高级别产品的过程中。因此，在讨论对比结果时，需要密切关注严格意义上每个数据产品的偏差和不确定性。根据反演算法的开发版本和类型，即使探测平台相同，数据产品通常也有明显差异。由于直接探测的局限性，特别是在地球卫星观测覆盖为全球范围，不可能对这种数据产品进行完全验证。例如，文献[27]中给出了由热带降雨测量任务(TRMM)核心卫星探测的不同版本和不同降雨类型的反演算法得出的逐月纬向平均降水率的差异。在版本 4 的数据(初步修正了发布算法)中，根据算法的选择，平均降水估差异的范围在热带地区最多，约为 40%。在进行了第一次实质性改进之后，在版本 5 中，这个范围已经缩小到约 24%。Liu[29]研究了 TRMM 多卫星降水分析(TMPA)产品版本 6 和版本 7 之间的差异，版本 7 显示小雨出现频率高于版本 6，大雨出现频率则相反。

10.3.2 后向散射辐射对比的信号模拟器的研制

在 21 世纪，特别是在 A-Train 卫星群[50]初步形成之后，研究界认识到了遥感反演的重大不确定性，其中包括地表降水、云液水和冰水含量以及由不同卫星仪器和平台获得的光学深度。原因是，卫星反演算法必须设置独特的微物理假设，根据卫星仪器类型和可用通道频率估计地球物理参数。这些微物理假设往往在不同算法之间有所不同，也不同于大气模式中云微物理参数化所采用的相应假设。为了避免这种不一致，提出了一种替代方法，即直接利用经过校准的卫星探测原始信号，如可见光辐亮度、红外微波亮温、雷达反射率和激光雷达后向散射。同时，研制开发了所谓的卫星信号模拟器，目的是将模式输出变量转换为与卫星可观测信号相同的物理量[5,9,13,31,33,34]。

为了在模拟器中精确模拟卫星信号，卫星信号模拟器与大气模式之间描述云层、降水和背景环境的物理量分布必须相同。模式输出的气压、温度、水汽和其他气体种类被用来表示分子辐射的消光和吸收。此外，模式输出的云和降水必须准确地表示遥感探测的发射和散射信号，包括水凝物粒子相态、大小和密度，以便表示仪器特定波长下辐射传输过程的尺寸积分消光系数 k，以及单次散射反照率 ω 和散射相位函数 p：

$$k = \frac{\pi}{4}\int D^2 Q_{\text{ext}}(D)N(D)\mathrm{d}D \tag{10.3}$$

$$\tilde{\omega} = \frac{\pi}{4k}\int D^2 Q_{\text{scat}}(D)N(D)\mathrm{d}D \tag{10.4}$$

$$P = \frac{\int D^2 p_{\text{scat}}(D)Q_{\text{scat}}(D)N(D)\mathrm{d}D}{\int D^2 Q_{\text{scat}}(D)N(D)\mathrm{d}D} \tag{10.5}$$

其中粒径分布 $N(D)$必须与大气模式中使用的云微物理参数化中的假设一致；Q_{ext}、Q_{scat}和 p_{scat}分别为直径为 D 的粒子的消光效率、散射效率和相位函数。估算这些单粒子要素时还必须与大气模式的云微物理参数化的假设相一致，包括粒子相位、密度及可能的形状。Q_{ext}、Q_{scat}一般来源于实验室的结果，而 p_{scat}既可以来源于解析解也可以来源于数值解。通常引入无量纲尺度参数 X 来计算 p_{scat}；用粒子直径 D 和归一化波长 λ 表示为：

$$X=\frac{\pi D}{\lambda} \tag{10.6}$$

当 X 小到可以忽略不计（<0.002）时，可以忽略来自该尺寸元素（如分子）的散射；当 X 在 0.002～0.2 时，将散射机制归类为瑞利散射，其中无偏振入射辐射的相位函数 P(θ)表示为：

$$P_{\text{scat}}(\theta)=\frac{3}{4}(1+\cos^2\theta) \tag{10.7}$$

公式中 θ 为散射角。上面的公式可以应用于任何粒子形状。单个粒子可视为有效密度为 ρ_e 的蓬松球。在这种散射机制下，通过复杂介质的解析解，使用空气、水、冰等不同特性混合后的介电常数，以表示不同的水凝物特征[2]。

当 X 大于 0.2 时，正向散射比瑞利散射强，散射相位函数更加复杂。这些散射机制分为 Mie 散射（$0.2<X<2000$）和几何散射（$2000<X$）。在这些机制中，Lorenz-Mie[11]和 T 矩阵方案（如[35]）被广泛用作估计球形和扁圆形粒子散射的解析解。然而，大多数冰晶及其聚合体的形状完全不同于球形或扁圆形，为了处理这些粒子，需要使用更多计算量的解决方案，如离散偶极近似[6]和时域有限差分方法[55]，这些数值解通常被用来精确地表示冰或混合相云的 W 波段雷达后向散射或微波亮度温度[39]。

一旦散射和吸收粒子元素在粒径和类型上被整合，就可以通过求解辐射传输来估计探测器的辐亮度。辐射传输计算基于各种数学方法、数值模式技术以及多种假设，这些假设取决于粒子的散射量级[49]和偏振大小[8]以及边界条件（例如太阳常数、地表温度）。在主动传感器的情况下，如雷达/激光雷达，后向散射信号由每个目标物产生辐射或后向散射则进一步在特定的卫星覆盖范围(即视场)上通过高斯加权函数进行卷积得到[33]。

大气模式与正演模拟器在水凝物粒子相态、PSD和其他背景环境假设方面的一致性必须尽可能一致，同样重要的是，需要采用适当的前向辐射传输和散射模型，其质量与实际反演中使用的模型相当。

图10.1给出示意图，用以说明基于传统反演方法和基于模拟器方法之间的区别。在基于反演的比较中，根据反演算法的不同，目标物理量可以有多个产品。通常情况下，反演算法中的假设和模式参数化不需要进行匹配比较。事实上，来自不同反演算法得到的物理量的差异可能是评估模型仿真结果的一个问题。相比之下，基于模拟器的比较遵循更直接的工作流程，以获得合成校准辐亮度数据。由于信号模拟器采用了与模型参数化相同的假设，因此在模型参数化中使用的假设存在固有的差异，从而使获得的合成数据存在差异；因此，综合校正数据相对于观测数据的误差与耦合参数化的模型模拟误差更密切相关[34]。

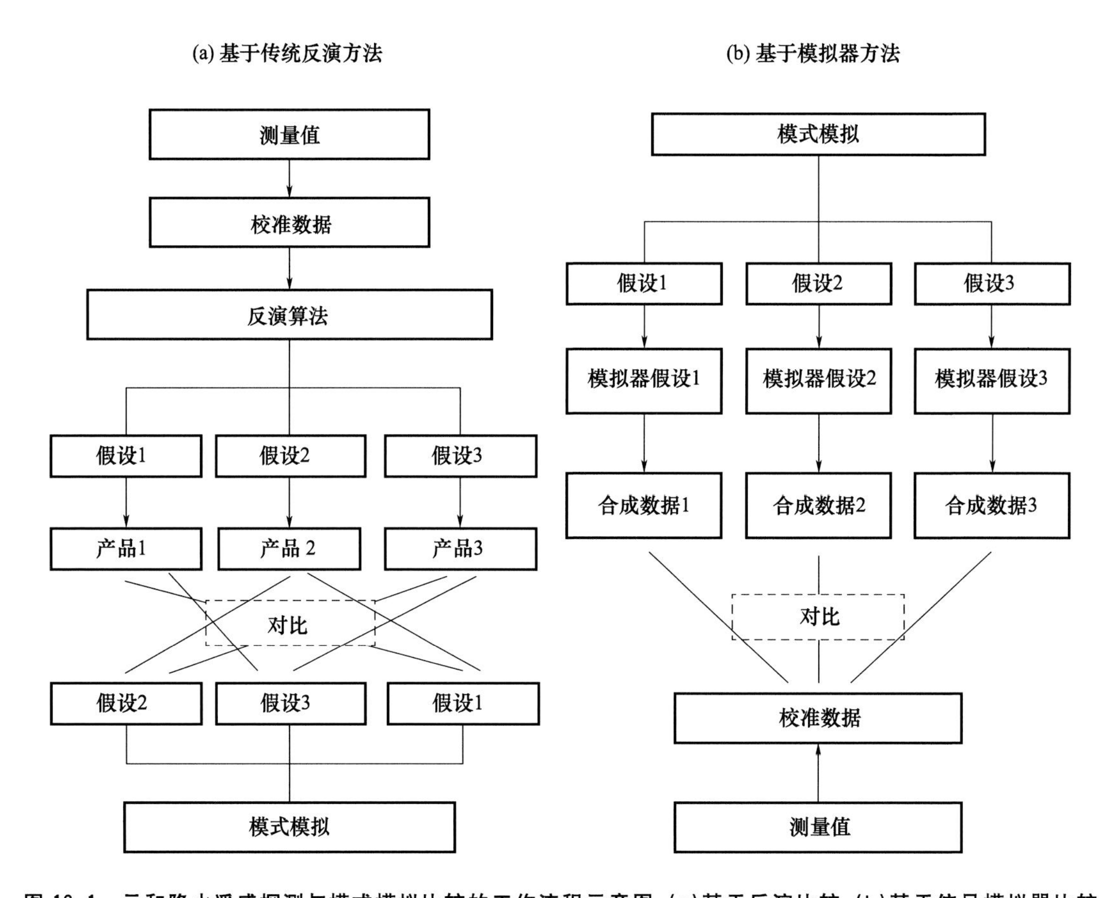

图10.1　云和降水遥感探测与模式模拟比较的工作流程示意图：(a)基于反演比较；(b)基于信号模拟器比较

与传统基于反演比较的方法相比，使用信号模拟器进行模式评估具有这些优点。然而，使用获得的辐射度或后向散射来解释评估结果需要一定的理论基础和理解能力，因为它们包含各种固有的物理量。例如，PSD、水凝物粒子密度和相态都是根据传感器通道频率所产生的微波发射和散射信号的相关参数得到的。

在通过此类模拟器模拟辐射信号时，谱分档微物理参数化在本质上优于体积水微物理参数化。水凝物的 PSD 信息是模拟受云和降水影响的辐射信号的基本要素。通过谱分档微物理模型模拟得到的水凝物 PSD 可以直接输入在信号模拟器中，因此模拟的信号具有显著的可变性，这可能是与自然界中各种 PSD 观测信息相一致所必需的。相比之下，来自体积水微物理参数化的相应模拟信号的变异性受到 PSD 模型假设的严格限制。

10.3.3 使用信号模拟器的案例研究回顾

本小节简要回顾了以往基于数值模拟和通过辐射信号模拟器进行的遥感探测比较的研究。由于对反演的物理量进行比较更为常见并且得到了广泛使用，因此本节特别介绍使用信号模拟器的研究。

Iguchi 等[18]通过与舰载或星载 W 波段雷达探测值的比较，分析了中纬度降水系统的后验模拟。日本气象厅非流体静力模型（JMA-NHM[44]）与希伯来大学云模型（HUCM[23]）和单参谱分档云微物理[7]用于水平网格间距为 3 km 的中尺度后验模拟。设计了一个离线雷达信号模拟器[37,38]，将后验模拟的输出转换为与实际舰载或星载雷达探测等效的合成雷达反射率因子和多普勒速度。模拟器中 PSD 谱和粒子下落速度等云微物理的假设与单参谱分档或体积微物理参数化的假设相一致，比较时参考未经过订正大气水凝物和气体衰减校正的观测雷达反射率因子，以便对衰减进行建模并纳入信号模拟器，从后验模拟计算相应的反射率因子。

图 10.2 显示了雷达反射率因子的时间-高度序列图，该反射率因子由舰载 W 波段雷达探测得到，并通过信号模拟器利用 JMA-NHM 与单参微物理耦合计算得到。信号模拟器每隔 1 分钟对水平网格中最接近科考船坐标的大气柱中的模型输出变量进行一次采样。这种配置旨在使模型输出的采样与实际测量尽可能一致。

然而，两图的比较看出在探测和后验模拟之间云的发生有些不一致，判断这不是由于云微物理参数化的问题。这种差异往往是由于通过初始和横向边界条件对后验进行确定性运行的强迫数据集控制下的预报误差，而不是由模型微物理方案的差异引起的。这些误差阻碍了对云微物理参数化中存在问题的讨论，不利于云微物理参数化的改进，这个问题在模型物理参数化验证的测量比较研究中很常见，不局限于使用信号模拟器的研究。

本研究引入了归一化等值线频率高度图（CFADs）分析等进行定量讨论，以避免预报误差对讨论的影响。结果表明，模拟结果高估了冰点高度以上冰相云和混合相云的反射率因子，这一结论与在文献[28]中使用几乎相同的谱-分档微物理参数化方案的研究结果一致。然而，针对反射率因子预测的高估问题，不同文献提出了不同的解决方法。Iguchi 等[18]从控制试验中测试了雪聚合物的下落速度增加和冰成核过程的不同参数化方案。另一方面，Li 等[28]测试了雪聚合物的体密度和下落速度的变化、温度对冰粒子收集效率的作用，同时也引入了大粒径雪聚合物的破碎。这两份出版物的这些变化在减少反射率的过度预测和减少观测和模拟之间的差异方面发挥了相似的作用。然而，这些选项中哪一种方法能有效地解决模式参数化问题是不确定的，与观测结果的比较缺乏足够的信息来确定云微物理计算中的误差来源。最终，HUCM 中谱分档微物理的最新版本[24]引入了大粒径雪聚合物的破碎过程，尽管没有直接的证据证明该更新是正确的。

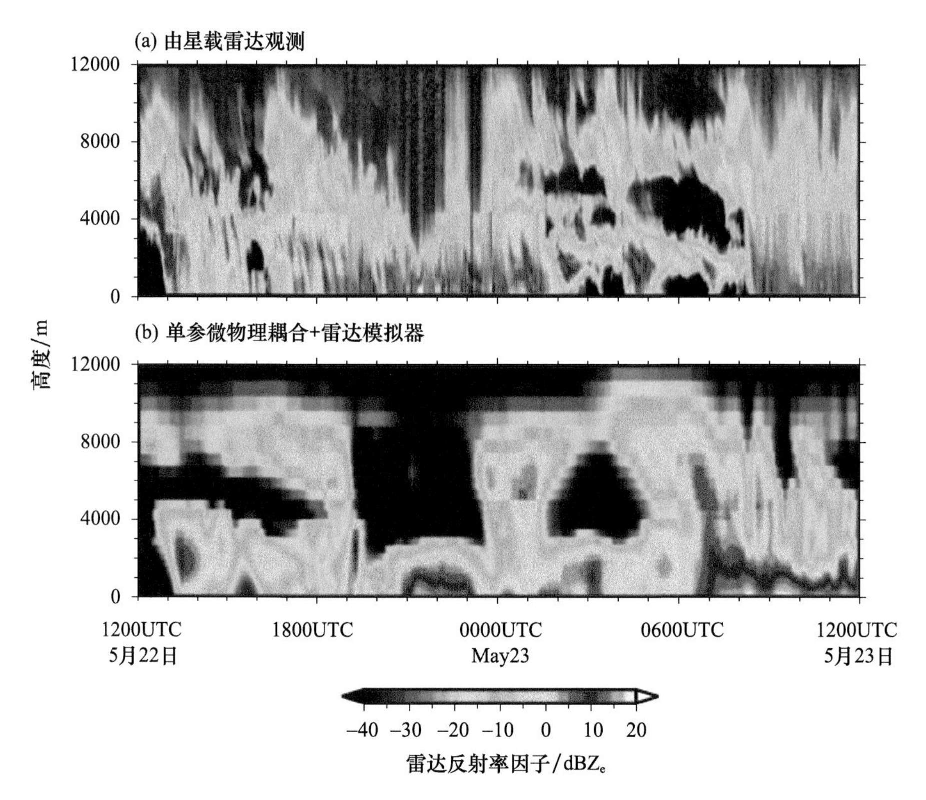

图 10.2 等效雷达反射率因子(dBZ_e)的时间高度剖面图(2001 年 5 月 22 日 1200 UTC—23 日 1200UTC)
(a)科考船上的 95-GHz 多普勒雷达探测结果;(b)使用 JMA-NHM 与谱-分档微物理方案耦合输出的雷达产品模拟器计算结果(源自文献[18],有变动)

Iguchi 等[18]使用这种信号模拟器对云微物理参数化进行了评估和改进的初步研究。在整个讨论过程中发现了许多限制和问题,特别是,谱分档微物理参数化比典型的体积微物理需要更多的方程和参数。因此,在敏感性试验中可以改变许多参数来检验模拟结果的改进。事实上,作者对冰相水凝物的凝华增长速度、雪聚合效率、冰成核参数化、水凝物粒子的下落速度等进行了修改,并对这些成分的复合进行了超过 30 种的敏感性试验,这种基于几乎随机猜测的方法在模拟和分析中增加了不必要的成本。尽管该报告得出的结论是,增加雪聚集物的下落速度是一种有效的解决方案,但这一结论并不是基于从案例研究中现场观测结果进行直接比较而获得的令人信服的科学证据。这个问题是目前科学讨论中的一个薄弱环节,在类似的研究中,可以通过与遥感测量的比较来改进模型物理参数化。研究界需要建立稳健和合理的方法来讨论云微物理参数化中每个组成部分产生误差的概率和可能性,而不是仅仅基于灵敏度测试来确定误差来源的传统方法。

Iguchi 等[19]尝试验证引入 HUCM 谱-分档微物理方案的冰粒子时间依赖性融化方案[40,24]。在大气中冰相水凝物下落过程中的融化会产生一个亮带,在雷达回波剖面中被识别为在 0℃温度层以下具有相对较大反射率的水平方向的薄层。

更好地了解冰粒子融化和随后形成的暗带是从雷达探测准确估计降水率的基本要素。对亮带结构的精确模拟对于理解微物理过程及其与云动力学的相互作用具有重要意义。然而,在典型的大气模式中，大多数云微物理参数化在模拟亮带结构时，由于简化了冰粒子融化的模型，需要增加主观假设。

文献[19]对2010年秋季在芬兰南部进行的弱降水验证试验(LPVEx)野外活动中的两个降水事件进行了分析，将具有时间依赖的HUCM谱-分档微物理融化方案引入到WRF(WRF-arw)模型的高级研究版本3.4[47]中。为了突出冰粒子融化对亮带结构的雷达反射率和多普勒速度分布的影响，分别进行了包含或不含时间依赖性融化方案的两种WRF后验模拟。戈达德卫星数据模拟器单元(G-SDSU)[32,33]中的离线雷达模拟器模块用于计算合成雷达反射率和多普勒速度廓线。WRF模拟所预测的融化冰粒子的液态水分数被包括在雷达模拟器的计算中，以便除了其他标准模式输出变量外，对亮带的模拟更加一致。

图10.3以归一化频率等值线温度图(CFTDs)的形式显示了第一次降水个例中观测和模拟的雷达反射率和多普勒速度廓线。观测到的雷达反射率和多普勒速度来自于部署在LPVEx试验现场的垂直指向微雨雷达(发射频率24.15 GHz)的探测结果，将这些数据与在一个现场观测到的探空数据的垂直温度廓线相结合绘制出CFTDs。与传统的CFADs相比，CFTDs在显示目标量的概率分布与温度之间的相关性方面更有用。这种相关性对于讨论冰粒子的融化以及随后根据周围大气温度形成的亮带非常重要，因为融化对温度的变化非常敏感。事实上，CFADs是在研究中绘制的，用于分析亮带结构，但没有在发表的出版物中出现。原因是真实大气的垂直温度廓线与模式模拟的不完全相同[19]，由于温度廓线的模拟误差，观测到的亮带与模拟的亮带的垂直位置可能不同；这个问题可能会掩盖通过分析亮带结构来模拟冰粒子融化的研究重点，引入时间依赖性融化方案对模拟亮带的影响在CFTD图中得到了展示。图10.3a中雷达反射率观测结果的CFTD显示表明反射率在0到3℃之间相对较大。在0℃层以下，反射率的增加似乎与亮带效应相对应。WRF-SBM与时间依赖性融化方案的模拟结果得到了类似的CFTD结构(图10.3b)，表明反射率在大约0到3℃的温度范围内增加。相比之下，使用过时的瞬时融化方案代替时间依赖性融化方案的模式模拟无法重现CFTD结构(图10.3c)；反射率的增加在温度范围内没有出现。从观测得到的多普勒速度CFTD(图10.3d)显示表明速度随温度在0到3℃范围内逐渐增加，采用时间依赖性融化方案的WRF-SBM模拟再次得到类似的CFTD结构(图10.3e)，而采用瞬时融化方案的模拟CFTD(图10.3f)没有出现这种速度逐渐变化的区域。

Iguchi等[18]试图对云和降水的整个结构进行整体验证，而文献[19]则专注于雷达测量中与特定云微物理过程相关的一个现象，即冰粒子的融化。尽管观测和模拟之间的分歧细节没有得到充分的讨论和解决，但后者能够提供关于下落的冰粒子逐渐融化对亮带形成的影响的更确凿的科学信息。

由于测量和大气模式模拟的多种不确定性和误差因素，即使是通过信号模拟器，也很难严格定量地讨论遥感测量和模式模拟之间的差异，以获得稳健的结论。其中一种解决方案是使用多传感器多频率方法，以避免微物理的局部调整[10,32,34]，更多的观测可以为微物理过程提供更好的约束。此外，新兴的偏振雷达可以更好地约束水凝物分布和风暴动力学[42,43]。

10.3.4　建立发展反演算法的测试平台

一旦基于辐射的模型评估甚至传统的基于反演的评估证明模拟的云和降水是有效的，就

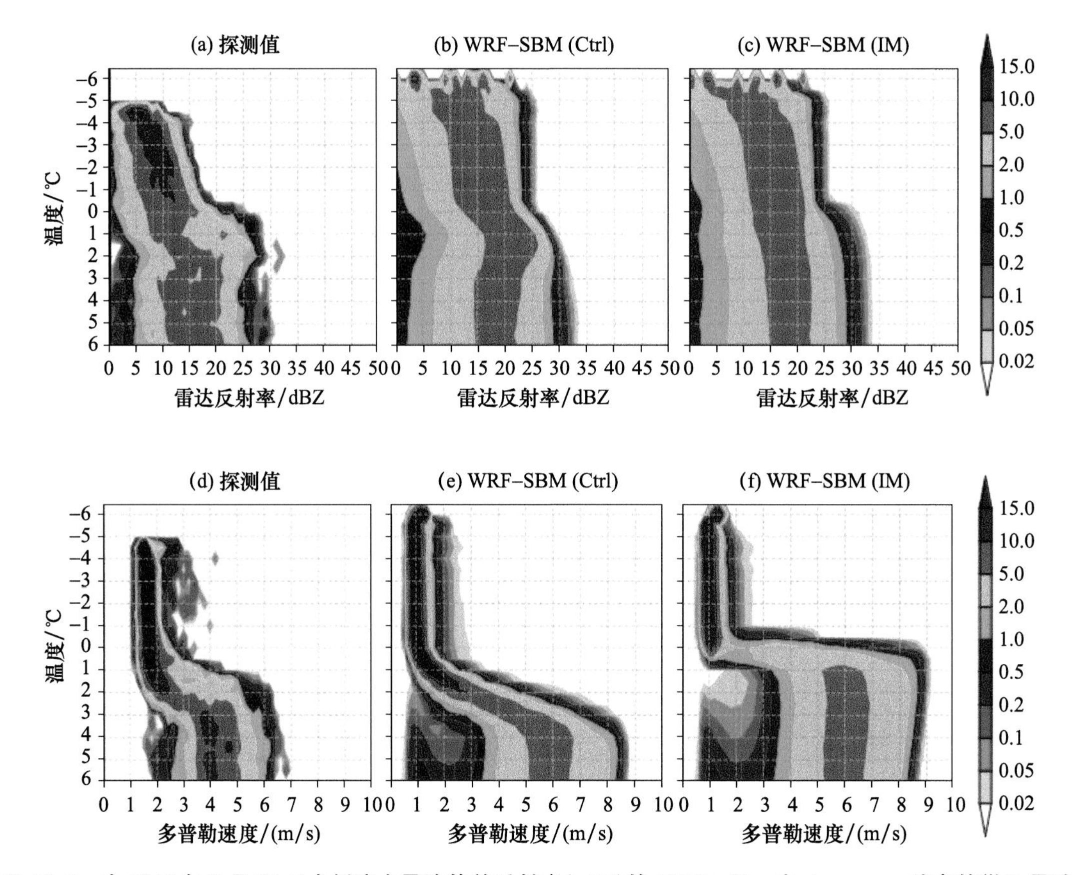

图 10.3 在 2010 年 9 月 21 日个例液水雷达等效反射率(dBZ)的 CFTDs 图。由 Jarvenpaa 站点的微雨雷达探测结果与 Jokioinen 站点的探空数据的垂直温度廓线相结合,并由 WRF 与谱分档微物理模拟相结合获得。(a)探测值,(b)WRF-SBM 模拟采用时间依赖性融化方案,(c)模拟采用瞬时融化方案。平均多普勒速度的 CFTDs(m/s):(d)探测值,(e)使用时间依赖性融化方案的模式模拟,和(f)使用瞬时融化方案的模拟(源自文献[19],有变化)

可以利用模拟的结果来测试和开发其他遥感测量反演算法,模型输出包含所有网格尺度的物理量,如压力、温度、水蒸气、液水和冰水含量等。首先,模式输出中的这些物理量通过信号模拟器转换为目标遥感仪器可以获得的辐射强度或后向散射;其次,通过反演算法将模拟器的辐射强度或后向散射转换为用于检验目标的物理量;最后,将反演所得物理量与原始模式输出中的相同物理量进行比较,以进行一致性检查。该过程的流程图见图 10.1[54]。

该研究的一个例子是全球降水测量(GPM)任务。GPM 是一个由多个地球观测卫星提供全球降水测量的国际任务[15]。GPM 核心卫星上的多通道雷达和辐射计在校准统一 GPM 卫星群测量结果的降水算法中发挥着核心作用。2014 年 2 月平台发射后,核心卫星开始从太空进行测量。然而,即使在核心卫星发射之前,第一个版本的 GPM 降水算法必须在现有数据库上建立和测试,主要针对现有 TRMM 核心卫星观测范围以外的中高纬度降水系统。所谓的合成 GPM 模拟器建立在各种 GPM 地面验证(GV)站点获得的外场试验探测数据集上[33]。基于地面和飞机的 GV 观测被用于验证和约束使用 WRF 模型与谱分档微物理参数化耦合的反演模拟[16,17,19]。随后,将 GV 约束的模式输出值输入合成 GPM 模拟器,生成 GPM 微波亮温和双频雷达反射率，以及模拟的降水率和含水量等地球物理参数。该 GPM 模拟器由详细的卫星-轨道和传感器扫描模块、微波辐射计和降水雷达的统一辐射传输模块以及对真实天线增益进行 WRF 模式仿真的模块组成。有了数据库，算法开发人员甚至可以在卫星发射之前测试他们的算法(图 10.4)。

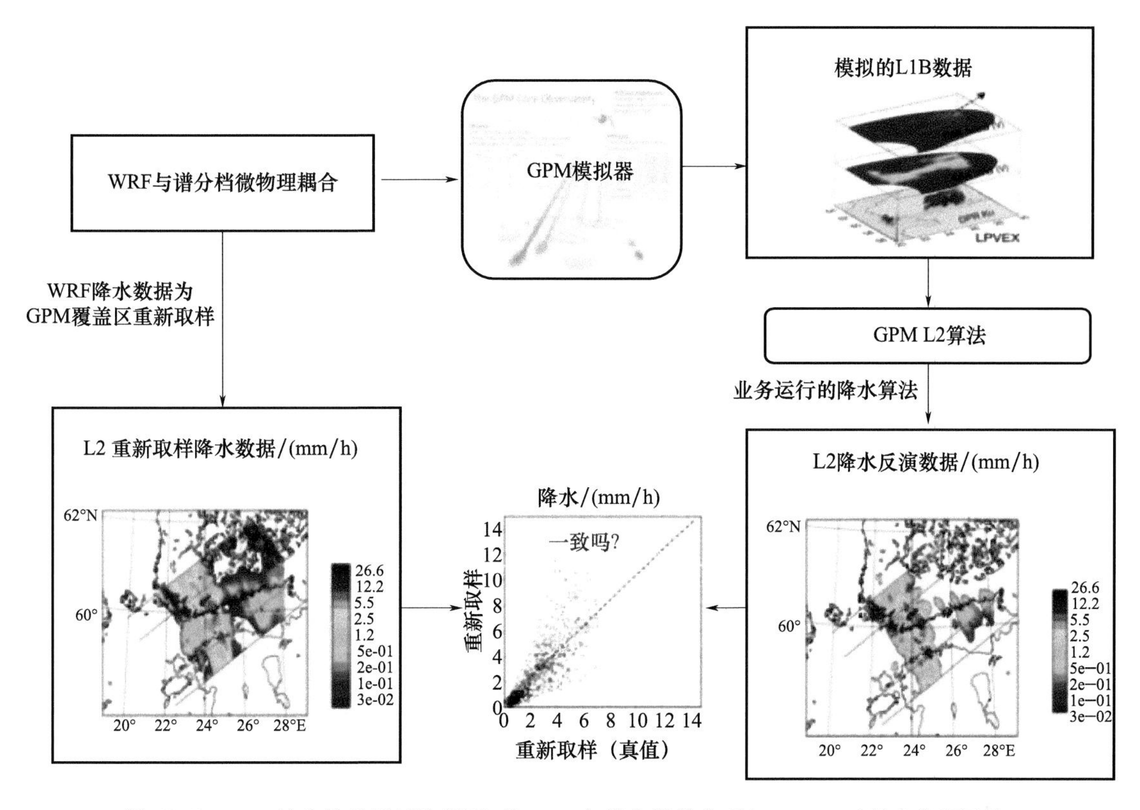

图 10.4　GPM 综合算法测试流程图，从 WRF 与谱分档微物理(WRF-SBM)耦合仿真开始，到利用合成信号数据和模拟降水率对 GPM 发射前降水算法进行评估

Kidd 等[26]建立了一种基于物理的反演方案，通过 GPM 任务卫星群上的交叉轨道(XT)被动微波传感器估算降水，该算法基于物理基础的戈达德廓线(GPROF)算法，使用从观测约束模型仿真中导出的数据库。该改进的 GPROF 方案利用 NASA 多尺度建模框架(MMF)数据库提供统一、全面的数据库，类似于锥形扫描(CS)反演方案使用的数据库。从一系列模拟中随机选择 MMF 内的大约 100 万列，即二维云解析模型廓线，然后应用到 GPM 模拟器[33]上，通过计算不同通道频率下传感器特定的覆盖区大小和视角，模拟来自前向辐射传输的合成信号或亮温，对模式数据库进行评价，并根据观测结果对模式数据库进行了改进，修改了 MMF 中水凝物的混合比和辐射传输的单次散射技术参数。该数据库通过提供一套更全面的全球微物理廓线，确保了气象条件和地表类型之间更好的一致性。数据库对 CS 数据库的偏差进行了修正，以确保最终产品的一致性。将该技术应用于微波湿度探测(MHS)传感器的观测，MMF-GPROF 算法在许多地区提供了与 CS 传感器相当或更好的降水估计。MMF-GPROF 算法还应用于先进微波技术探空仪(ATMS)和用辐射测量法测定热带湿度廓线的大气探测仪(SAPHIR)，提供了卫星群 XT 降水估算的第三版。

10.4 总结和结论

本章介绍了云和降水的遥感探测和后验模拟的比较方法，并回顾了几种大气模式的模拟结果，用以验证和改进云微物理参数化方案。传统的比较是基于通过反演算法从探测值中估计的物理量，与从模式模拟中计算出的相应量进行比较，这种方法仍然被广泛接受，作为一种简单直接的方法来识别目标量的误差，对天气和气候研究（如地表降水率）有很大的意义。然而，人们逐渐认识到从遥感探测中反演的物理量有很大的不确定性，这可能会降低验证的可靠性。此外，还提出了信号模拟器的概念，研究界已经开发了与各种遥感仪器相匹配的模拟器模块。利用这种模拟器可以在探测结果和大气模式模拟结果之间进行基于信号的比较，而不会存在反演算法中固有的不确定性。然而，探测和模拟得到的信号辐射的一致性并不能保证对所有预测和诊断变量（如地表降水率）的良好模拟。

尽管在信号模拟器上比较的不确定性较小，但由于强迫数据驱动模拟、动力学核心和其他物理参数化以及测量中的误差，云微物理参数化的评估受到误差和不确定性的阻碍。蒙特卡罗实验（即数值天气预报中的集合预报）是一种广泛使用的通过扰动初始条件来研究模拟结果中概率和误差的方法。另一方面，数据同化[20]旨在通过将观测数据纳入同化算法中，对模拟场进行系统校正，提高模式预报的准确度和精度。基于辐射的同化系统可以利用遥感探测数据，甚至不需要通过反演算法[56]将探测结果进行物理量转换。然而，传统的同化方法不能直接用于模式参数化的改进。

最近的一些研究直接将贝叶斯分析应用于探究模式输出中目标物理参数对物理参数化中特定参数的具体敏感性。Posselt[41]在单参体积云微物理参数化中的参数进行了扰动，以检验它们对代表云结构、动力学以及理想云分辨模拟中潜热和辐射加热的模式输出参数的影响，采用贝叶斯马尔可夫链蒙特卡（MCMC）算法进行了大量的集成模拟。研究得出的结论是，同化平均降水率、垂直积分液水含量和冰水含量以及辐射通量不能唯一地约束研究中所调查的一组参数。它还表明，包含唯一约束水凝物 PSD 信息的观测系统可能是产生参数值的唯一估计值的必要条件。

基于贝叶斯推论的这类分析有可能为利用遥感探测在云和降水建模方面的进展提供新的视角，超越基于推测的传统敏感性试验方法。通过讨论辐射/后向散射信号与目标物理量之间的参数依赖关系，也可能为使用信号模拟器的研究提供了重要的下一步。

致谢 作者对 NASA 降水测量任务（PMM）、NASA 建模分析预测（MAP）和美国能源部（DOE）大气系统研究（ASR）项目的支持表示感谢。

参考文献*

[1] Atlas,D. 1953. Optical extinction by rainfall. *Journal of Meteorology* 10:486-488.

[2] Bohren,C. F. ,and L. J. Battan. 1980. Radar backscattering by inhomogeneous precipitation particles. *Journal of the Atmospheric Sciences* 37:1821-1827.

[3] Borys,R. D. ,and M. A. Wetzel. 1997. Storm Peak Laboratory:a research,teaching,and service facility for the atmospheric sciences. *Bulletin of the American Meteorological Society* 78(10):2115-2123.

[4] Chen,J. -P. ,and D. Lamb. 1994. Simulation of cloud microphysical and chemical processes using a multicomponent framework. Part I:Description of the microphysical model. *Journal of the Atmospheric Sciences*51:2613-2630.

[5] Donovan,D. P. ,N. Schutgens,J. P. V. Baptista,H. Barker,J. P. Blanchet,A. Belaulne,and J. Testud. 2004. The EarthCARE Simulator. Users Guide and Final Report. KNMI,ESA/ESTEC,MSC,UQAM,GKSS,University of Kiel,CETP-UVSQ and Pennsylvania State University,28,48-49.

[6] Draine,B. T. ,and P. J. Flatau. 1994. Discrete-dipole approximation for scattering calculations. *Journal of the Optical Society of America* A11:1491-1499.

[7] Eito,H. ,and K. Aonashi. 2009. Verification of hydrometeor properties simulated by a cloudresolving model using a passive microwave satellite and ground-based radar observations for a rainfall system associated with theBaiu front. *Journal of the Meteorological Society of Japan. Series II*87:425-446.

[8] Evans,K. F. ,and G. L. Stephens. 1991. A new polarized atmospheric radiative transfer model. *Journal of Quantitative Spectroscopy and Radiative Transfer*46:413-423.

[9] Han,Y. ,P. vanDelst,Q. Liu,F. Weng,B. Yan,R. Treadon,and J. Derber. 2006. JCSDA communityradiative transfer model(CRTM)-Version 1. NOAA Technical Report NESDIS,122,33.

[10] Han,M. ,S. A. Braun,T. Matsui,and C. R. Williams. 2013. Evaluation of cloud microphysics schemes in simulations of a winter storm using radar and radiometer measurements. *Journal of Geophysical Research:Atmospheres*118:1401-1419.

[11] Hansen,J. E. ,and L. D. Travis. 1974. Light scattering in planetary atmospheres. *Space Science Reviews*16:527-610.

[12] Hashino,T. ,and G. J. Tripoli. 2007. The Spectral Ice Habit Prediction System(SHIPS). Part I:Model description and simulation of the vapor deposition process. *Journal of the Atmospheric Sciences*64:2210-2237.

[13] Haynes,J. M. ,Z. Luo,G. L. Stephens,R. T. Marchand,and A. Bodas-Salcedo. 2007. Amultipurpose radar simulation package:QuickBeam. *Bulletin of the American Meteorological Society* 88:1723-1727.

[14] Heymsfield,A. J. 2007. On measurements of small ice particles in clouds. *Geophysical Research Letters*34(23). https://doi. org/10. 1029/2007GL030951.

[15] Hou,A. Y. ,R. K. Kakar,S. Neeck,A. A. Azarbarzin,C. D. Kummerow,M. Kojima,R. Oki,K. Nakamura,and T. Iguchi. 2014. The global precipitation measurement mission. *Bulletin of the American Meteorological Society*95:701-722.

[16] Iguchi,T. ,T. Matsui,J. J. Shi,W. K. Tao,A. P. Khain,A. Hou,R. Cifelli,A. Heymsfield,and A. Tokay.

* 参考文献沿用原版书中内容,未改动

2012. Numerical analysis using WRF-SBM for the cloud microphysical structures in the C3VP field campaign: impacts of supercooled droplets and resultant riming on snow microphysics. *Journal of Geophysical Research-Atmospheres* 117. https://doi.org/10.1029/2012JD018101.

[17] Iguchi, T., T. Matsui, A. Tokay, P. Kollias, andW. K. Tao. 2012. Two distinct modes in one-day rainfall event during MC3E field campaign: analyses ofdisdrometer observations and WRFSBM simulation. *Geophysical Research Letters*, 39. https://doi.org/10.1029/2012GL053329.

[18] Iguchi, T., T. Nakajima, A. P. Khain, K. Saito, T. Takemura, H. Okamoto, T. Nishizawa, and W. K. Tao. 2012. Evaluation of cloud microphysics in JMA-NHM simulations using bin or bulk microphysical schemes through comparison with cloud radar observations. *Journal of the Atmospheric Sciences* 69: 2566-2586.

[19] Iguchi, T., T. Matsui, W. K. Tao, A. P. Khain, V. T. J. Phillips, C. Kidd, T. L'Ecuyer, S. A. Braun, and A. Hou. 2014. WRF-SBM simulations of melting-layer structure in mixed-phase precipitation events observed duringLPVEx. *Journal of Applied Meteorology and Climatology* 53: 2710-2731.

[20] Kalnay, E. 2003. *Atmospheric Modeling, Data Assimilation and Predictability*. Cambridge: Cambridge University Press.

[21] Kessler, E. 1969. *On the Distribution and Continuity of Water Substance in Atmospheric Circulations*. Meteorological Monographs. Vol. 32. Berlin: Springer.

[22] Kessler, E. 1995. On the continuity and distribution of water substance in atmospheric circulations. *Atmospheric Research* 38: 109-145.

[23] Khain, A., M. Ovtchinnikov, M. Pinsky, A. Pokrovsky, and H. Krugliak. 2000. Notes on the state-of-the-art numerical modeling of cloud microphysics. *Atmospheric Research* 55: 159-224.

[24] Khain, A. P., V. Phillips, N. Benmoshe, and A. Pokrovsky. 2012. The role of small soluble aerosols in the microphysics of deep maritime clouds. *Journal of the Atmospheric Sciences* 69: 2787-2807.

[25] Khain, A. P., K. D. Beheng, A. Heymsfield, A. Korolev, S. O. Krichak, Z. Levin, M. Pinsky, V. Phillips, T. Prabhakaran, and A. Teller. 2015. Representation of microphysical processes in cloud-resolving models: spectral(bin)microphysics versus bulk parameterization. *Reviews of Geophysics* 53: 247-322.

[26] Kidd, C., T. Matsui, J. Chern, K. Mohr, C. Kummerow, and D. Randel. 2016. Global precipitation estimates from cross-track passive microwave observations using a physically based retrieval scheme. *Journal of Hydrometeorology* 17: 383-400.

[27] Kummerow, C., J. Simpson, O. Thiele, W. Barnes, A. T. C. Chang, E. Stocker, R. F. Adler, A. Hou, R. Kakar, and F. Wentz. 2000. The status of the Tropical Rainfall Measuring Mission (TRMM) after two years in orbit. *Journal of Applied Meteorology* 39: 1965-1982.

[28] Li, X., W.-K. Tao, T. Matsui, C. Liu, and H. Masunaga. 2010. Improving a spectral bin microphysical scheme using TRMM satellite observations. *Quarterly Journal of the Royal Meteorological Society* 136: 382-399.

[29] Liu, Z. 2015. Comparison of precipitation estimates between Version 7 3-hourly TRMM Multi-Satellite Precipitation Analysis(TMPA)near-real-time and research products. *Atmospheric Research* 153: 119-133.

[30] Marshall, J. S., and W. M. Palmer. 1948. The distribution of raindrops with size. *Journal of Meteorology* 5: 165-166.

[31] Masunaga, H., T. Matsui, W.-K. Tao, A. Y. Hou, C. D. Kummerow, T. Nakajima, P. Bauer, W. S. Olson, M. Sekiguchi, and T. Y. Nakajima. 2010. Satellite data simulator unit: amultisensor, multispectral satellite simulator package. *Bulletin of the American Meteorological Society* 91: 1625-1632.

[32] Matsui, T., X. Zeng, W.-K. Tao, H. Masunaga, W. S. Olson, and S. Lang. 2009. Evaluation of long-term cloud-resolving model simulations using satellite radiance observations and multifrequency satellite simulators. *Journal of Atmospheric and Oceanic Technology* 26:1261-1274.

[33] Matsui, T., T. Iguchi, X. Li, M. Han, W.-K. Tao, W. Petersen, T. L'Ecuyer, R. Meneghini, W. Olson, and C. D. Kummerow. 2013. GPM satellite simulator over ground validation sites. *Bulletin of the American Meteorological Society*94:1653-1660.

[34] Matsui, T., J. Santanello, J. J. Shi, W.-K. Tao, D. Wu, C. Peters-Lidard, E. Kemp, M. Chin, D. Starr, and M. Sekiguchi. 2014. Introducingmultisensor satellite radiance-based evaluation for regional Earth System modeling. *Journal of Geophysical Research: Atmospheres* 119:8450-8475.

[35] Mishchenko, M. I., L. D. Travis, and D. W. Mackowski. 1996. T-matrix computations of light scattering bynonspherical particles: a review. *Journal of Quantitative Spectroscopy and Radiative Transfer* 55: 535-575.

[36] Murakami, M., and T. Matsuo. 1990. Development of the hydrometeorvideosonde. *Journal of Atmospheric and Oceanic Technology*7(5):613-620.

[37] Okamoto, H., T. Nishizawa, T. Takemura, H. Kumagai, H. Kuroiwa, N. Sugimoto, I. Matsui, A. Shimizu, S. Emori, and A. Kamei. 2007. Vertical cloud structure observed from shipborne radar and lidar: midlatitude case study during the MR01/K02 cruise of the research vessel Mirai. *Journal of Geophysical Research: Atmospheres* 112:D08216.

[38] Okamoto, H., T. Nishizawa, T. Takemura, K. Sato, H. Kumagai, Y. Ohno, N. Sugimoto, A. Shimizu, I. Matsui, and T. Nakajima. 2008. Vertical cloud properties in the tropical western Pacific Ocean: validation of the CCSR/NIES/FRCGC GCM by shipborne radar and lidar. *Journal of Geophysical Research: Atmospheres* 113. https://doi.org/10.1029/2008JD009812.

[39] Olson, W. S., L. Tian, M. Grecu, K.-S. Kuo, B. T. Johnson, A. J. Heymsfield, A. Bansemer, G. M. Heymsfield, J. R. Wang, and R. Meneghini. 2016. The microwave radiative properties of falling snow derived fromnonspherical ice particle models. Part II: Initial testing using radar, radiometer and in situ observations. *Journal of Applied Meteorology and Climatology* 55:709-722.

[40] Phillips, V. T. J., A. Pokrovsky, and A. Khain. 2007. The influence of time-dependent melting on the dynamics and precipitation production in maritime and continental storm clouds. *Journal of the Atmospheric Sciences*64:338-359.

[41] Posselt, D. J. 2016. A Bayesian examination of deep convective squall-line sensitivity to changes in cloud microphysical parameters. *Journal of the Atmospheric Sciences* 73:637-665.

[42] Putnam, B. J., M. Xue, Y. Jung, G. Zhang, and F. Kong. 2017. Simulation of polarimetric radar variables from 2013 CAPS spring experiment storm-scale ensemble forecasts and evaluation of microphysics schemes. *Monthly Weather Review* 145:49-73.

[43] Ryzhkov, A., M. Pinsky, A. Pokrovsky, and A. Khain. 2011. Polarimetric radar observation operator for a cloud model with spectral microphysics. *Journal of Applied Meteorology and Climatology*50:873-894.

[44] Saito, K., T. Fujita, Y. Yamada, J.-I. Ishida, Y. Kumagai, K. Aranami, S. Ohmori, R. Nagasawa, S. Kumagai, and C. Muroi. 2006. The operational JMA nonhydrostatic mesoscale model. *Monthly Weather Review*134:1266-1298.

[45] Seifert, A., and K. D. Beheng. 2001. A double-moment parameterization for simulating autoconversion, accretion and selfcollection. *Atmospheric Research* 59:265-281.

[46] Seifert, A., and K. D. Beheng. 2006. A two-moment cloud microphysics parameterization for mixed-phase clouds. Part 1: Model description. *Meteorology and Atmospheric Physics* 92:45-66.

[47] Skamarock, W. C., J. B. Klemp, J. Dudhia, D. O. Gill, D. M. Barker, M. G. Duda, X. Y. Huang, W. Wang, and J. G. Powers. 2008. A description of the Advanced Research WRF Version 3, NCAR technical note, Mesoscale and Microscale Meteorology Division. National Center for Atmospheric Research, Boulder, CO.

[48] Srivastava, R. C. 1967. A study of the effect of precipitation on cumulus dynamics. *Journal of the Atmospheric Sciences* 24:36-45.

[49] Stamnes, K., S.-C. Tsay, W. Wiscombe, and K. Jayaweera. 1988. Numerically stable algorithm for discrete-ordinate-method radiative transfer in multiple scattering and emitting layered media. *Applied Optics* 27:2502-2509.

[50] Stephens, G. L., D. G. Vane, R. J. Boain, G. G. Mace, K. Sassen, Z. Wang, A. J. Illingworth, E. J. O'Connor, W. B. Rossow, and S. L. Durden. 2002. The CloudSat mission and the A-Train: a new dimension of space-based observations of clouds and precipitation. *Bulletin of the American Meteorological Society* 83:1771-1790.

[51] Takahashi, T. 2010. Thevideosonde system and its use in the study of East Asian monsoon rain. *Bulletin of the American Meteorological Society* 91(9):1231-1246.

[52] Tobo, Y., J. Uetake, Y. Uji, Y. Iwamoto, K. Miura, and R. Misumi. 2017. Routine measurements of atmospheric ice nucleating particles on the TokyoSkytree, Atmospheric Ice Nucleation Conference-Focus Meeting 9, Leeds, UK. https://aerosol-soc.com/abstracts/routine-measurements-atmospheric-ice-nucleating-particles-tokyo-skytree.

[53] Ulbrich, C. W. 1983. Natural variations in the analytical form of the raindrop size distribution. *Journal of Climate and Applied Meteorology* 22:1764-1775.

[54] Woods, C. P., D. E. Waliser, J.-L. Li, R. T. Austin, G. L. Stephens, and D. G. Vane. 2008. EvaluatingCloudSat ice water content retrievals using a cloud-Rˇ resolving model: sensitivities to frozen particle properties. *Journal of Geophysical Research: Atmospheres* 113:D00A11.

[55] Yang, P., H. Wei, H.-L. Huang, B. A. Baum, Y. X. Hu, G. W. Kattawar, M. I. Mishchenko, and Q. Fu. 2005. Scattering and absorption property database fornonspherical ice particles in the near-through far-infrared spectral region. *Applied Optics* 44:5512-5523.

[56] Zhang, S. Q., T. Matsui, S. Cheung, M. Zupanski, and C. Peters-Lidard. 2017. Impact of assimilated precipitation-sensitive radiances on the NU-WRF simulation of the West African monsoon. *Monthly Weather Review*. https://doi.org/10.1175/MWR-D-16-0389.1.

术　语

A-矩阵(A-Train)
阿卡(Aqua)
暴雨(heavy rainfall)
贝叶斯估计理论(Bayesian estimation theory)
被动辐射计(passive radiometer)
冰雹(hail)
冰雹探测(hail detection)
冰粒(ice particles)
冰水含量(Ice Water Content (IWC))
波束宽度(beamwidth)
布拉格散射(Bragg scattering)
参数化(parameterization)
层云降水(stratiform precipitation)
差分传播相移(specific differential phase)
差分反射率(differential reflectivity)
超级单体(supercell)
尺寸参数(size parameter)
传播差分相移(propagation differential phase shift)
垂直指向雷达(vertically pointing radar)
磁控管(magnetron)
大气辐射(atmospheric radiation)
大气辐射测量(Atmospheric Radiation Measurement (ARM))
大气模型(atmospheric model)
大气透射率(atmospheric transmissivity)
大涡模拟(Large Eddy Simulation)
单散射反照率(single-scattering albedo)
等效反射率(effective reflectivity)
等效体积球形直径(equivalent volume spherical diameter)
低地球轨道(Low Earth Orbit (LEO))
低空旋转(low-level rotation)

低噪声放大器(low-noise amplifier)
地面验证(ground validation)
地球观测系统(Earth Observing System (EOS))
地物杂波目标(ground clutter targets)
电磁波(electromagnetic waves)
电磁辐射(electromagnetic radiation)
电荷分离(charge separation)
电活动(electrical activity)
定量分析(quantitative analysis)
定量降水估计(Quantitative Precipitation Estimation (QPE))
冬季风暴(winter storms)
对流单体(convective cell)
对流风暴(convective storms)
对流降水(convective precipitation)
多频雷达(multiple-frequency radar)
多普勒雷达(Doppler radar)
多普勒频谱(Doppler spectrum)
多普勒频谱宽度(Doppler spectrum width)
多普勒频移(Doppler shift)
多普勒效应(Doppler effect)
二次冻结特征(refreezing signature)
二次制冰(secondary ice production)
反射率(reflectivity)
反向散射(backscattering)
返回功率(returned power)
范围(range)
方位(azimuth)
非球形粒子(nonspherical particles)
菲涅耳(Fresnel)
峰值功率(peak power)
辐射传输(radiative transfer)
辐射计(radiometer)
傅里叶分析(Fourier analysis)
工作频率(operating frequency)
功率放大器(power amplifier)
共极相关系数(co-polar correlation coeffcient)
固态发射机(solid-state transmitters)

光环(Aura)
光学厚度(optical thickness)
广角照相机(wide-field-camera)
国家极轨业务环境卫星系统(NPOESS)
过冷水(supercooled liquid water)
海拔(elevation)
红外辐射成像计(infrared imager radiometer)
洪水(flood)
后向散射截面(backscattering cross-section)
后向散射相位差分(backscatter differential phase)
回波(echoes)
混合相态降水(mixed-phase precipitation)
极化(polarization)
极化雷达(polarimetric radar)
监控扫描(surveillance scan)
降水(precipitation)
降水测量(precipitation measurement)
降水分类方案(precipitation classification scheme)
降水雷达(precipitation radar)
降水强度(precipitation intensity)
结冰危险(icing hazard)
介电常数(dielectric constant)
近场相互作用(near-field interactions)
距离一高度指示(range-height indicator (RHI))
飓风(hurricane)
卡塞格伦天线(Cassegrain antenna)
科罗拉多州立大学 S 波段双极化可移动气象雷达(CSU-CHILL radar)
可见光红外成像辐射仪(VIIRS)
空间分辨率(spatial resolution)
快速扫描雷达(rapid-scan radar)
Ka 波段雷达(Ka-band radar)
馈源喇叭(feed horn)
拉曼激光雷达(Raman lidar)
雷暴(thunderstorm)
雷达波长(radar wavelength)
雷达波束宽度(radar beamwidth)
雷达采样体积(radar sampling volume)

雷达等效反射率系数(equivalent radar reflectivity factor)
雷达反射率(radar reflectivity)
雷达反射系数(radar reflectivity factor)
雷达方程(radar equation)
雷达观测体积(radar observing volume)
雷达回波(radar echoes)
雷达偏振变量(polarimetric radar variables)
雷达偏振测量(radar polarimetry)
雷达气象学(radar meteorology)
雷达散射截面(radar cross section)
雷达图像(radar imaging)
雷达校准(radar calibration)
离散偶极近似(Discrete Dipole Approximation (DDA))
粒子谱分布(Particle Size Distribution (PSD))
亮带(bright band)
亮度温度(brightness temperature)
龙卷(tornado)
龙卷爆发(tornado outbreak)
龙卷残骸(tornadic debris)
龙卷超级单体(tornadic supercell)
龙卷动力学(tornado dynamics)
龙卷警报(tornado warning)
龙卷碎片特征(tornado debris signature)
龙卷探测(tornado detection)
龙卷形成(tornadogenesis)
龙卷预报(tornado forecast)
龙卷涡旋特征(Tornado Vortex Signature (TVS))
脉冲长度(pulse length)
脉冲调制器(pulse modulator)
脉冲多普勒雷达(pulsed-Doppler radar)
脉冲宽度(pulse width)
脉冲重复频率(pulse repetition frequency)
脉冲重复时间(pulse repetition time)
美国国家航空航天局(NASA)
蒙特卡洛(Monte Carlo)
米氏理论(Mie theory)
模糊逻辑(fuzzy logic)

奈奎斯特速度(Nyquist velocity)
内部电场(internal electric field)
欧洲极轨卫星系列(MetOp)
欧洲中期天气预报中心(ECMWF)
偶极矩(dipole moment)
抛物面(parabolic dish)
频率(frequency)
平面位置显示(plan position indicator (PPI))
气候模型(climate model)
气溶胶和辐射监测卫星(EarthCARE)
气象(meteorology)
气象雷达(weather radar)
气象雷达网(weather radar network)
气象卫星(meteorological satellite)
气象卫星(weather satellite)
气象研究与预报模式(Weather Research and Forecasting (WRF))
强对流风暴(severe convective storms)
倾斜角度(canting angle)
球形粒子(spherical particles)
去极化(depolarization)
去极化比(depolarization ratio)
全球降水测量(Global Precipitation Measurement (GPM))
热带降雨测量任务卫星(TRMM)
热带气旋(tropical cyclone)
日本宇宙航空研究开发机构(Japan Aerospace Exploration Agency (JAXA))
融化层(melting layer)
瑞利近似(Rayleigh approximation)
瑞利散射(Rayleigh scattering)
散射(scattering)
散射截面(scattering cross section)
散射相位函数(scattering phase function)
扫视(sweep)
闪电(lightning)
神经网络(neural networks)
时域有限差分法(Finite Difference Time Domain Method)
数据同化(data assimilation)
数值模型(numerical model)

数值天气预报(Numerical Weather Prediction (NWP))
数字波束形成(digital beam forming)
衰减(attenuation)
衰减校正(attenuation correction)
双偏振(dual-polarization)
双偏振雷达(dual-polarization radar)
水凝物(hydrometeor)
水凝物分类(hydrometeor classification)
水文循环(hydrological cycle)
水循环(water cycle)
水蒸气(water vapor)
凇附(riming)
速调管(klystorn)
TRMM 微波成像仪(TRMM microwave Imager (TMI))
体积扫描模式(Volume Coverage Pattern (VCP))
天线(antenna)
天线反射直径(antenna reflector diameter)
天线转换开关(duplexer)
跳频(frequency hopping)
湍流动能(Turbulent Kinetic Energy (TKE))
湍流运动(turbulent motion)
W 波段雷达(W-band radar)
微波成像仪(microwave imager)
微波辐射计(microwave radiometer)
微物理过程(microphysical process)
卫星辐射计(satellite radiometer)
卫星估计降水(satellite precipitation estimate)
卫星相互校准(inter-satellite calibration)
卫星遥感(satellite remote sensing)
稳定本机振荡器(stabilized local oscillator (STALO))
涡流比(swirl ratio)
无线电波(radio waves)
X 波段雷达(X-band radar)
吸收(absorption)
下落速度(terminal velocity)
线性退极化比(linear depolarization ratio)
霰(graupel)

相对介电常数(relative permittivity)
相干振荡器(coherent oscillator (COHO))
相控阵雷达(phased-array radar)
相同高度平面位置指示(constant-altitude plan-position indicator(CAPPI))
相位(phase)
相位检测器(phase detector)
相移(phase shift)
消光系数(extinction coefficient)
新一代天气雷达(NEXRAD)
信号处理(signal processing)
信号模拟器(signal simulator)
信息内容分析(information content analysis)
信噪比(signal to noise ratio)
星载激光雷达(CALIPSO)
行波管(travelling wave tube (TWT))
雪的凝结(snow aggregate)
遥感(remote sensing)
液态水路径(liquid water path)
移动雷达(mobile radar)
有效视场(Effective Field of View (EOV))
雨滴(raindrop)
雨滴谱测量仪(disdrometer)
雨量估计(rainfall estimation)
雨量计(rain gauge)
原位测量(in-situ measurement)
圆偏振(circularly polarized)
云(clouds)
云滴(cloud droplet)
云解析模型(cloud resolving model)
云雷达(cloud radar)
云网络(Cloudnet)
云探测卫星(CloudSat)
云微物理(cloud microphysics)
云掩码(cloud mask)
载波频率(carrier frequency)
折射率(refractive index)
振幅(amplitude)

正向运算(forward operators)
中分辨率成像光谱仪(MODIS)
中频(intermediate frequency (IF))
中气旋(mesocyclone)
中气旋探测(mesocyclone detection)
主成分分析(principal component analysis)
自然灾害(natural hazard)
总差分相移(total differential phase shift)
最大不模糊速度(maximum unambiguous velocity)